Teaching Computational Thinking and Coding to Young Children

Marina Bers
Tufts University, USA

IGI Global
PUBLISHER of TIMELY KNOWLEDGE

A volume in the Advances in Early Childhood and K-12 Education (AECKE) Book Series

Published in the United States of America by
IGI Global
Information Science Reference (an imprint of IGI Global)
701 E. Chocolate Avenue
Hershey PA, USA 17033
Tel: 717-533-8845
Fax: 717-533-8661
E-mail: cust@igi-global.com
Web site: http://www.igi-global.com

Library of Congress Cataloging-in-Publication Data

Names: Ber, Marina, 1970- editor.
Title: Teaching computational thinking and coding to young children / Marina Ber, editor.
Description: Hershey, PA : Information Science Reference, 2022. | Includes bibliographical references and index. | Summary: "This book starts with an introduction to the topic of computational thinking and young children and then presents chapters of different aspects to consider in teaching computational thinking to young children,"-- Provided by publisher.
Identifiers: LCCN 2021023096 (print) | LCCN 2021023097 (ebook) | ISBN 9781799873082 (hardcover) | ISBN 9781799873099 (paperback) | ISBN 9781799873105 (ebook)
Subjects: LCSH: Computer science--Study and teaching (Elementary) | Computer programming--Study and teaching (Elementary) | Logic--Study and teaching (Elementary) | Problem solving--Study and teaching (Elementary) | Computer literacy.
Classification: LCC QA76.27 .T423 2022 (print) | LCC QA76.27 (ebook) | DDC 004.071--dc23
LC record available at https://lccn.loc.gov/2021023096
LC ebook record available at https://lccn.loc.gov/2021023097

This book is published in the IGI Global book series Advances in Early Childhood and K-12 Education (AECKE) (ISSN: 2329-5929; eISSN: 2329-5937)

British Cataloguing in Publication Data
A Cataloguing in Publication record for this book is available from the British Library.

All work contributed to this book is new, previously-unpublished material.
The views expressed in this book are those of the authors, but not necessarily of the publisher.

For electronic access to this publication, please contact: eresources@igi-global.com.

Advances in Early Childhood and K-12 Education (AECKE) Book Series

ISSN:2329-5929
EISSN:2329-5937

MISSION

Early childhood and K-12 education is always evolving as new methods and tools are developed through which to shape the minds of today's youth. Globally, educational approaches vary allowing for new discussions on the best methods to not only educate, but also measure and analyze the learning process as well as an individual's intellectual development. New research in these fields is necessary to improve the current state of education and ensure that future generations are presented with quality learning opportunities.

The **Advances in Early Childhood and K-12 Education (AECKE)** series aims to present the latest research on trends, pedagogies, tools, and methodologies regarding all facets of early childhood and K-12 education.

COVERAGE

- Curriculum Development
- Literacy Development
- Head Start and Pre-K Programs
- STEM Education
- Special Education
- Learning Outcomes
- Standardized Testing
- Common Core State Standards
- Early Childhood Education
- Performance Assessment

IGI Global is currently accepting manuscripts for publication within this series. To submit a proposal for a volume in this series, please contact our Acquisition Editors at Acquisitions@igi-global.com or visit: http://www.igi-global.com/publish/.

The Advances in Early Childhood and K-12 Education (AECKE) Book Series (ISSN 2329-5929) is published by IGI Global, 701 E. Chocolate Avenue, Hershey, PA 17033-1240, USA, www.igi-global.com. This series is composed of titles available for purchase individually; each title is edited to be contextually exclusive from any other title within the series. For pricing and ordering information please visit http://www.igi-global.com/book-series/advances-early-childhood-education/76699. Postmaster: Send all address changes to above address.

Titles in this Series

For a list of additional titles in this series, please visit:
http://www.igi-global.com/book-series/advances-early-childhood-education/76699

Handbook of Research on Critical Issues in Special Education for School Rehabilitation Practices
Ajay Singh (Texas A&M International University, USA) Chia Jung Yeh (East Carolina University, USA) Sheresa Blanchard (East Carolina University, USA) and Luis Anunciação (Pontifical Catholic University of Rio de Janeiro, Brazil)
Information Science Reference • © 2021 • 590pp • H/C (ISBN: 9781799876304) • US $245.00

Physical Education Initiatives for Early Childhood Learners
Pedro Gil-Madrona (University of Castilla-La Mancha, Spain)
Information Science Reference • © 2021 • 431pp • H/C (ISBN: 9781799875857) • US $195.00

Teaching Practices and Equitable Learning in Children's Language Education
Christina Nicole Giannikas (Cyprus University of Technology, Cyprus)
Information Science Reference • © 2021 • 273pp • H/C (ISBN: 9781799864875) • US $195.00

Connecting Disciplinary Literacy and Digital Storytelling in K-12 Education
Leslie Haas (Buena Vista University, USA) and Jill Tussey (Buena Vista University, USA)
Information Science Reference • © 2021 • 378pp • H/C (ISBN: 9781799857709) • US $195.00

Promoting Positive Learning Experiences in Middle School Education
Cherie Barnett Gaines (Lincoln Memorial University, USA) and Kristy M. Hutson (Lincoln Memorial University, USA)
Information Science Reference • © 2021 • 335pp • H/C (ISBN: 9781799870579) • US $195.00

For an entire list of titles in this series, please visit:
http://www.igi-global.com/book-series/advances-early-childhood-education/76699

IGI Global
PUBLISHER of TIMELY KNOWLEDGE

701 East Chocolate Avenue, Hershey, PA 17033, USA
Tel: 717-533-8845 x100 • Fax: 717-533-8661
E-Mail: cust@igi-global.com • www.igi-global.com

Table of Contents

Detailed Table of Contents

Computer programming is becoming an essential skill in the 21st century, and in order to best prepare future generations, the promotion of computational thinking and literacy must begin in early childhood education. Computational thinking can be defined in many ways. The broad definition offered in this chapter is that computational thinking practices refer to techniques applied by humans to express themselves by designing and constructing computation. This chapter claims that one of the fundamental ways in which computational thinking can be supported and augmented is by providing children with opportunities to code and to create their own interactive computational media. Thus, computational literacy will allow children to become producers and not only consumers of digital artifacts and systems.

As coding and computer science become established domains in K-2 education, researchers and educators understand that children are learning more than skills when they learn to code – they are learning a new way of thinking and organizing thought. While these new skills are beneficial to future programming tasks, they also support the development of other crucial skills in early childhood education. This chapter explores the ways that coding supports computational thinking in

The chapter begins with an exploration of computational thinking (CT) and its relationship to computational literacy, followed by a summary of theoretical and empirical work that aims to elucidate the connections among coding, CT, and literacy. The authors argue that these connections thus far have been predominantly one of support (i.e., unidirectional) and motivated by technological and policy advances, as opposed to considering the connections as mutually reinforcing and developmentally coaligned. The authors discuss the coding as another language (CAL) pedagogical approach, a pedagogy that presents learning to program as akin to learning how to use a new language for communicative and expressive functions, emphasizing the bidirectional connections between the two domains. Finally, the authors detail various curricula that use the CAL approach and discuss the implications of CAL for teaching and learning in early childhood.

Life science and computer science share the educational goals of fostering students to engage in inquiry-based learning and solve problems through similar practices of discovery, design, and experimentation. This chapter outlines the pedagogical links among traditional life science and emerging computer science domains in early childhood education, and describes an educational intervention using the CRISPEE technological prototype. CRISPEE, designed by a research team of developmentalists, biologists, educators, and computer scientists, invites young children to use computational logic to model design processes with biological materials. Findings are discussed as they relate to new understandings about how young children leverage computational thinking when engaged in design-based life science, or biodesign.

In the past two decades, STEM education has been slowly replaced by "STEAM," which refers to learning that integrates science, technology, engineering, arts, and mathematics. The added "Arts" portion of this pedagogical approach, although an important step towards integrated 21st century learning, has long confused policymakers, with definitions ranging from visual arts to humanities to art education and more. The authors take the position that Arts can be broadly interpreted to mean any approach that brings interpretive and expressive perspectives to STEM activities. In this chapter, they present illustrative cases inspired by work in real learning

Section 3
Contexts

This chapter examines the relationship between coding, computational thinking, and the contexts in which those concepts are learned. It recounts a pilot study where a 12-week robotics curriculum was taught in kindergarten classrooms at eight interfaith and secular schools in Boston, United States of America and Buenos Aires, Argentina. In this chapter, the authors explore how teachers and students drew from their socio-cultural environments to inform the language of computational thinking and support the internalization of computational concepts and, in turn, how computational thinking was used as a tool for deeper exploration of cultural traditions and beliefs, meaning-making, and creative expression.

The representation of women in technical fields such as computer science and engineering continues to be an issue in the United States, despite decades of research and interventions. According to the most recent Bureau of Labor Statistics reports, only 21.1% of computer programmers are women, and only 16.5% of engineering and architecture positions are filled by women. This chapter discusses the long-term importance of exposing girls to computational thinking during their formative early childhood years (Kindergarten through second grade) in order to set them up for equal opportunities in technical fields throughout their later educational and career years. This chapter presents a case example of a K-2nd grade robotics and coding curriculum in order to highlight examples of developmentally appropriate technologies, activities, and strategies that educators can implement to foster young girls' computational thinking skills. Best practices and instructional strategies to support girls—as well as young children of any gender identity—are discussed.

This chapter discusses understandings of coding and computational thinking education for students with disabilities. The chapter describes the special education system in the United States, including limitations in how computer science education is made

Preface

Thinking About Computational Thinking in Early Childhood

My mentor, Seymour Papert, used to say that one cannot “think about thinking without thinking about something”. Behind this playful phrase hides a deep concern about epistemology: how do we get to know what we know? How do we gain new insights about our own cognitive processes and the world around us?

This book is inspired by that quest for gaining new knowledge. In September 2001, I started as a young assistant professor at Tufts University, and I created an interdisciplinary research group called “Developmental Technologies”, DevTech. Students with diverse backgrounds: from child development to computer science, from education to engineering, from cognitive science to anthropology, joined the group. For the last twenty years, DevTech has focused on understanding how new technologies that engage in coding, robotics and making can play a positive role in children’s development and learning. Early on, DevTech’s work focused on older children, teens and pre-teens, and coding opportunities embedded in virtual worlds. Later on, as I had my own children, I realized there was a lack of tools for them. Thus, DevTech’s focus shifted to early childhood.

This fascinating period of development invites us to re-think how we design interfaces for children who cannot yet read and write, who have short attention span and working memory, who are honest in expressing engagement and frustration, who are just learning how to work with others and who are eager to explore the world by touching, making and breaking. At DevTech we focused on designing and creating novel programming environments, such as KIBO robotics and ScratchJr, to be developmentally appropriate. We collaborated with others, such as KinderLab Robotics, the LifeLong Kindergarten group at the MIT Media Lab and the Scratch Foundation to make sure these technologies could go out into the world and become products used by millions, and not only research prototypes. We developed teaching materials and pedagogical strategies for professional development of early childhood educators and opportunities for family and community engagement.

All of this work is based on the theoretical frameworks that I have developed over the years, such as Positive Technological Development (Bers, 2012); Coding as a Playground (Bers, 2018; 2020) and Coding as Another Language (Bers, 2019). We are conducting studies all over the world to understand how diverse children learn with and about computer science, how this discipline can help them make connections to more traditional domains of learning and how it can support the development of positive human values (Bers, 2022).

Our research involves three dimensions: theoretical contributions, design of new technologies, and empirical work to test and evaluate the theory and the technologies. Our long-time commitment is to inspire sustainable and scalable evidence-based programs for young children that promote the learning of programming with a playful and developmentally appropriate approach.

As we were busy with our work, a new term in the field started to gain traction: "computational thinking". Jeannette Wing popularized it in the mid 2000's. However, those of us working within a Constructionist framework (Papert, 1980; Resnick, 2017) recognized the concept immediately from observing the kinds of questions children engaged with when programming, the kind of problems they encountered when creating their projects, and the multiple creative solutions they proposed.

Nowadays, there is a push to embed computational thinking throughout the national and international educational frameworks, to conduct research to better understand what it means, to package it through curriculum, media and games, and to assess it. This brings new opportunities for the field of computer science education to reach mainstream.

In this book, we focus on a particular segment of the population: early childhood. The fifteen chapters in this book were all written by current or former students in the DevTech research lab. Thus, all chapters share a similar theoretical and pedagogical framework, but they each focus on a particular aspect of computational thinking and early childhood education. The book is organized in four parts: "Deep Dive," "Connections," "Contexts." and "Evaluation."

The three chapters in "Deep Dive" set the stage for thinking about computational thinking and its relationship to coding and unplugged activities. Chapter 1, "From Computational Thinking to Computational Doing," by Marina Bers, provides a broad literature review and positions computational thinking practices as techniques applied by humans to express themselves by designing and constructing computational artifacts. This chapter claims that one of the fundamental ways in which computational thinking can be supported and augmented is by providing children with opportunities to code and to create their own interactive computational media. Chapter 2, "Why Teach Coding to Early Elementary Learners," by Claudia Mihm, explores the ways in which coding supports computational thinking in young children, and connects the core concepts of computational thinking to other crucial skills in early childhood education – such as literacy, numeracy, and organization skills. Chapter 3, "Unplugged Learning: Recognizing Computational Thinking in Everyday Life," by Emily Relkin and Amanda Strawhacker, explores both plugged and unplugged opportunities that, building on traditional early education experiences and activities, can engage young children in computational thinking in both formal and informal learning settings.

The four chapters in "Connections" provide examples of the associations and relationships that come about when computational thinking is explored through the life sciences, literacy, dramatic arts, and the development of executive functions. Chapter 4, "The Role of Executive Function and Self-Regulation in the Development of Computational Thinking," by Elizabeth Kazakoff-Myers, explores theoretical connections between computational thinking, coding, self-regulation, and executive function and presents findings from an early study. Chapter 5, "Rhyme and Reason: The Connections Among Coding, Computational Thinking, and Literacy," by Madhu Govind, Ziva Hassenfeld, and Laura de Ruiter, discusses theoretical and empirical work to elucidate the connections among coding, computational thinking, literacy, and language. The authors argue that these connections thus far have been predominantly one of support (i.e., unidirectional) and motivated by technological and policy advances, as opposed to considering the connections as mutually reinforcing and developmentally coaligned. The authors present the Coding as Another Language (CAL) pedagogical approach and curricula, which addresses the bidirectional connection between computer science and literacy. Chapter 6, "Computational Thinking and Life Science: Thinking About the Code of Life," by Amanda Strawhacker, outlines the pedagogical links among traditional life science and emerging computer science domains in early childhood education, and describes an educational intervention using the CRISPEE prototype that invites young children to leverage computational thinking when engaged in design-based life science, or biodesign. Chapter 7, "Computational Expression: How Performance Arts Support Computational Thinking," by Amanda Strawhacker and Amanda Sullivan, explores how in the past two decades, STEM education has been slowly replaced by "STEAM", which refers to learning that integrates Science, Technology, Engineering, Arts, and Mathematics and presents case studies in which painting, sculpture, architecture, poetry, music, dance, and drama supported the teaching of computational thinking skills.

The five chapters in "Contexts" present examples of different experiences in which children and families learn to code by making their own computationally rich projects and by thinking in computational ways. The diversity of the five chapters in this part span from a focus on girls, children with disabilities and families, to learning environments explicitly designed to promote making activities and exploration of cultural and religious identities. Chapter 8, "Fostering Computational Thinking in Homes and Other Informal Learning Spaces," by Madhu Govind, explores the different kinds of roles, interactions, and opportunities afforded by parents, caregivers, and siblings when engaging in collaborative coding activities. Chapter 9, "Makerspaces as Learning Environments to Support Computational Thinking," by Amanda Strawhacker and Miki Vizner, explores the ways in which tools, facilitation, and the physical environment of informal makerspaces can support children's engagement with powerful ideas of computer science and the maker movement. Chapter 10, "Coding, Computational Thinking, and Cultural Contexts," by Libby Hunt and Marina Bers, describes a pilot study in eight interfaith kindergarten classrooms in Boston, United States of America and Buenos Aires, Argentina that set out to explore different ways a robotics curriculum could promote computational thinking, and, in turn, how computational thinking was used as a tool for deeper exploration of cultural traditions and beliefs, meaning-making, and creative expression. Chapter 11, "Supporting Girls' Computational Thinking Skillsets: Why Early Exposure Is Critical to Success," by Amanda Sullivan, discusses the long-term importance of exposing girls to computational thinking during their formative early childhood years in order to set them up for equal opportunities in technical fields throughout their later educational and career years and presents a case study of a K-2nd grade robotics and coding curriculum that illuminates best practices and instructional strategies. Chapter 12, "Including Students With Disabilities in the Coding Classroom," by Tess Levinson, Libby Hunt, and Ziva Hassenfeld, describes the special education system in the United States, and how computer science education is made available to students receiving special education services. It presents a case study of a student with a mild disability. Implications for future research directions are discussed.

Lastly, "Evaluation" presents three chapters focused on how to evaluate computational thinking using projects created by children, validated unplugged assessments and data analytics. Chapter 13, "TechCheck: Creation of an Unplugged Computational Thinking Assessment for Young Children," by Emily Relkin, describes the development and validation of TechCheck, a novel instrument for rapidly assessing Computational Thinking (CT) skills in 5-9 years old children. Chapter 14, "Examining Young Children's Computational Artifacts," by Apittha Unahalekhaka and Madhu Govind, examines various rubrics and assessment tools used to measure the levels of programming competency, creativity, and purposefulness displayed in students' coding artifacts, and discusses the ScratchJr and KIBO Project Rubrics for researchers and educators to evaluate diverse projects. Chapter 15, "Insights Into Young Children's Coding With Data Analytics," by Apittha Unahalekhaka, Jessica Blake-West, and XuanKhanh Nguyen, discusses the potential of learning analytics for computational thinking assessment in early childhood education and provides examples of its use to deepen understanding of computational thinking through observing young children's engagement with ScratchJr.

Together, these 15 articles divided in four parts presents a snapshot of some of the work done over two decades by members of the DevTech research group. As the field progresses, and more labs and researchers around the country and the world engage with computer science and early childhood, it is my hope that we will grow our collective ability to "think about thinking by thinking about computational thinking".

REFERENCES

Bers, M. (2006). The role of new technologies to foster positive youth development. *Applied Developmental Science, 10*(4), 200–219.

Bers, M. (2020). *Coding as a Playground: Programming and Computational Thinking in the Early Childhood Classroom* (2nd ed.). Routledge Press.

Bers, M. (in press). *Beyond Coding: How Children learn Values through programming*. MIT Press.

Bers, M. U. (2008). *Blocks to robots learning with technology in the early childhood classroom*. Teachers College Press.

Bers, M. U. (2010). Beyond computer literacy: Supporting youth's positive development through technology. *New Directions for Youth Development, 128*, 13–23.

Bers, M. U. (2012). Designing Digital Experiences for Positive Youth Development: From Playpen to Playground. Cary, NC: Oxford.

Bers, M. U. (2017). The Seymour test: Powerful ideas in early childhood education. *International Journal of Child-Computer Interaction*.

Bers, M. U. (2018). *Coding as a Playground: Programming and Computational Thinking in the Early Childhood Classroom*. Routledge Press.

Bers, M. U. (2019). Coding as another language: A pedagogical approach for teaching computer science in early childhood. *Journal of Computers in Education, 6*(4), 499–528.

Bers, M. U., & Resnick, M. (2015). *The Official ScratchJr Book*. No Starch Press.

Papert, S. (1980). *Mindstorms: Children, computers, and powerful ideas*. Basic Books, Inc.

Resnick, M. (2017). *Lifelong Kindergarten: Cultivating Creativity through Projects, Passion, Peers, and Play*. MIT Press.

Sullivan, A., & Bers, M. U. (2017). Dancing robots: Integrating art, music, and robotics in Singapore's early childhood centers. *International Journal of Technology and Design Education*.

Acknowledgment

The research presented here is based on the past and present contributions of members of the DevTech research group directed by Marina Bers at Tufts University. This book took shape during group meetings, but it would not have been possible without the careful coordination of Madhu Govind. Thank you, Madhu for your professionalism in all aspects involved in making this book possible.

Section 1
Deep Dive

Chapter 1
From Computational Thinking to Computational Doing

Marina Umaschi Bers
Tufts University, USA

ABSTRACT

Computer programming is becoming an essential skill in the 21st century, and in order to best prepare future generations, the promotion of computational thinking and literacy must begin in early childhood education. Computational thinking can be defined in many ways. The broad definition offered in this chapter is that computational thinking practices refer to techniques applied by humans to express themselves by designing and constructing computation. This chapter claims that one of the fundamental ways in which computational thinking can be supported and augmented is by providing children with opportunities to code and to create their own interactive computational media. Thus, computational literacy will allow children to become producers and not only consumers of digital artifacts and systems.

A SCENARIO

Henry, 6 years old, is working with the free ScratchJr introductory programming language on an iPad loaded in his kindergarten class. He is focused on making an animation of a train. Every so often, he wiggles. Other times, he is frustrated and watches over his friend Liana's project to ask her a question. "How did you make your cat appear and disappear on the screen so many times?" Henry is trying to program a train that travels into a tunnel. He drew the train and the tunnel with the paint tool in ScratchJr. He is happy with how they look, but now comes the hardest part: he needs to program the train to move forward, while making a "choo-choo-

DOI: 10.4018/978-1-7998-7308-2.ch001

choo" noise. He has to time it perfectly, so the train disappears as it travels into the tunnel, but the noise continues to play out.

"Look at my cat! Look at my cat!" Liana is excited to show Henry how to use new programming blocks, purple blocks to be more specific, called Looks blocks. She has programmed her ScratchJr kitten to appear and disappear on the screen ten different times. She has put together a long sequence. Although Liana cannot read yet, she knows that these programming blocks can make her ScratchJr kitten show and hide.

Henry wants to do the same thing, except that he wants the train to always hide while it is inside the tunnel. Slowly, by trial and error, he figures that he will need to put together a sequence with ten hide blocks. And then, he could put one show block and the train will become visible again. The problem is that all of this needs to happen while the train keeps sounding its "choo choo choo" horn. He is not sure how to do this.

Henry's teacher hears his call and walks over to him. Henry explains what he wants. The teacher shows him how to create parallel programs, so two different events can happen at the same time. Henry is happy to try it. He records the "choo choo choo" sound with his voice and he creates a sequence that makes the sound start with the green flag. The same green flag that starts the train moving forward to go through the tunnel. "It works!" exclaims Henry while jumping up and down. However, after watching the animation for a few seconds, Henry notices that the train doesn't hide for long enough. Self-confident, Henry decides to add a few more blocks to the hide sequence until he runs out of space in the screen. He is about to ask Liana again, when he suddenly remembers about the new programming block they learned a few days ago, a long orange block, called "Repeat." This block allows for other blocks to be inserted inside its "loop". The repeat block then runs the blocks inside its loop as many times as the programmer decides.

After some trial and error, in which Henry plays with inserting different numbers of hide blocks inside the "Repeat" block, he figures it out. He can put just one hide block inside the "Repeat" block and set the number of repetition times to the highest he needs for the train to be inside the tunnel. He chooses the number 20 and clicks the "Green flag" to see the animation. The train moves forward on its tracks and goes into a tunnel while the "choo choo choo" sound plays in the background. Then, the train disappears and comes out on the other side. After watching the animation, Henry realizes it is boring to wait for so long for the train to appear again. He goes back to his code and reduces the number of repetitions to 5. Figure 1 shows Henry's code for the train.

During this experience, Henry had fun. He also put his coding skills to work making a project he cared about. He learned that a programming language has a syntax in which symbols represent actions. He understood that his choices had an impact on what was happening on the screen. While programming, Henry encountered some of the most powerful ideas of computer science that are accessible for a young child. He developed computational thinking. He was able to create a sequence of programming blocks to represent a complex behavior (e.g., appearing and disappearing), as well as create parallel sequences so two different events could happen at the same time (e.g., parallel programming). He used logic in a systematic way to correctly order the blocks in a sequence and he problem-solved. He exercised his tenacity and learned how to ask for help from peers and his teacher. Finally, Henry was able to create a project from his own original idea and turn it into a final product, a project he chose and to which he was personally attached. He was happy to revise it when the final outcome did not meet his expectations (i.e., it ended up being so long that it was boring to watch). He also engaged with mathematical ideas of estimation and number sense.

Figure 1. The ScratchJr interface with Henry's train. In this photo, the train is programmed with a repeat loop to disappear 5 times while it goes in the tunnel and then appear at the end.
Source: IGI, 2021

To code, Henry used ScratchJr, a programming language specifically designed for young children and available for free on touchscreen tablets. ScratchJr was designed by the DevTech Research Group at Tufts University in collaboration with MIT's Lifelong Kindergarten group at the MIT Media Lab and the Playful Invention Company (PICO) company. To date, over 20 million young children all over the world are using ScratchJr to create their own projects. As children make computational projects, they develop computational thinking. This involves more than problem-solving or logical thinking; it means gaining the concepts, skills, and habits of mind to express themselves through coding. In this approach, doing and thinking come together, echoing decades of research done by developmental scientists and educational researchers.

This chapter explores the recent construct of computational thinking as it applies to young children. First, the chapter provides an overview about computational thinking, and then it focuses on a limited set of seven powerful ideas of computer science that are developmentally appropriate. Finally, the chapter discusses the relationship between computational thinking and coding. Computational thinking is often thought of as a cognitive activity that involves problem-solving through both unplugged activities and computer programming. In this chapter, though, the definition is framed in a broader context. Computational thinking is conceptualized as an expressive process that involves problem solving. In this perspective, problem solving is not an end in itself, it is also a means for expression, for making projects.

COMPUTATIONAL THINKING: THEORETICAL FOUNDATIONS

The idea that the "theory of computation" is for everyone, and not only for computer scientists, dates back to the 1960's (Perlis, 1962). As technology progressed, computers became more accessible, but programming languages' syntax and grammar were still too difficult to understand and manipulate. In 1982, Perlis, one of the pioneers in developing the ALGOL programming language wrote "most people find the *concept of programming* obvious, but the doing impossible" (Perlis, 1982, p. 10). What Perlis referred to as the "concept of programming" is close to our current understanding of computational thinking. Furthermore, Perlis distinguished the cognitive processes associated with thinking in abstract and logical ways from the mastery of a programming language. Perlis also observed that, since computer programming requires logical and creative thought, its teaching needs to start early in life and become part of everyone's education. The intuition that even young children could grasp concepts associated with computational thinking such as sequencing, patterns, modularity, cause and effect, and problem-solving when presented with them in a way that made sense, has now been confirmed by extensive research (Bers, 2018, 2020).

These insights echoed Seymour Papert's work, who, at the time, was working on the development of a programming language that children could use, so the "doing" would not be impossible, as Perlis had argued. Building on knowledge about human development from Jean Piaget, Papert collaborated with Wally Feurzeig and others to create LOGO, the first programming language designed for children to think in computational ways. Papert argued that this new way of thinking could happened at its best when children were given tools to create personally meaningful projects. That is, computational thinking and computational doing could happen hand-in-hand (Bers, 2010; Papert, 1980). Children who could think like a computer were children who could use a computer to express themselves in a fluent way to create computational media, and children who could develop habits of mind such as persistence (Bers, 2021).

In the process of learning how to use a programming language, one learns to think and act in different ways. In his writing, Papert did not use the term computational thinking. Instead, he defined these new ways of thinking by referring to powerful ideas, central concepts and skills within a discipline, that children could encounter while using programming languages, such as sequencing, abstraction, modularization, problem solving and logical thinking (Bers, 2008, 2017, 2020). Those cognitive mechanisms are associated with what researchers call now computational thinking (Barr, Harrison, & Conery, 2011; Barr & Stephenson, 2011; Computer Science Teachers Association, 2020; Lee et al., 2011; Wing, 2006;).

In 2006, Jeannette Wing's influential article "Computational Thinking" appeared in the *Communications of the ACM* (Wing, 2006). Echoing Perlis and Papert, Wing argued that computational thinking, a broad set of analytic and problem-solving skills, dispositions, and habits, rooted in computer science, is universally applicable and therefore should be part of every child's analytical ability. Wing defined computational thinking as "solving problems, designing systems, and understanding human behaviour, by drawing on the concepts fundamental to computer science" (p. 33). At the heart of computational thinking is abstraction (Kramer, 2007), that is the ability to identify salient pieces of a problem or model and ignore inessential details.

Computational thinking includes mental tools such as thinking recursively, using abstraction when figuring out a complex task, and applying heuristic reasoning to discover a solution and to identify potential "bugs" or problems. Wing asserts that just as the printing press facilitated the spread of the three Rs (reading, writing, and arithmetic), computers facilitate the spread of computational thinking. The question is: are computers, per se, facilitating the spread of computational thinking or is it the ability to program that allows for the development of computational thinking? This chapter proposes that, although using computers opens the door to the world of computation, it is through programming them that we develop new ways of thinking associated with the discipline of computer science. Consuming technology is not the same as producing it. Using a computer doesn't require the same kind of thinking as programming a computer. This perspective seems to be in alignment with Wings later writing in which she defined computational thinking as a "thought processes involved in formulating problems and their solutions so that the solutions are represented in a form that can be effectively carried out by an information-processing agent" (Wing, 2011).

Wing's writing brought newfound light to the importance of computer science education. However, it also limited the discourse around computational thinking to a problem-solving process that complements mathematical and engineering thinking. It does not incorporate the relationship between thinking and doing as a way of personal expression. Mainstream computational thinking came to represent a type of analytical thinking that shares similarities with mathematical thinking (e.g., problem-solving), engineering thinking (designing and evaluating processes), and scientific thinking (systematic analysis; Bers, 2010). This perspective easily found a niche in the K-12 curriculum as it enabled the teaching of computational thinking outside the context of computer science courses and programming languages.

However, for some researchers, this decoupling is problematic. For example, Bers (2018) claims the thought processes involved in computational thinking need to support the creation of computational projects, and vice versa. While Wing describes the computational thinker as "an information-processing agent," Bers refers to this thinker as an "expressive agent." (Bers, 2020). An expressive agent is someone who has the internal and external resources and the required fluency with technological environments to be able to translate ideas into computational projects to share with others (Bers, 2021). Along this line of thinking, Brennan and Resnick (2012) broke down computational thinking into a three-dimensional framework that comprises *concepts*, *practices*, and *perspectives*. At a higher level, computational thinking practices refer to techniques applied by humans to express themselves by designing and constructing computation. The ideas presented in this chapter are aligned with this understanding.

Although computational thinking has received considerable attention over the past several years, there is little agreement on what a definition for computational thinking might encompass (Allan et al., 2010; Barr & Stephenson, 2011; Grover & Pea, 2013; National Academies of Science, 2010; Relkin, 2018; Relkin & Bers, 2019; Shute, Sun, & Asbell-Clarke, 2017; Grover & Pea, 2013; Guzdial, 2008). However, there is consensus on the fact that the science of computation must be available to thinkers of all disciplines, regardless of their ability to program (Guzdial, 2008; Yadav, 2011). Thus, motivated by a shortage of software engineers and programmers and the need of diversity in the industry, frameworks and initiatives have been put in place to promote the teaching of computational thinking in K-12, so to attract a wider pool of interested students before college (Barr, Harrison, & Conery, 2011).

This decision is informed by the need to broaden access to the field of computer science, which has traditionally been populated by white males. Extensive prior work demonstrates the importance of piquing the interest of girls during their formative early childhood years before gender stereotypes regarding these traditionally masculine fields are ingrained in later years (Sullivan, 2019; Metz, 2007; Steele, 1997). In addition, work is done to extend opportunities for participation in computing communities to black and Latino students (Erete et al., 2017). To this end, computer science educators have a unique responsibility to promote social justice and combat systemic inequities in the field of computer science (Vogel et al., 2017). Starting to teach computer science in the early years is one way to do so. But it must be done with an age-appropriate pedagogical approach.

Playgrounds for Thinking

Researchers, practitioners, funding institutions, and policy makers have traditionally associated computer programming and computational thinking with problem-solving. Thus, when translated into the educational curriculum, computer science is grouped with science, technology, engineering, and math disciplines: STEM. When integrated with these programs, computational thinking is defined as a set of cognitive skills for identifying patterns, breaking apart complex problems into smaller steps, organizing and creating a series of steps to provide solutions, and building a representation of data through simulations (Barr & Stephenson, 2011). When extended to the arts, STEM becomes STEAM and a design component is usually added. Children are invited to create their own projects. This addition facilitates the understanding of computational thinking as involved in the process of creation. However, not everyone agrees on the materials for creation. While some researchers believe that computational tools and learning how to code are essential, others pose that it is possible to engage in computational thinking without working with computers of any kind.

Recently, a low-tech or unplugged approach to computational thinking has been growing (e.g., Bell, Witten, & Fellows, 1998). For example, computational thinking can occur in everyday activities, including: sorting LEGO (using the concept of "hashing" to sort by color, shape, and size), cooking a meal (using "parallel processing" to manage cooking different types of food at different temperatures for different amounts of time), and looking up a name in an alphabetical list (linear: starting at the beginning of the list; binary: starting at the middle of the list). Traditional board games have also been designed to explicitly support computational thinking. For example, Robot Turtles (Shapiro, 2015) was designed for young children aged 3–8 years to start thinking in computational ways while playing a traditional turn-based board game, in which there are multiple paths for reaching a goal and for solving a problem successfully.

The DevTech Research Group at Tufts University has also developed low-tech strategies to promote computational thinking through the Coding as Another Language (CAL) curriculum: singing and dancing, card games, bingo, and the Simon Says game (Bers, 2019). The powerful ideas that children encounter while playing these games, such as sequencing and debugging, breaking one big problem into smaller steps, and planning and testing a strategy, all tap into the core of computational thinking. However, DevTech's approach, which is presented through the different chapters in this book, claims that although unplugged activities can support, enable and augment the development of computational thinking, in order to fully engage with it, children must experience the activity of computer programming.

Programming languages involve problem-solving, while supporting personal expression through the making of computational projects. Coding, then, becomes a vehicle for new forms of thinking and for the expression of the resulting thoughts.

There is a constant interplay between making new things in the world and making new ideas in our heads. As you make new things and get feedback from others (and from yourself), you can revise, modify, and improve your ideas. And based on these new ideas, you are inspired to make new things. (Resnick, 2001, p.3)

In order for children to make computational projects, there is a need for programming languages. When those are designed in developmentally appropriate ways, they can become coding playgrounds (Bers, 2012; 2020; 2018). In the coding playgrounds, children can learn to think computationally, while also making their own projects. For example, with tools such as KIBO robotics, shown in figure 2, and the free ScratchJr introductory language, children can create different forms of computational projects, from screen animations to dancing robots. (Kazakoff, Sullivan, & Bers, 2013; Portelance & Bers, 2015; Sullivan & Bers, 2015).

Programming languages such as KIBO and ScratchJr are coding playgrounds that promote problem-solving, imagination, cognitive challenges, social interactions, motor skills development, emotional exploration, and making different choices. They provide tools to create projects to express our thinking, to communicate who we are and what we love. In the process, computational thinking develops. Early childhood is a wonderful time for discovering new ways of thinking.

Figure 2. A KIBO robotics project programmed with wooden blocks
Source: IGI, 2021

Research shows the economic and developmental impact of interventions that begin in early childhood. These are associated with lower costs and more durable effects than interventions that begin later on (e.g., Cunha & Heckman, 2007; Heckman & Masterov, 2007; National Research Council Committee on Early Childhood Pedagogy, 2001; Shonkoff & National Research Council, 2000). Thus, if promoting computational thinking is important in our information age, it needs to be introduced in early childhood, given the plasticity of young children. However, pedagogical approaches and programming languages must be consistent with developmentally appropriate practice (Bredekamp, S, 1987) and must embrace the maturational stages of children by inviting play and discovery, socialization and creativity (Bers, 2018a). In addition, the powerful ideas of the discipline of computer science must be developmentally appropriate. It is not enough to copy models used in later schooling.

Research shows that, when beginning in prekindergarten, learning to program can significantly improve a child's ability to logically sequence picture stories (Kazakoff, Sullivan, & Bers, 2013) and to improve executive functioning (Arfé et al., 2019). These findings are consistent with other research that shows the positive impact that learning computer programming and computational thinking can have on skills such as reflectivity, divergent thinking, and cognitive, social, and emotional development (Clements & Gullo, 1984; Clements & Meredith, 1992; Flannery & Bers, 2013). In early childhood, the playground approach to coding provides opportunities to encounter a complex system of ideas that is logically organized and utilizes abstraction and representation. In addition, it enables the development of skills and habits of mind to put those powerful ideas to use—by making personally meaningful projects (Bers, 2020).

THE POWERFUL IDEAS OF COMPUTATIONAL THINKING

Seymour Papert coined the term "powerful ideas" to refer to a central concept and skill within a domain (i.e., computer science) that is at once personally useful, interconnected with other disciplines, and has roots in intuitive knowledge that a child has internalized over a long period. According to Papert, powerful ideas afford new ways of thinking, new ways of putting knowledge to use, and new ways of making personal and epistemological connections with other domains of knowledge (Papert, 2000).

Papert envisioned the computer as a carrier of powerful ideas and as an agent for educational change. While school reform is complex, he proposed that the teaching of computer science could help children encounter powerful ideas about new disciplines, such as computer science, old disciplines, such as math, and learning itself. For example, he showed how when children learned how to program a turtle to create geometrical shapes on the screen with LOGO, they were not only exploring abstract thinking, modularity, problem solving, and recursion, but also powerful mathematical ideas such as angles as well as thinking about their own thinking (Abelson & DiSessa, 1981).

Over the years, a growing community of researchers and educators has used the term "powerful ideas" to refer to a set of intellectual tools worth learning, as decided by a community of experts in each of the fields of study (Bers, 2008). However, different people have used the term in diverse ways and amongst the powerful ideas community there are divergent opinions about the benefits and dangers of presenting a unified definition (Papert & Resnick, 1996).

When exploring the concept of computational thinking, it is useful to do it in such a way to identify the powerful ideas we hope children will encounter and develop. Powerful ideas of computer science are not tied to a particular programming environment but to the discipline of computer science and its associated habits of mind. While most of these ideas can be encountered when engaging in low-tech or unplugged activities, it is in the activity of programming that they can be further explored. However, the challenge for early childhood education is that powerful ideas need to be defined in a developmentally appropriate way and described at different levels of depth, in a spiral manner, in the sequence of PreK-2. For example, understanding algorithmic thinking in PreK might focus on linear sequencing, while in second grade it extends to loops. Children will understand that within a sequence there are patterns that repeat themselves. Inspired by already existing computational thinking curriculum such as the framework and resources for educators launched by Google in 2010 (Google for Education, 2010), the work of Karen Brennan and Mitchel Resnick with Scratch (2012), and previous work with the KIBO robotic system (Sullivan & Bers, 2015) and with ScratchJr (Portelance, Strawhacker, & Bers, 2015), I propose a framework involving seven developmentally appropriate powerful ideas for early childhood computer science education: algorithms, modularity, control structures, representation, hardware/software, design process, and debugging.

- **Algorithms** refer to a series of ordered steps taken in a sequence to solve a problem or achieve an end goal. Sequencing is an important skill in early childhood; it is a component of planning and involves putting objects or actions in the correct order. For example, retelling a story in a logical way or ordering numbers in a line is sequencing. Understanding algorithms involves understanding abstraction (i.e., identifying relevant information to define what constitutes a step in the sequence) and representation (i.e., depicting and organizing information in an appropriate form).
- **Modularity** involves breaking down tasks or procedures into simpler, manageable units that can be combined to create a more complex process. This process of decomposition involves subdividing jobs. In early childhood, decomposition can be taught anytime a complex task needs to be broken down into smaller units.
- **Control Structures** determine the order (or sequence) in which instructions are followed or executed within an algorithm or program. Control structures provide a window into understanding the computational concept of making decisions based on conditions (e.g., variable values, branching, etc.). Children learn about sequential execution first, and later they become familiar with multiple control structures that involve repeat functions, loops, conditionals, events, and nested structures. Loops can be used to repeat patterns of instructions, conditionals to skip instructions, and events to initiate an instruction. Understanding control structures in early childhood requires familiarity with patterns.
- **Representation** is related to the way computers store and manipulate data and values in a variety of ways. These data need to be made accessible through different representations. Early on, children learn that concepts can be represented by symbols. For example, letters represent sounds, numbers represent quantities, and programming instructions represent behaviors. As children grow and advance to more complex programming languages, they learn about other data types, such as variables. The notion that concepts can

be represented using symbols is foundational in early childhood and has strong links to both math and literacy.

- **Hardware/Software Systems** refer to the fact that computing systems need both hardware and software to operate and to accomplish tasks, such as receiving, processing, and sending information. The relationship between hardware and software becomes increasingly important in understanding the ways that components affect a system. As children grow they will encounter the need to understand the complexities of different system.
- **Design Process** is an iterative process used to develop programs and tangible artifacts that involves several steps and has similarities with the engineering design cycle (Ertas & Jones, 1996). The design process starts with asking a question, planning an approach, proposing prototypes, testing and re-testing them, revising and sharing the result. Children can begin at any step, move back and forth between steps, or repeat the cycle over and over (Bers, 2018). As children become more familiar with the design process, they become instilled with the ability to iteratively create and refine their work, to give and receive feedback to others, and to continually improve a project through experimenting and testing. This leads to iterative improvement, involves perseverance, and has strong associations with some aspects of executive functions, such as self-control, planning and prioritizing, and organization (Bers, 2020)
- **Debugging** refers to the systematic analysis and evaluation that allows us to fix problems and involves using skills such as testing, logical thinking, and problem-solving in an intentional, iterative step-by-step way. As children learn how to debug their systems, they start to develop common troubleshooting strategies that can be used on a variety of computing systems. Debugging teaches the powerful lesson that things do not just happen to work on the first try, but, in fact, that many iterations are usually necessary to get it right.

In the chapters in this book, these seven powerful ideas of computer science that are developmentally appropriate will be further explored and examples and case studies will be presented.

CONCLUSION: FROM THINKING TO DOING

There is debate amongst researchers and educators regarding whether computational thinking can be classified as a unique category of thought (Gadanidis, 2017; Pei, Weintrop, & Wilensky, 2018). However, the term has grown popular at a time when schools are starting to incorporate the teaching of computer science in more massive ways (K–12 Computer Science Framework Steering Committee, 2016).

Thinking is at the core of why we introduce coding in early childhood. Coding can not only help children think in systematic and sequential ways, but also create and express themselves in new ways. That is, in computational ways. The power of computational thinking extends beyond thinking like computer scientists in two significant ways. First, by engaging children to think about their own thinking, they can develop metacognition. And that is a useful skill, regardless of the discipline of study or the job of the future. Second, when children engage in coding or in creating their own computational projects, they put their thinking at the service of making by engaging in an activity that requires the application of their abstract thinking in very concrete ways.

This chapter claims that one of the fundamental ways in which computational thinking can be supported and augmented is by providing children with opportunities to code and to create their own interactive computational media. It is in this process, that the seven powerful ideas described earlier are best encountered.

Thinking implies the ability to make sense, interpret, represent, model, predict and invent our experiences in the world. Research shows that children learn to think with and through language (Vygotsky, 1978). Thus, by learning to use a programming language that involves logical sequencing, abstraction, and problem solving, children can learn how to think in analytical ways. Wittgenstein argued that the language we speak determines the thoughts we are able to have. In other words, learning a new language can make new patterns of thought, new conceptual frameworks, and new ways of using language (Wittgenstein, 1997). Wittgenstein's philosophy echoes Vygotsky's developmental perspective in terms of the relationship between language and thinking at the individual level. Programming languages provide opportunities for new ways of thinking that involve a problem-solving dimension as well as the use and manipulation of a language, a symbolic representational system, to create a sharable product that others can interpret (Bers, 2020).

Computer programming is becoming an essential skill in the 21st century. Each month, there are an estimated 500,000 openings for computing jobs nationwide, and a lack of adequately trained people to fill them (Code.org, 2018; Fayer, Lacey, & Watson, 2017). However, the rationale for supporting the introduction of computer science and computational thinking starting in kindergarten is not the creation of the future workforce, but the future citizenry (Bers, 2020). Without understanding the fundamentals of what an algorithm is and how it works, people might not understand why and how certain data is displayed and become illiterate in the information age.

We understand by doing. Therefore, in this chapter I argue that in order to promote computational thinking, we should expose young children to learning a developmentally appropriate programming language with a playful approach. Coding is a new literacy, and as such, those who learn how to code from a young age, will not only participate in the automated economy, but will also be able to create new opportunities for civic participation and for living in healthy social communities (Bers, 2022). Computational literacy will allow children to become producers, and not only consumers of digital artifacts and systems. They will be able to do, to change the world, and not only to think about it.

ACKNOWLEDGMENT

This work was supported by the U.S. Department of Education [grant U411C190006].

REFERENCES

Abelson, H., & DiSessa, A. (1981). Turtle geometry: The computer as a medium for exploring mathematics (The MIT press series in artificial intelligence). Cambridge, MA: MIT Press.

Arfé, B., Vardanega, T., Montuori, C., & Lavanga, M. (2019). Coding in Primary Grades Boosts Children's Executive Functions. *Frontiers in Psychology*, *10*, 2713. doi:10.3389/fpsyg.2019.02713 PMID:31920786

Barrouillet, P., & Lecas, J. (1999). Mental Models in Conditional Reasoning and Working Memory. *Thinking & Reasoning*, *5*(4), 289–302. doi:10.1080/135467899393940

Bers, M. U. (2008). *Blocks to Robots: Learning with Technology in the Early Childhood Classroom*. Teachers College Press.

Bers, M. U. (2012). *Designing Digital Experiences for Positive Youth Development: From playpen to playground*. Oxford University Press. doi:10.1093/acprof:oso/9780199757022.001.0001

Bers, M. U. (2017). The Seymour Test: Powerful Ideas in early childhood education. *International Journal of Child-Computer Interaction*, *14*, 10–14. doi:10.1016/j.ijcci.2017.06.004

Bers, M. U. (2018a). *Coding as a playground: Computational thinking and programming in early childhood*. Routledge.

Bers, M. U. (2018b). Coding, Playgrounds and Literacy in Early Childhood Education: The Development of KIBO Robotics and ScratchJr. *IEEE Global Engineering Education Conference (EDUCON)*, 2100. 10.1109/EDUCON.2018.8363498

Bers, M. U. (2019). Coding as Another Language: "Why Computer Science in Early Childhood Should Not Be STEM. In C. Donohue (Ed.), *Key Issues in Technology and Early Childhood*. Routledge. doi:10.4324/9780429457425-11

Bers, M. U. (2020). Playgrounds and Microworlds: Learning to Code in Early Childhood. In Designing Constructionist Futures: The Art, Theory, and Practices of Learning Designs. Academic Press.

Bers, M. U. (2022). *Beyond Coding: How Children Learn Human Values through Programming*. The MIT Press.

Bers, M. U., & Resnick, M. (2015). *The Official ScratchJr Book: Help your Kids Learn to Code*. No Starch Press.

Blikstein, P. (2013). Digital Fabrication and 'Making' in Education: The Democratization of Invention. In J. Walter-Herrmann & C. Büching (Eds.), *FabLabs: Of Machines, Makers and Inventors*. Transcript Publishers. doi:10.14361/transcript.9783839423820.203

Bowman, B., Donovan, S., & Burns, M. (2001). Eager to learn: Educating our preschoolers. Washington, DC: National Academy Press.

Bredekamp, S. (1987). *Developmentally appropriate practice in early childhood pro- grams serving children from birth through age 8*. National Association for the Education of Young Children.

Clements, D. H. (2007). Curriculum Research: Toward a Framework for "Research-based Curricula". *Journal for Research in Mathematics Education*, *38*(1), 35–70.

Clements, D. H., & Sarama, J. (2004). Learning trajectories in mathematics education. *Mathematical Thinking and Learning*, *6*(2), 81–89. doi:10.120715327833mtl0602_1

Code.org. (2018). *2018 Annual Report*. https://code.org/files/annual-report-2018.pdf

Cunha, F., & Heckman, J. (2007). The Technology of Skill Formation. *The American Economic Review*, *97*(2), 31–47. doi:10.1257/aer.97.2.31

Dalbey, J., & Linn, M. C. (1985). The demands and requirements of computer programming: A literature review. *Journal of Educational Computing Research*, *1*(3), 253–274. doi:10.2190/BC76-8479-YM0X-7FUA

de Strulle, A., & Shen, C. (n.d.). *STEM + Computing K-12 Education (STEM+C)*. https://wwwnsf.gov/funding/pgm_summ.jsp?pims_id=505006

DiSessa, A. A. (2000). *Changing minds: Computers, learning, and literacy*. MIT Press. doi:10.7551/mitpress/1786.001.0001

Elkin, M., Sullivan, A., & Bers, M. U. (2016). Programming with the KIBO Robotics Kit in Preschool Classrooms. *Computers in the Schools*, *33*(3), 169–186. doi:10.1080/07380569.2016.1216251

Erete, S., Martin, C. K., & Pinkard, N. (2017). Digital Youth Divas: A program model for increasing knowledge, confidence, and perceptions of fit in STEM amongst black and brown middle school girls. In Moving students of color from consumers to producers of technology (pp. 152-173). IGI Global. doi:10.4018/978-1-5225-2005-4.ch008

Fayer, S., Lacey, A., & Watson, A. (2017). *BLS Spotlight on Statistics: STEM Occupations-Past, Present, and Future*. U.S. Department of Labor, Bureau of Labor Statistics.

Fedorenko, E., Ivanova, A., Dhamala, R., & Bers, M. U. (2019). The Language of Programming: A Cognitive Perspective. *Trends in Cognitive Sciences*, *23*(7), 525–528. doi:10.1016/j.tics.2019.04.010 PMID:31153775

Gadanidis, G. (2017). Five affordances of computational thinking to support elementary mathematics education. *Journal of Computers in Mathematics and Science Teaching*, *36*(2), 143–151.

Grover, S., & Pea, R. (2013). Computational Thinking in K–12: A Review of the State of the Field. *Educational Researcher*, *42*(1), 38–43. doi:10.3102/0013189X12463051

Guzdial, M. (2008). Education: Paving the way for computational thinking. *Communications of the ACM*, *51*(8), 25–27. doi:10.1145/1378704.1378713

Guzdial, M., & Morrison, B. (2016). Seeking to making computing education as available as mathematics or science education. *Communications of the ACM*, *59*(11), 31–33. doi:10.1145/3000612

Heckman, J., & Masterov, D. (2007). The Productivity Argument for Investing in Young Children. *Review of Agricultural Economics*, *29*(3), 446–493. doi:10.1111/j.1467-9353.2007.00359.x

Hubwieser, P., Armoni, M., Giannakos, M. N., & Mittermeir, R. T. (2014). Perspectives and Visions of Computer Science Education in Primary and Secondary (K-12) Schools. *ACM Transactions on Computing Education, 14*(2).

Janveau-Brennan, G., & Markovits, H. (1999). The Development of Reasoning with Causal Conditionals. *Developmental Psychology*, *35*(4), 904–911. doi:10.1037/0012-1649.35.4.904 PMID:10442860

Jenkins, T. (2002). *On the difficulty of learning to program.* https://www.psy.gla.ac.uk/~steve/localed/jenkins.html

K-12 Computer Science Framework Steering Committee. (2016). *K–12 computer science framework.* https://k12cs.org

Kafai, Y. B., & Resnick, M. (1996). *Constructionism in practice: Designing, thinking, and learning in a digital world.* Erlbaum.

Kramer, J. (2007). Is abstraction the key to computing? *Communications of the ACM*, *50*(4), 36–42.

Lockwood, J., & Mooney, A. (2018). Computational thinking in education: Where does it fit? A systematic literary review. *International Journal of Computer Science Education in Schools*, *2*(1), 41–60.

Madill, H., Campbell, R. G., Cullen, D. M., Armour, M. A., Einsiedel, A. A., Ciccocioppo, A. L., & Coffin, W. L. (2007). Developing career commitment in STEM-related fields: Myth versus reality. In R. J. Burke, M. C. Mattis, & E. Elgar (Eds.), *Women and Minorities in Science, Technology, Engineering and Mathematics: Upping the Numbers* (pp. 210–244). Edward Elgar Publishing.

Markert, L. R. (1996). Gender related to success in science and technology. *The Journal of Technology Studies*, *22*(2), 21–29.

National Research Council. (2011). *Report of a Workshop of Pedagogical Aspects of Computational Thinking.* National Academy Press.

National Research Council. (2012). *A Framework for K-12 Science Education: Practices, Crosscutting Concepts, and Core Ideas. Committee on a Conceptual Framework for New K-12 Science Education Standards. Board on Science Education, Division of Behavioral and Social Sciences and Education.* The National Academies Press.

Papert, S. (1980). *Mindstorms: Children, computers, and powerful ideas.* Basic Books, Inc.

Papert, S. (1987). Computer Criticism vs. Technocentric Thinking. *Educational Researcher*, *16*(1).

Pea, R. D., & Kurland, D. M. (1984). On the cognitive effects of learning computer programming. *New Ideas in Psychology*, *2*, 137–168.

Pei, C., Weintrop, D., & Wilensky, U. (2018). Cultivating computational thinking practices and mathematical habits of mind in lattice land. *Mathematical Thinking and Learning*, *20*(1), 75–89.

Perlis, A. J. (1962). The computer in the university. In M. Greenberger (Ed.), *Computers and the world of the future* (pp. 180–219). MIT Press.

Piaget, J. (1952). *The origins of intelligence in children*. International Universities Press.

Resnick, M. (2017). *Lifelong Kindergarten: Cultivating Creativity through Projects, Passion, Peers, and Play*. MIT Press.

Resnick, M., Maloney, J., Monroy-Hernández, A., Rusk, N., Eastmond, E., Brennan, K., ... Kafai, Y. (2009). Scratch: Programming for all. *Communications of the ACM*, *52*(11), 60–67.

Resnick, M., & Siegel, D. (2015). A Different Approach to Coding. *International Journal of People-Oriented Programming*, *4*(1), 1–4.

STEM Education Act of 2015, House of Representatives 1020, 114th Congress. (2015). https://www.congress.gov/bill/114th-congress/house-bill/1020

Strawhacker, A. L., & Bers, M. U. (2015). "I want my robot to look for food": Comparing children's programming comprehension using tangible, graphical, and hybrid user interfaces. *International Journal of Technology and Design Education*, *25*(3), 293–319.

Sullivan, A. (2019). *Breaking the STEM stereotype: reaching girls in early childhood.* Rowman & Littlefield.

Vee, A. (2013). Understanding computer programming as a literacy. *Literacy in Composition Studies*, *1*(2), 42–64.

Vogel, S., Santo, R., & Ching, D. (2017, March). Visions of computer science education: Unpacking arguments for and projected impacts of CS4All initiatives. In *Proceedings of the 2017 ACM SIGCSE Technical Symposium on Computer Science Education* (pp. 609-614). ACM.

Vygotsky, L. S. (1978). *Mind in society: The Development of higher psychological processes*. Harvard University Press.

Vygotsky, L. S. (1987). Thinking and speech (N. Minick, Trans.). In R. W. Rieber & A. S. Carton (Eds.), The collected works of L. S. Vygotsky (Vol. 1., pp. 39-285). New York: Plenum Press. (Original work published 1934)

Wilson, C., Sudol, L. A., Stephenson, C., & Stehlik, M. (2010). *Running on empty: The failure to teach K-12 computer science in the digital age*. The Association for Computing Machinery and the Computer Science Teachers Association.

Wing, J. (2006). *Computational thinking. Communications of Advancing Computing Machinery, 49 (3), 33-36*. Association for Computing Machinery.

Wing, J. (2011). *Research notebook: Computational thinking—What and why?* https://www.cs.cmu.edu/link/research-notebookcomputational-thinking-what-and-why

Wing, J. M. (2006). Computational thinking. *Communications of the ACM, 49*(3), 33-35. through programming robots in early childhood. *Journal of Educational Computing Research, 50*(4), 553–573.

Wittgenstein, L. (1997). Philosophical Investigations (2nd ed.). Cambridge: Blackwell.

ADDITIONAL READING

Bruner, J. (1983). *Child's Talk: Learning to Use Language*. W. W. Norton & Company.

Lave, J., & Wenger, E. (1991). *Situated learning. Legitimate peripheral participation*. Cambridge University Press. doi:10.1017/CBO9780511815355

Papert, S. (1993). *The children's machine: Rethinking school in the age of the computer*. Basic Books.

Resnick, M. (2017). *Lifelong Kindergarten: Cultivating Creativity through Projects, Passions, Peers, and Play*. MIT Press. doi:10.7551/mitpress/11017.001.0001

Turkle, S. (1984). *The Second Self: Computers and the Human Spirit*. Basic Books.

Vee, A. (2017). *Coding Literacy: How Computer Programming is Changing Writing*. The MIT Press. doi:10.7551/mitpress/10655.001.0001

Vygotsky, L. S. (1978). *Mind in society: The development of higher psychological processes*. Harvard University Press.

KEY TERMS AND DEFINITIONS

Coding: The process or activity of writing computer programs.

Computational Thinking: Techniques applied by humans to express themselves by designing and constructing computation.

Computer Science: The study of the principles and use of computers.

Constructionism: The theory that learning should be done through student-centered discovery.

Early Childhood: Period of time between birth and age eight.

KIBO: A screen-free programmable robotics kit for young children with blocks, sensors, modules, and art platforms.

Learning: The acquisition of knowledge or skills through experience, study or being taught.

ScratchJr: A free block-based programming application for young children.

Chapter 2
Why Teach Coding to Early Elementary Learners

Claudia M. Mihm
Tufts University, USA

ABSTRACT

As coding and computer science become established domains in K-2 education, researchers and educators understand that children are learning more than skills when they learn to code – they are learning a new way of thinking and organizing thought. While these new skills are beneficial to future programming tasks, they also support the development of other crucial skills in early childhood education. This chapter explores the ways that coding supports computational thinking in young children and connects the core concepts of computational thinking to the broader K-2 context.

INTRODUCTION

Steve Jobs once said, "I think everybody in this country should learn how to program a computer…because it teaches you how to think." He said this over twenty years ago, and schools are starting to catch up to his vision (Code.org, 2013). 47% of public high schools offer coding courses, and there were over 1 million teacher and 36 million student accounts on Code.org at the end of 2018, up from 10,000 and 500,000, respectively, in 2013 (Code.org, 2021a; Code.org, 2021b). In the employment realm, computer science is projected to make up 67% of new STEM jobs in the United States by 2028, totaling nearly 600,000 jobs (Code.org, 2021b). However, only 25.2% of people employed in computing fields in 2020 identified as women, and less than 35% were non-white (U.S. Bureau of Labor Statistics, 2021). Even

DOI: 10.4018/978-1-7998-7308-2.ch002

earlier in the computer science career pipeline, only 33% of students who took the AP Computer Science exam were female in 2020, showing that the gender divide is already present in high school (Code.org, 2021b). Despite these trends, research shows that providing students early exposure to computer science helps improve the gender gaps and overall diversity in tech fields (Bers, 2018; Jungert et al., 2018; Sullivan & Bers, 2016). By providing students with early exposure to computational thinking, schools can create opportunities for them to enter this growing field.

Beyond career preparedness, there are several reasons why it is important for early elementary students to learn how to code. Early coding exposure can help attract students from historically underrepresented backgrounds to engage in STEM, by building comfort with engineering concepts and helping young children, and especially girls, to picture themselves as future engineers and software developers (Jungert et al., 2018; Sullivan & Bers, 2018). Additionally, engaging with computers can help children develop a deeper understanding of their own learning, by asking them to examine their thought processes as they are instructing the computer (Papert, 1980). Studying computer science has also been shown to improve elementary student's performance in other subjects and strengthen their problem-solving abilities (Code.org, 2020). Finally, learning to code builds up computational thinking skills, which are transferrable far beyond programming, and include practices that are especially crucial for early elementary students (Bers, 2018; Yadav et al., 2016). I will spend this chapter exploring how to teach computational thinking to young children through programming and discuss the importance of these skills outside the realm of code.

Background

Before continuing to explore computational thinking, I will establish what I mean by programming. I came to understand how young children can code in my time working at the DevTech research group, when I was an undergraduate student at Tufts University. I was majoring in Computer Science, so I was learning conventional, text-based programming languages like C++. While the K-2 students I was teaching would have needed significant help to use C++ to program anything, they were able to create complex programs using developmentally appropriate tools created at DevTech: ScratchJr and KIBO. ScratchJr is an iPad app where users program with image-based blocks (see Figure 1), and KIBO is a robotics kit where users give instructions to a robot by scanning wooden blocks which represent different actions (see Figure 2) (Portelance et al., 2015; Sullivan et al., 2017). Programming at its core is the act of giving instructions to a computer or computational device in a form that both the user and the computer can understand. Both ScratchJr and KIBO facilitate this through picture-based blocks, which can be put together into sequences to make the on-screen characters or robot execute different actions (Portelance et al., 2015; Sullivan et al., 2017).

Figure 1. A ScratchJr screen displaying a cat character and movement blocks

While the tools are different, both the software engineer and the student are engaging in the same cognitive processes. Both use the tools available to go through an iterative problem-solving process, leveraging the power of technology and a programming language, to create a solution. While ScratchJr and KIBO remove certain barriers to conventional programming, such as literacy requirements, syntax errors and top-to-bottom sequencing, they still provide access to the core fundamentals of computer science and computational thinking – such as sequencing, looping, debugging, and more. As I will demonstrate throughout this chapter, these tools facilitate mastery of computational thinking concepts, just as traditionally designed programming languages do. I will explore each powerful idea of computational thinking, and how it can be taught by using either KIBO or ScratchJr. As someone who has a degree in computer science and works in the technology field, it has always been important to me that the ways I teach computational thinking are aligned with the profession and can set students up to continue succeeding in computational activities. As I explore each idea, I will pull from my own experience teaching these ideas, and also share the broader relevance of the idea for foundational skill building, beyond becoming an expert programmer.

Figure 2. KIBO robot with programming blocks and accessories

Powerful Ideas of Computational Thinking

As Bers explains in the first chapter of this book, computational thinking is a way of thinking that grows out of computer science. At its core, it is a collection of problem-solving and analytic skills, grounded in practices necessary to program successfully, but that have far-reaching applications. Bers' seven powerful ideas of computational thinking help to illustrate the core concepts that students learn when they learn to code, with a particular focus on developmentally relevant skills for early childhood learners (Bers, 2018). The powerful ideas are: Algorithms, Modularity, Control Structures, Representation, Hardware & Software, Design Process and Debugging (Bers, 2018). These powerful ideas illustrate skills that are not only important for mastering programming, but also for navigating the challenges of 21st century, with its constantly evolving technological landscape, and that are particularly useful for early elementary learners. By developing these seven areas, learners are equipping themselves with intellectual tools that have broad applications – from developing literacy to problem-solving skills, and much in between (Hassenfeld & Bers, 2020; Kewalramani et al., 2016; Wilson-Lopez et al., 2017).

The seven powerful ideas, rooted in learning to program, have historically been challenging to teach to young learners. As with LOGO, Scratch, and other programming languages designed for elementary and middle-grade children, ScratchJr and KIBO are programming languages that help younger students engage with powerful ideas of computational thinking in developmentally appropriate ways (Flannery et al., 2013; Papert, 1980; Portelance et al., 2015; Sullivan et al., 2017). Previously, programming – even through block-based platforms like Scratch – was inaccessible to young children, due to the baseline of motor and cognitive abilities necessary to utilize those platforms (Flannery et al., 2013). Scratch, for instance, requires students to be able to read the text on the blocks, use a mouse to move the blocks around a screen, and understand mathematical concepts like degrees and percentages. Recently, there have been new approaches to programming technology that remove those barriers, allowing access for early elementary students (Jacobson, 2016; Portelance et al., 2015; Sullivan et al., 2017). By engaging with these tools, designed with their motor and cognitive abilities in mind, young learners have access to a whole realm of knowledge that has previously been inaccessible. The value of this is not just to get them started earlier as coders, but also to give them access to a whole new set of skills that may serve useful to them in other areas of learning. In this chapter, I will explore how we teach these powerful ideas of computational thinking through KIBO and ScratchJr, illustrate strategies and approaches that facilitate computational thinking, and discuss why it is important to start in early childhood.

TEACHING THE POWERFUL IDEAS OF COMPUTATIONAL THINKING

Algorithms

Bers (2018) defines algorithms as “A series of ordered instructional steps taken in a sequence to solve a problem or achieve an end goal.” By looking at this definition, it becomes clear that the term algorithm, often attributed to the world of engineering and mathematics, is simply another word for a sequential plan. We create algorithms every day, from making a sandwich to planning a walking route. Learning to program is a concrete way to practice building algorithms, given the logical structure and clear steps required to instruct a non-sentient machine to carry out a task. ScratchJr and KIBO are especially helpful tools for young learners to master planning because the blocks allow for quick iteration, a visual representation of the steps your program is taking, and a structured environment in which to execute on a plan.

One successful way that I have found to teach algorithms to young learners is to ask them to create a story in ScratchJr. While there are many ways to approach creating algorithms, I have found that asking students to tell a story helps them build personal connection and relevance. To create the algorithm, they first have to think of the story they would like to tell, and the conclusion it will reach. By doing this, it creates a foundation from which to build the algorithm. In my experience, if there is not a clear end goal it can be incredibly challenging to create the story – and thus craft an algorithm – without it petering out or continuing beyond the scope of a meaningful narrative.

Once they have decided on their plot, they must identify the different characters in the story, and the actions that the characters must take. Only then do I ask them to start building a sequence, by articulating the order in which these actions take place, as not everything can happen at once. I often ask the students to write down or speak the order of actions, and who is performing them, to ensure that they are creating a structured sequence. By articulating each step, they are learning how to break something whole – a story – down into component parts, as well as how to put those parts together into an ordered sequence. By nature of building this story in ScratchJr, the sequence that they design is reinforced by the block-based, ordered structure inherent to the platform.

Learning to sequence is an important skill supported by developing algorithms – if you don't establish sequence, your algorithm is just an unordered collection. Sequencing is not just important for developing algorithms, though. It has applications in reading comprehension – understanding the order that events take place in a story – writing, and numeracy (Chase et al., 2014). Sequencing is included in reading and mathematics curricular frameworks for kindergarten and first grade in several states (e.g. Massachusetts, California) and in the US Common Core framework (California State Board of Education, 2013; Massachusetts Department of Elementary and Secondary Education, 2017; National Governors Association Center for Best Practices, Council of Chief State School Officers, 2010a). Research has found that engaging with block-based programming also improves early childhood learner's scores on evaluations of story-based sequencing ability (Kazakoff et al., 2013). The evaluations were not tied to programming knowledge, but rather to sequencing knowledge more broadly. This indicates that while learning sequencing in the context of programming, young learners are developing skills that will aid them in reading comprehension and writing production.

Creating algorithms also provides opportunities to practice planning. By learning how to move from a high-level goal (tell a story) to a step-by-step plan, students are learning to break down goals into achievable steps, to understand how individual components work together to create a whole, and to grasp order and sequence. Whether learning to read, write a story, plan their day, or tackle a word problem, understanding how to create and act on algorithms is something that students will use with throughout their lives.

Modularity

Modularity is defined as "The breaking down of tasks or procedures into simpler, manageable units that can be combined or re-used to create a more complex process" (Bers, 2018). Modularity focuses on breaking a problem down into achievable tasks, and often takes it one step further by repurposing the practice to achieve something new. In text-based programming languages, functions are a great example of modularity. It is common practice to package a task as a function, which can be referenced throughout the program. For example, a "print" function will initiate a series of smaller sub-tasks to print whatever input it is given on the screen, and can be used throughout a program without having to write the same sub-task code again.

Let's look at an example with ScratchJr: say a learner wants to have their character move diagonally across the screen. There is no "diagonal" block in ScratchJr, so they will need to combine multiple blocks in order to create the action – in other words, they will need to create a function. In order to create the function, students must understand that a diagonal move can be broken down to moving right and moving up at the same time. Through trial and error, they will eventually learn that they need to have two parallel programs, or programs that run at the same time, to achieve a straight diagonal line. One program instructs their character to move right, and the other instructs their character to move up, starting at the same time (see Figure 3). In order to create this diagonal function, they breaking down a complex task – moving diagonally – into two component parts – moving right, and moving up. They are then repurposing their knowledge of the move right and move up blocks to create a new action. In order to teach modularity through ScratchJr, I like to make sure students are grounded in the ScratchJr blocks they can use– if there is not a block for it, then they'll need to break the task down another level. Similarly, if they have already solved the problem somewhere else in the program, it can be helpful to ask them if they've created a solution with blocks elsewhere already.

Figure 3. ScratchJr blocks demonstrating parallel programs to make a character move diagonally

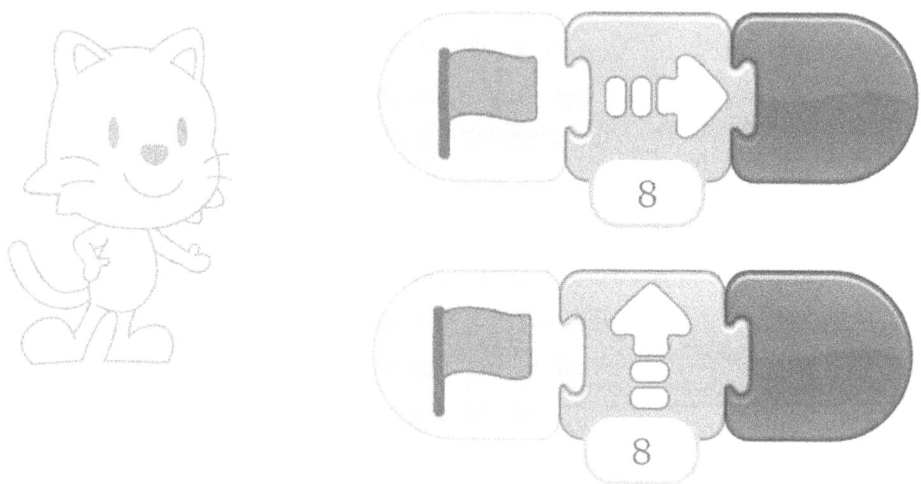

An example of modularity outside of programming is understanding base-ten number concepts, including number decomposition. Number decomposition is a primary focus in K-2 Common Core Mathematics guidelines – for example, the understanding that "the numbers from 11 to 19 are composed of a ten and one, two, three, four, five, six, seven, eight, or nine ones," or "the numbers 10, 20, 30, 40, 50, 60, 70, 80, 90 refer to one, two, three, four, five, six, seven, eight, or nine tens (and 0 ones)" (National Governors Association Center for Best Practices, Council of Chief State School Officers, 2010b). As both of these examples show, it is not just a matter of breaking down a number into units of one, but understanding that both one and ten are units that all numbers can be broken down to. Just as students decompose a diagonal move until it takes the form of familiar blocks – in this case, 'Move Right' and 'Move Up' – they can apply the same approach decomposing a number into a collection of ones and tens. By equipping them with the tangible experience of breaking a task down, thinking modularly helps expose them to an important approach to knowledge they will use for the rest of their lives.

Control Structures

"Control structures determine the order (or sequence) in which instructions are followed or executed within an algorithm or program" (Bers, 2018). Control structures are some of the most complex programming features supported by KIBO and ScratchJr, and so many teachers and curricula will leave them as some of the last blocks to introduce. However, they are incredibly powerful, and open up new ways to create and express oneself when they are mastered.

In KIBO, control structures take the form of blocks, such as repeat blocks, conditionals, and events (e.g., “start” blocks). They play an important role in sequencing, as they disrupt the otherwise linear sequencing pattern that KIBO adopts. To teach these concepts, it can be helpful to return back to the algorithm. Going back to the idea of creating a story, which can be just as easily done with KIBO, if a child wants to code a character to take the same action more than once (say, they pace back and forth three times), then that can be a great moment to introduce control structures, specifically the repeat block. While you do not necessarily need the repeat block in this case – you can just use the same blocks three times in a row – it simplifies the action, and allows the learner to practice using control structures.

Similarly, learning to leverage control structures can lead to more complex programs. Take the case of a student who wants to program their dog to bump into their cat and have the cat say “Ow!” when they are bumped. Their first instinct will probably be to use a “Wait” block to try to have the cat wait enough time to let the dog move to them. I have seen many do this, and while it is one solution, it is tedious and requires a lot of trial and error, especially if the child later decides to add something to the project that alters the timing (for example, dog saying “Hello!” before walking). It also does not actually mimic the interactivity that the original story implies, since the cat and the dog are operating completely independently within the program. It would make much more sense to say “when the cat is bumped by the dog, then have the cat say ‘Ow!’.” There is, in fact, an event block in ScratchJr called “Start on Bump” that allows this exact interaction (see Figure 4). Event blocks allow for interrelated programs, which opens up the possibility for the logical sequencing of events. This can also be challenging to teach, but I’ve found it helpful to act out the event blocks with people – having two students (gently!) reenact the scene helps them understand the role of the block. In addition, this is another example where grounding it in a story helps students grasp the concept. Because they are able to act it out and understand what should happen, it gives them a stronger foundation for translating actions to code.

Figure 4. A ScratchJr program showcasing the "Start on Bump" block

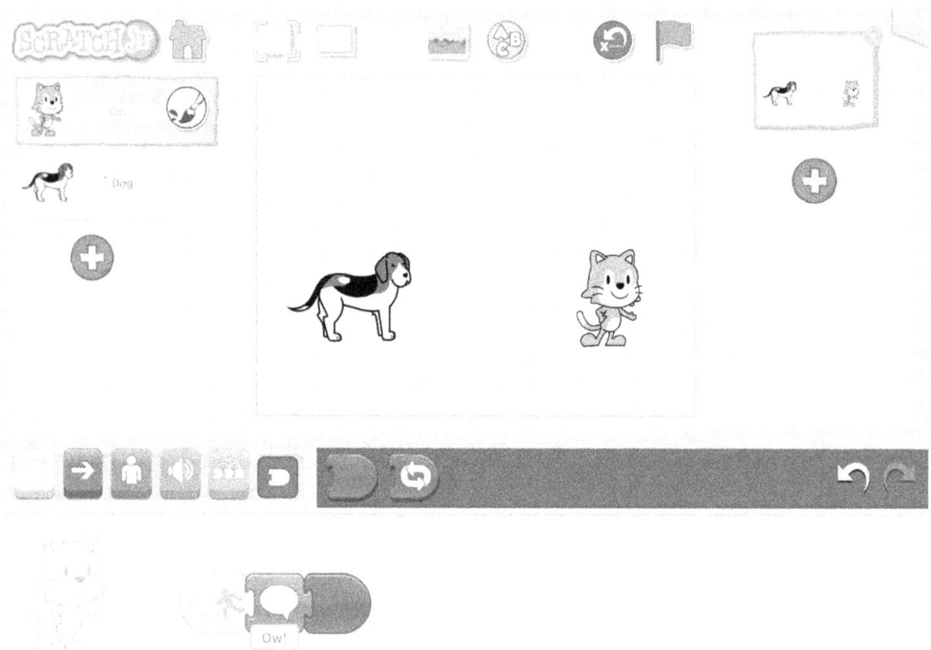

Control structures are relevant outside of computing because they help students visualize cause and effect. In the example with the dog and the cat, we can see the contrast between the first and second solutions, and how they can help demonstrate cause and effect. In the first solution, the student uses a "Wait" block, which means there is no relationship between the dog's actions and the cat saying "Ow!" However, in the second solution, there is a relationship between the two, because the cat will only say "Ow!" (the effect) when it is bumped by the dog (the cause). While understanding cause and effect has relevance in many subjects, such as understanding the plot of a story or color mixing in art, it is also a foundational skill that has implications for future science learning. Cause and effect is listed as a cross-cutting subject in the *Next Generation Science Standards*, indicating its importance as a foundational science skill, and appears in the standards as early as kindergarten (NGSS Lead States, 2013). While there has been evidence to show that younger children can master cause and effect in specific circumstances, despite developmental claims that they are not able to, it is still a challenging concept to grasp and teach (Goddu et al., 2020; Springer & Keil, 1991). By engaging with control structures in a visible way, where you can immediately see the output of an action, children have the opportunity to build their understanding of cause and effect in a visual environment, with a finite amount of possible causes and possible effects. When children grasp control structures, they are opened up to a whole new way of looking at programming and logical thinking.

Representation

Bers (2018) introduces representation by saying "programming languages represent information through the use of a symbol system." With ScratchJr and KIBO, as with all programming languages, symbols are used to represent instructions to the tool. KIBO has blocks that represent movement (e.g., move forward, turn right, etc.), sound (sing), colors (light up blue), and more. These blocks, when scanned, translate into an action that the robot takes.

To teach representation, I ask the students to explain or show me what a block tells the machine to do. For example, I would hand them a "Move Forward" block and ask them what it does. Often, I will ask them to act it out with their body, to make sure they really understand what action the block represents. In order to teach the connection between the symbol (the block) and its meaning (the outcome), I encourage students to test out what a block does. Because KIBO has such visual outputs, it is easier for the programmers to see the connection – they can scan the block, and see what the output is. Through this practice, they learn to connect the block to the corresponding KIBO action that it represents.

Learning KIBO blocks helps to build a mental model of representation, which can be hugely beneficial when it comes to equipping students with skills to express themselves. Tinkering with language and storytelling gives children new, novel ways to express themselves, and can expand the tools they have for communication (Maureen, et al., 2020). By learning a new representative system and new storytelling tool, children are given new ways to communicate their ideas. As we see with some young people preferring to communicate via spoken word, written work, drawing, etc., introducing programming as a novel form of communication allows for creative forms of expression, collaboration and meaning-making.

Additionally, symbolic representation is the foundation of building any sort of human communication system – be that a programming language or a natural spoken and written one (Jones, et al., 2012). Numbers and letters, like KIBO blocks, are objects that represent a larger concept. Building a strong foundation of symbolic understanding is crucial for future literacy and numeracy endeavors (Berninger et al., 2002). Therefore, helping students understand how to connect symbols to what they represent to others in their society is a hugely important skill in developing literacy. Understanding the connection between a KIBO block and a KIBO movement is a similar practice to understanding the connection between a letter and the sound it represents. By learning with KIBO, where they can test out the outcome of a symbol as many times as they need, coding can be a helpful way to build an understanding of representation.

Hardware and Software

Understanding hardware and software, and the relationship between them, is crucial for anyone navigating our increasingly digital world. As Bers (2018) states, "Computing systems need hardware and software to operate. The software provides instructions to the hardware, which might or might not be visible. Hardware and software work together as a system to accomplish tasks, such as receiving, processing, and sending information." KIBO is an incredibly powerful tool for helping illustrate the relationship between hardware and software, because of the tangible nature of the software. With KIBO, the blocks provide the instructions to the robot, and thus serve as the software. Users can see the hardware of the robot by looking at the clear bottom, which reveals the wires and other internal workings of how the robot's hardware operates.

To introduce the terms of hardware and software, I like to use the metaphor of the body and the brain. While this is an oversimplification, it helps to ground abstract mechanical relationships in something the students understand. In this metaphor, the brain is the software, giving instructions to your body, which acts on those instructions, making it the hardware. To translate this relationship to KIBO, I ask the learner to press the "Go" button on KIBO. Nothing happens, because we have not yet given it any instructions – so they start to understand that the hardware is not functional without the software. I then ask them to pick a KIBO block and tell me how to use it, which usually leads to an explanation that you have to scan the block in order to use it. This illustrates the opposite point, that software is not useful unless it has hardware to execute the commands. Especially because KIBO is a tangible tool, this is a helpful way to start the conversation about software and hardware.

We often hear people assume that because young learners are "digital natives," things like hardware and software are concepts that they innately understand (Prensky, 2001). However, while young children might have a level of comfort and familiarity with a variety of technological devices, this does not necessarily mean that they have a deeper understanding of how they actually work. It is the difference between knowing *what* you can do with your tablet and actually understanding *how* your tablet works. Familiarity with the interaction between hardware and software can serve as a powerful foundation to create interesting, meaningful artifacts, and to shift from being a consumer of technology to a creator with technology (Kafai, Fields, & Searle, 2014).

It is important to teach the relationship between hardware and software, in order to help students begin to view technology as a human-made object, rather than a magical one. The clear base on KIBO helps them to see that it is in fact a collection of hardware that makes up a robot, rather than a mysterious collection of things they would never hope to understand. This process leads to an improved relationship with any sort of technology in their life, and an increased ability to do more interesting things with it. Additionally, our workforce is requiring increasingly higher levels of digital abilities and understanding – as of 2016, an analysis of 545 occupations in the U.S. revealed that 23% of the country's jobs required high digital ability, and 48% required medium digital ability (Muro et al., 2017). By pairing use of technology with a deep understanding of how it works, we are equipping young learners with the ability to meaningfully engage with and understand the technologies they will need to use in the future.

Design Process

While there are many different definitions of a design process, the general sentiment is the same. It is an iterative process that helps move from concept to implementation, and then cycles through the same steps making improvements. Bers (2018) outlines the steps as: "ask, imagine, plan, create, test, improve, and share. The process is open-ended, in that a problem may have many possible solutions." Below, I explain each step more.

Ask: you must define a question that you want to answer, or a problem that you want to solve.

Imagine: in order to answer your question or challenge, you must create multiple potential solutions to the question that you are asking.

Plan: you move from high-level concept to executable plan, which means getting more specific about how you will use the tools at your disposal to build the solution. The plan stage is when you create an algorithm.

Create: in this step, you actually start building. This is not the end of the design process – while creation is of course a key step, the important iteration happens after the first round of creation is completed.

Test: once you have created something, you must test if it solves the initial question you asked yourself in the first step.

Improve: you iterate on your design based on the results of your testing. This also requires going back to your plan, and adapting it based on what you have learned.

Share: you share your solution and process with others. Sharing is a great way to synthesize your learnings, get more feedback, and learn from others as they share their creations with you.

An important part of all design processes is that they are not linear. While there are certain steps to move through, a designer will probably loop through them multiple times and in different orders. It is helpful to understand the steps of the design process before beginning, because then there is an established vocabulary for referring to the steps that you are moving through. It also introduces the concept of iteration.

To teach the design process, it is important to actually name the steps of it. At the DevTech research group, we had a poster that illustrates the process. This helps to provide a shared vocabulary at the beginning, that can be referenced back to. For instance, a student I worked with was particularly stuck in the creation phase, and she kept losing sight of what she was trying to build. In the planning phase, we had discussed the story that she was trying to tell and mapped out the different characters and events needed. By revisiting the plan she had put together, she was able to remind herself of the vision that she had already created. Because we had a shared vocabulary, and understanding of a progression of steps, it was easier to discover what her challenge was, and land on a solution.

Learning the design process has been shown to build similar strategies to reading comprehension (Wilson-Lopez et al., 2017). By becoming familiar with the design process, students also are developing problem-solving strategies that will be useful to them in other disciplines. Another important facet of the design process is iteration. By learning to iterate, one builds comfort with the idea that they are not going to get it right on the first try. With this approach, failure to achieve a goal becomes not a failure, but a step in the right direction, and an opportunity to try again with a new approach. Through programming, students develop emotional intelligence through overcoming their fear of something not working, which can help build a willingness to continue through adversity (Kewalramani et al., 2016). While this mental shift is crucially important for learning to program, iteration is also an important process outside the realm of programming – any product benefits from multiple rounds of design, creation, and feedback. Learning to go through the design process with a structured tool like ScratchJr or KIBO can help students internalize the process and apply it to other undertakings – such as writing a story, building a block tower, or solving a math problem.

Debugging

As Bers (2018) defines it, debugging is: “Fixing problems through systematic analysis and evaluation, while developing troubleshooting strategies.” As you can see from this definition, while debugging is often relegated to the world of programming, it is really about creating practices for problem solving. Teaching debugging inherently comes with teaching programming, as problems will always arise, especially if students are new to programming. It is especially helpful if the programmer has made a plan – such as an algorithm – because that provides a foundation of what you are hoping to achieve, which can be compared to what the program is actually doing. When a student's program is not doing what is expected, I ask them to start by comparing the plan of what they want the program to be doing with what it is actually doing. The goal is to build up a strategy for understanding where things went wrong, not just that it is not working. One of the strengths of tools like ScratchJr and KIBO is that they allow for immediate output and rapid tweaking. So, if someone is stuck, they do not just have to think about what is wrong. They can actually start changing the program and see what changes.

One of my favorite debugging strategies is to take the sequence apart, and test it out step by step. For instance, if you are trying to navigate KIBO through a maze, you take all the blocks off your sequence but the first one. Then, you run KIBO to make sure that the first block is achieving what you want it to do. You can then continue to add your blocks back into your sequence, systematically working until you identify where it is going wrong. This is not only an effective strategy for this particular instance, but also helps demonstrate a methodical approach to problem solving more broadly. Rather than giving up, or starting completely over, it models a step-by-step problem-solving approach that can be used in other challenging situations.

Students have shown more comfort with debugging in programming than in writing endeavors – in other words, they are more comfortable with editing their program than with editing their written work (Hassenfeld & Bers, 2020). This has exciting implications for building up student's comfort with editing and revision in other areas of their life. Additionally, Seymour Papert writes about the power of programming to teach metacognition, and debugging is a wonderful example of this (Papert, 1980). By going through debugging processes, learners are finding the gaps in their own knowledge or understanding of their programming language. This metacognition serves to make them stronger learners, and more self-aware people (Erdmann & Hertel, 2019). Metacognition is not just useful for debugging – their awareness of their learning approaches is not just tied to programming languages, and thus can help them become stronger learners across domains.

CONCLUSION

In this chapter, I show how the benefit of learning to code at an early age far surpasses any programming skills that may be mastered. Learning to code teaches computational thinking skills, which in turn, have wide-ranging benefits for students in K-2 classrooms. Computational thinking is not only a helpful cognitive tool, but a necessary one to equip our young learners with the skills they need to tackle the challenges of the 21st Century. The seven powerful ideas developed by Bers and further explored in this chapter, important on their own, boil down to an understanding of creative problem solving and digital competency that can set students up for success as they move forward.

It is my hope that in the coming years, we will see an incorporation of programming into classroom activities. Rather than a separate STEM lesson, I have seen coding bring a reading or science lesson to life in new ways, engaging learners who felt isolated or allowing others a new way to understand a concept. Learning to program will be most powerful if they are aligned with core teaching goals and incorporated into classroom routines. In short, in order to best leverage the strengths of computational thinking skills, we must incorporate computational thinking activities into the core of our classrooms. Computational thinking is a powerful approach to thinking, and the earlier we can expose our students to it, the better served they will be.

REFERENCES

An, S., Tinajero, J., Tillman, D., & Kim, S. (2019). Preservice Teachers' Development of Literacy-Themed Mathematics Instruction for Early Childhood Classrooms. *International Journal of Early Childhood, 51*(1), 41–57. doi:10.100713158-019-00232-9

Barr, V., & Stephenson, C. (2011). Bringing computational thinking to K-12. *ACM Inroads, 2*(1), 48–54. doi:10.1145/1929887.1929905

Berninger, V. W., Abbott, R. D., Vermeulen, K., Ogier, S., Brooksher, R., Zook, D., & Lemos, Z. (2002). Comparison of Faster and Slower Responders to Early Intervention in Reading: Differentiating Features of Their Language Profile. *Learning Disability Quarterly, 25*(1), 59–76. doi:10.2307/1511191

Bers, M. U. (2018). *Coding as a Playground: Programming and Computational Thinking in the Early Childhood Classroom*. Routledge Press.

California State Board of Education. (2013). *California Common Core State Standards: English Language Arts & Literacy in History/Social Studies, Science, and Technical Subjects*. Retrieved from California Department of Education: https://www.cde.ca.gov/

Chase, M., Son, E. H., & Steiner, S. (2014). Sequencing and Graphic Novels with Primary-Grade Students. *The Reading Teacher, 67*(6), 435–443. doi:10.1002/trtr.1242

Code.org. (2013). *Steve Jobs on Computer Science*. Academic Press.

Code.org. (2020a, April 15). *CS helps students outperform in school, college, and workplace*. codeorg.medium.com

Code.org. (2020b). *CS helps students outperform in school, college, and workplace*. Retrieved from codeorg.medium.com

Code.org. (2021a). *Code.org Statistics*. Retrieved from Code.org: code.org/statistics

Code.org. (2021b). *Why Computer Science?* Retrieved from code.org: code.org/promote

Erdmann, K. A., & Hertel, S. (2019). Self-regulation and co-regulation in early childhood – development, assessment and supporting factors. *Metacognition and Learning*, *14*(3), 229–238. doi:10.100711409-019-09211-w

Flannery, L. P., Silverman, B., Kazakoff, E. R., Bers, M. U., Bontá, P., & Resnick, M. (2013). Designing ScratchJr: support for early childhood learning through computer programming. In *Proceedings of the 12th International Conference on Interaction Design and Children*. New York, NY: Association for Computing Machinery. 10.1145/2485760.2485785

Goddu, M. K., Lombrozo, T., & Gopnik, A. (2020). Transformations and Transfer: Preschool Children Understand Abstract Relations and Reason Analogically in a Causal Task. *Child Development*, *91*(6), 1898–1915. doi:10.1111/cdev.13412 PMID:32880903

Hassenfeld, Z. R., & Bers, M. U. (2020). Debugging the Writing Process: Lessons From a Comparison of Students' Coding and Writing Practices. *The Reading Teacher*, *73*(6), 735–746. doi:10.1002/trtr.1885

Jacobson, L. (2016). The Codemakers: J is for Javascript. *School Library Journal*, *62*(4).

Jones, C. D., Clark, S. K., & Reutzel, D. (2012). Enhancing Alphabet Knowledge Instruction: Research Implications and Practical Strategies for Early Childhood Educators. *Early Childhood Education*, *41*(2), 81–89. doi:10.100710643-012-0534-9

Jungert, T., Hubbard, K., Dedic, H., & Rosenfield, S. (2018). Systemizing and the gender gap: Examining academic achievement and perseverance in STEM. *European Journal of Psychology of Education*, 479–500.

Kafai, Y. B., Fields, D. A., & Searle, K. A. (2014). Electronic Textiles as Disruptive Designs: Supporting and Challenging Maker Activities in Schools. *Harvard Educational Review*, *84*(4), 532–557. doi:10.17763/haer.84.4.46m7372370214783

Kazakoff, E., Sullivan, A., & Bers, M. U. (2013). The effect of a classroom-based intensive robotics and programming workshop on sequencing ability in early childhood. *Early Childhood Education Journal*, *41*(4), 245–255. doi:10.100710643-012-0554-5

Kewalramani, S., Palaiologou, I., & Dardanou, M. (2016). Children's Engineering Design Thinking Processes: The Magic of the ROBOTS and the Power of BLOCKS (Electronics). *Eurasia Journal of Mathematics, Science and Technology Education*, *16*(3). Advance online publication. doi:10.29333/ejmste/113247

Massachusetts Department of Elementary and Secondary Education. (2017). *English Language Arts and Literacy*. Retrieved from Massachusetts Department of Education: https://www.doe.mass.edu/

Maureen, I. Y., van der Meij, H., & de Jong, T. (2020). Enhancing Storytelling Activities to Support Early (Digital) Literacy Development in Early Childhood Education. *International Journal of Early Childhood*, *52*(1), 55–76. doi:10.100713158-020-00263-7

Muro, M., Liu, S., Whiton, J., & Kulkarni, S. (2017). *Digitalization and the American workforce*. Brookings Institute.

National Governors Association Center for Best Practices, Council of Chief State School Officers. (2010a). *Common Core State Standards: English Language Arts Standards: Writing, Grade 1*. Washington, DC: National Governors Association Center for Best Practices, Council of Chief State School Officers. Retrieved from Common Core State Standards Initiative: http://www.corestandards.org/

National Governors Association Center for Best Practices, Council of Chief State School Officers. (2010b). Common Core State Standards: Mathematics Standards: Number & Operations in Base Ten, Grade 1. National Governors Association Center for Best Practices, Council of Chief State School Officers.

NGSS Lead States. (2013). *Next Generation Science Standards: For States By States*. Author.

Papert, S. (1980). *Mindstorms: Children, computers and powerful ideas*. Basic Books.

Portelance, D. J., Strawhacker, A., & Bers, M. U. (2015). Constructing the ScratchJr programming language in the early childhood classroom. *International Journal of Technology and Design Education*, •••, 1–16.

Prensky, M. (2001). Digital Natives, Digital Immigrants Part 2: Do They Really Think Differently? *On the Horizon*, *9*(6), 1–6. doi:10.1108/10748120110424843

Springer, K., & Keil, F. (1991). Early Differentiation of Causal Mechanisms Appropriate to Biological and Nonbiological Kinds. *Child Development*, *62*(4), 767–781. doi:10.2307/1131176 PMID:1935342

Sullivan, A., & Bers, M. U. (2016). Girls, boys, and bots: Gender differences in young children's performance on robotics and programming tasks. *Journal of Information Technology Education: Innovations in Practice*, *15*, 145–165. doi:10.28945/3547

Sullivan, A., & Bers, M. U. (2018). Investigating the use of robotics to increase girls' interest in engineering during early elementary school. *International Journal of Technology and Design Education*, *29*(5), 1033–1051. doi:10.100710798-018-9483-y

Sullivan, A. A., Bers, M. U., & Mihm, C. (2017). Imagining, Playing, and Coding with KIBO: Using Robotics to Foster Computational Thinking in Young Children. *Proceedings of the International Conference on Computational Thinking*.

U.S. Bureau of Labor Statistics. (2021). *Labor Force Statistics from the Current Population Survey CPS CPS Program Links*. Author.

Wilson-Lopez, A., Larsen, V., & Gregory, S. (2017). Reading and Engineering: Elementary Students' Co-Application of Comprehension Strategies and Engineering Design Processes. *Journal of Pre-College Engineering Education Research*, *6*(2), 39–57. doi:10.7771/2157-9288.1116

Yadav, A., Hong, H., & Stephenson, C. (2016). Computational Thinking for All: Pedagogical Approaches to Embedding 21st Century Problem Solving in K-12 Classrooms. *TechTrends*, *60*(6), 565–568. doi:10.100711528-016-0087-7

ADDITIONAL READING

Barr, V., & Stephenson, C. (2011). Bringing computational thinking to K-12. *ACM Inroads*, *2*(1), 48–54. doi:10.1145/1929887.1929905

Bers, M. U., & Sullivan, A. (2019). Computer science education in early childhood: The case of ScratchJr. *Journal of Information Technology Education: Innovations in Practice*, *18*, 113–138.

Education Design Center. (2020). Broadening Participation of Elementary Students and Teachers in Computer Science. Retrieved from https://www.edc.org/broadening-participation-elementary-students-and-teachers-computer-science

Elkin, M., Sullivan, A., & Bers, M. U. (2018). Books, Butterflies, and 'Bots: Integrating Engineering and Robotics into Early Childhood Curricula. In Early Engineering Learning (pp. 225-248). Singapore: Springer Singapore.

Grover, S., & Pea, R. (2013). Computational Thinking in K—12: A Review of the State of the Field. *Educational Researcher*, *42*(1), 38–43. doi:10.3102/0013189X12463051

Resnick, M., & Robinson, K. (2017). *Lifelong kindergarten (The MIT Press)*. The MIT Press. doi:10.7551/mitpress/11017.001.0001

Sáez-López, J., Román-González, M., & Vázquez-Cano, E. (2016). Visual programming languages integrated across the curriculum in elementary school: A two year case study using "Scratch" in five schools. *Computers & Education*, *97*, 129–141. doi:10.1016/j.compedu.2016.03.003

Sullivan, A., & Bers, M. U. (2016). Girls, boys, and bots: Gender differences in young children's performance on robotics and programming tasks. *Journal of Information Technology Education: Innovations in Practice*, *15*, 145–165. doi:10.28945/3547

KEY TERMS AND DEFINITIONS

Code: To create a program in a specific programming language.

Conditionals: A programming command that executes different actions based on evaluating a condition.

Programmer: One who writes code in a specific programming language.

Programming Language: A formal language consisting of commands that can be interpreted by a computer.

Robot: A machine that is programmable via a computer and is capable of carrying out actions automatically.

Sequencing: Putting events, steps, or other individual items in a specific order in relation to each other.

STEM: The collective field consisting of science, technology, engineering, and mathematics.

Chapter 3
Unplugged Learning:
Recognizing Computational Thinking in Everyday Life

Emily Relkin
Tufts University, USA

Amanda Strawhacker
Tufts University, USA

ABSTRACT

This chapter explores perspectives on unplugged coding and computational thinking (CT) in early childhood. Concepts, definitions, and research on unplugged learning and its relationship to computer science are considered. Several examples illustrate how young children can encounter powerful ideas of CT in both formal educational settings and in the process of everyday life. Resources are provided that aid in the identification and integration of unplugged activities into early childhood settings. Finally, the authors advocate for further research on teaching CT concepts to children that includes both coding and unplugged approaches.

INTRODUCTION

In the years before writing this chapter, we have had numerous conversations with members of the DevTech research group and colleagues about whether computers are necessary for children to learn computational thinking (CT). Many expressed the view that computers are an integral and inseparable part of computer science (CS) education. Although the precise definition of CT and its implications for education are still debated, in this chapter we attempt to unpack the distinctions between CS,

DOI: 10.4018/978-1-7998-7308-2.ch003

CT and unplugged learning for the purpose of envisioning a more equitable and inclusive computational pedagogy.

While it is hard to imagine CS education without computers, the logic, algorithms and other processes that contribute to CT have always been an important part of everyday life even before computers came into existence. After all, if it had been impossible to think "computationally" before the advent of computers, how would humans have invented them in the first place?

These questions mirror the rise of a national debate about the place of technology in computer science education. Critics of unplugged learning (notably, Huang & Looi, 2020) have argued that some CS concepts may be impossible to divorce from the technological medium of computers, and that school-based curricula that seek to teach CS concepts without computers are under-researched. We wholeheartedly agree with Huang and Looi's conclusion that more research is needed to understand whether and how "unplugged" learning that takes place without computers or other technology can support engagement with CS.

However, we argue that the debate is far from settled about whether unplugged learning can effectively support the development of computational thinking concepts. In this chapter, we use Bers' definition of CT concepts as those discipline-agnostic ideas, such as algorithmic logic and iterative design, that are foundational to CS, but not necessarily exclusive or unique to it. While education research is still exploring the relationships among tech tools, unplugged learning, and CS education, we propose that unplugged activities may engage children in CT skills practice, which may be beneficial for general development even beyond CS learning. For example, CS education has long faced criticism for lack of inclusion among girls, minorities, learners with disabilities, and learners unable to access tech tools and experiences. These inequities cause early gaps in CS achievement that last throughout a child's academic life and even long into adulthood, impacting career opportunities and contributing to a stratified CS workforce (Brackmann et al., 2017; Margolis et al., 2017; Wang & Hejazi Moghadam, 2017). By engaging children in unplugged CT learning, beginning with familiar early childhood activities, children may be able to build a foundation of CT awareness to mitigate gaps in CS readiness, and even the playing field for later pre-professional training.

Throughout this chapter we will be referring to concepts of Computer Science (CS) and Computational Thinking (CT). We will discuss methods and practices of exploring these domains such as unplugged learning and coding/programming. Although the meaning of these concepts are still evolving, for the purposes of this chapter, we will use the following definitions and conceptions: (1) CS is a field of study explicitly about understanding and utilizing computer technology. The Computer Science Teachers Association (CSTA) in the US has adopted the Association for Computing Machinery's (ACM) definition of CS, as "the study of computers and algorithms processes, their principles, their designs, their applications, and their impact on society" (CSTA, 2021; Tucker et al., 2006). (2) CT is often described as drawing upon computer science concepts, but CT skills are not exclusive to CS (Basu et al., 2016; Chen et al., 2017; Wing, 2006). CT broadly represents the set of thought processes that are required to engage in a range of analytic skills. These skills include, among other things, the ability to think recursively, apply abstraction, and use heuristic reasoning to solve a problem or complete a task. The process of engaging in CT is related to the field of CS but is also applicable to everyday life (Relkin et al., 2021; Wing, 2006, 2011). (3) We define "unplugged" as any activity or experience that does not require the child to be actively manipulating a smart device or computer, but that still promotes engagement with computer science concepts, skills, and practices. (see table 1). There are many tools and technologies that are available to introduce learners to CS and CT. New resources for unplugged computational learning are still emerging. In the following sections we outline the importance of unplugged opportunities to engage a diverse range of students in computational learning.

Unplugged Coding vs. Unplugged CT: Not All Unplugged Learning is Created Equal

One of the original intentions launching CS unplugged initiatives was to teach CS concepts in a way that made them accessible regardless of the student's access to computers. Coding (programming) involves using a set of instructions to solve problems and tasks with computers and other technologies (McLennan, 2017; Metin, 2020). Some activities that are described as "unplugged" are essentially coding exercises carried out offline using some of the same symbols and syntax as actual programming. For example, ScratchJr.org allows print out of large programming block cards that can be used to play a game called "Programmer Says". This game is similar to "Simon Says" but uses ScratchJr programming language instead of the usual verbal instructions to help students gain familiarity with coding commands. For the purpose of the present discussion, we will refer to these types of exercises as *unplugged coding activities*.

Other resources teach CT-related principles without directly invoking coding commands. For example, CSunplugged.org's "Divide and Conquer?" uses animal playing cards to teach about algorithms and related concepts. We will refer to activities such as these as *unplugged CT activities*. While coding is recognized as a means of acquiring CT skills, unplugged activities provide another route that may be more accessible to young children, especially those who are pre-literate or without ready access to computer hardware.

New research is constantly leading to revisions and refinements in educational practices. This includes how unplugged coding activities (e.g. board game or paper-based coding) versus unplugged CT activities (e.g. non-coding sorting and pattern matching) are employed in early childhood education (e.g., see Barr & Stephenson, 2011; Bell & Lodi, 2019; Bell & Vahrenhold, 2018; Upadhyaya et al., 2020). In the following section we discuss what the evidence currently suggests about how unplugged activities may contribute to children's future CS success.

Unplugged Activities as a Foundation for Computer Science Learning

Unplugged activities were created as a means of teaching Computer Science (CS) concepts and skills without requiring use of computers or programming (Bell & Vahrenhold, 2018). This approach was popularized in the early 1990s by a group of collaborating educators and researchers who coined the term "CS Unplugged" and shared a free collection of unplugged resources (Bell & Vahrenhold, 2018). CS unplugged concepts were taken from advanced CS courses and translated into physical activities where information could be more easily understood (Bell & Lodi, 2019). Since then, engaging students in unplugged activities have become widely recognized as beneficial for teaching CS concepts to young children. A variety of newer unplugged activities and curricula are found on websites such as csunplugged.org, in mainstream coding education initiatives such as www.code.org and https://www.barefootcomputing.org/ and in books and other resources (Caldwell & Smith, 2016).

Unplugged CS in Early Childhood

Limits to young children's literacy, numeracy, and abstract reasoning skills place some constraints on the CS concepts that can be mastered by young children (Bell et al., 2016). Young children are capable of learning to code but typically require developmentally appropriate programming platforms that use tangible blocks and graphical coding interfaces (Strawhacker et al., 2017; Sullivan et al., 2015; Resnick & Silverman, 2005). Unplugged activities can be used to introduce CS concepts to young children regardless of their ability to read, write, or count. An unplugged activity typically involves a set of artifacts and procedures that are well-known to most children and adults from their everyday lives. By presenting a readily understood analogy, or challenging students to formulate questions and find solutions, unplugged activities exercise some of the same skills that are involved in computer programming without the use of computers.

Past research on the impact of CS unplugged on young children has provided some contradictory messages. A few researchers have reported that unplugged activities do not increase interest or knowledge in CS/CT as much as traditional coding activities (e.g., Black et al., 2013). Other investigators have found that unplugged lessons alone are just as effective if not better at promoting CT as traditional coding (Hermans & Aivaloglou, 2017; Metin, 2020; Wohl et al., 2015). Yet another group of authors have suggested that the most powerful way to promote CT in young children is to integrate unplugged exercises and coding activities together (Metin, 2020; Huang & Looi, 2020; Bers, 2020; Thies & Vahrenhold, 2012, 2013).

How Does Computational Thinking Fit with Unplugged Learning?

The concept of CT was mentioned by Seymour Papert in his 1980 book *Mindstorms,* but the concept was more widely popularized by Jeanette Wing in 2006. Wing proposed that teaching CT should be part of every child's education. However, long before Wing (2006) popularized the term "computational thinking", young children have practiced sequencing, deconstructing problems, symbolic representation, and other CT skills in various science and non-technical classes as well as in their daily lives.

There have been several attempts to identify the subdomains of CT in a manner that conceptually aligned with child development. For example, the Computer Science Teachers Association (CSTA) and the International Society for Technology in Education (ISTE) developed a framework for grades K-12 that includes nine core concepts of CT (CSTA & ISTE, 2011). Dr. Marina Bers has identified seven developmentally appropriate powerful ideas of Computational Thinking for children in the range of 4-7 years of age (Bers, 2018; 2020). CSunplugged.org identified 6 CT subdomains for K-12. Table 1 shows the relationship among these three frameworks. In this chapter, we primarily employ Bers' seven powerful ideas because they are specifically developmentally appropriate for early childhood (ages 4-7).

Table 1. Computational Thinking domains as described by Bers (2018), CSunplugged.org and CSTA & ISTE (2011) respectively. The "CT Concept or Learning Goal" column highlights the overlap between these frameworks

Bers' CT Powerful Ideas	CS Unplugged Framework	ISTE & CSTA Standards	CT Concept or Learning Goal
Algorithms	Algorithmic Thinking	Algorithms & Procedures	Step-by Step sequencing/order used to achieve a task or solve problems
Modularity	Decomposition	Problem Decomposition	Breaking up larger task into smaller more manageable parts
Control Structures	Generalizing and Patterns	Parallelization	Recognizing patterns and repetition, cause and effect, determining the order of events, Organize resources
Representation	Abstraction	Abstraction, Data Representation	Symbolic Representation, Filtering to make less complex
Hardware / Software	*N/A*	Automation	Smart objects are not magical, objects are human engineered, understanding technology
Design Process	Evaluation	Simulation	Creative problem solving, gathering information, identifying possible solutions, iterative editing/ revision
Debugging	Logic	Data Analysis	Identifying problems, Making sense of data, Error correction

Source: (Bers, 2018; Csunplugged.org; CSTA & ISTE, 2011)

Unplugged Learning as a Pathway to Equitable Computational Thinking

CT has gained traction among education practitioners and researchers as a foundational skill set for CS education. However, research has identified that gaps in children's access to devices and technology instruction predict lasting and lowered outcomes for those students with lowered access to technology (Fraillon et al. 2020). Across schools, districts, and even nations, gaps are common between students from families of high and low socioeconomic status (SES), with lower SES correlated to lowered performance on CT assessments (Karpiński et al. 2021).

In an analysis of data from the 2018 IEA International Computer and Information Literacy Study (ICILS), which tested over 46,000 students from 14 countries, researchers found that persistent gaps among students' CT performance were linked to their family's socioeconomic backgrounds (Fraillon et al. 2020; Karpiński et al. 2021). Specifically, results "consistently showed that students from less advantaged backgrounds had lower levels of computer skills than those from more advantaged backgrounds, especially in CT" (Karpiński et al. 2021, p. 1).

Importantly, preliminary research has shown that unplugged activities may be useful for addressing these gaps by laying a foundation for later technology-mediated CS learning (Bers, 2020; del Olmo-Muñoz et al. 2020). Young children aged 4-8 years, with their developmental need for physical, hands-on play and limited screen engagement, may benefit the most from foundational unplugged CT experiences (Przybylski & Weinstein, 2019; Saxena et al. 2020). Further, CT already permeates children's daily activities and experiences as a cognitive skill set, in and out of school. It is important to look beyond formal education and explore the possibilities of democratizing CT by broadening awareness of its impact in everyday life. Given mounting evidence, we posit that one critical step toward addressing the known gap in CT performance among children from different backgrounds and with differential access to technology is to research and promote educational initiatives on aspects of CT that can be introduced in an unplugged format in formal, informal, and home learning settings.

By acknowledging and highlighting opportunities for everyday unplugged CT experiences, there is hope for democratizing foundational CS skills for all children, not just those whose families can afford it. In this chapter, instead of emphasizing novel activities and strategies to introduce CT into early childhood learning, we focus on recognizing how CT already exists in children's everyday learning. To help highlight the many opportunities to integrate CT throughout children's activities at home and school, we draw on existing definitions of computational thinking concepts, taken from Bers' (2020) Powerful Ideas of Computational Thinking. This reference was chosen for its evidence-based recommendations, developed from empirical research specifically relating to early childhood (Bers, 2020)

SUPPORTING THE ACQUISITION OF YOUNG CHILDREN'S CT THROUGH UNPLUGGED LEARNING

In the following sections, we follow a day in the life of a fictional child, Ava, who is a typically-developing 5-year-old girl. Like most kids her age, she explores and practices critical cognitive skills in her everyday life, preparing her for foundational computational thinking tasks before ever working with a computer. The examples below focus on typical "unplugged" experiences and daily practices that require problem solving and CT strategies. In each example, we offer evidence-based practice suggestions for further supporting unplugged engagement with CT.

Waking up With Algorithms

Ava's father knocks on her bedroom door and calls, "Rise and shine, Sleepyhead!" Ava stumbles into the bathroom, rubbing her eyes and yawning, and reaches for her toothbrush. Sometimes she is so sleepy in the morning, she has to stop and remember what to do first. She checks the list with sticker pictures that she made with her mother to remind her of the steps:

Figure 1.

1. Open the toothpaste 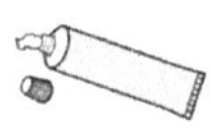	2. Squirt toothpaste on brush 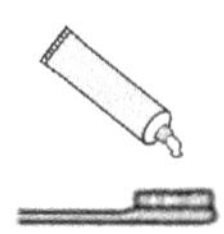	3. Brush teeth all over: left, right, middle, bottom & top 
4. Rinse out mouth 	5. Rinse the toothbrush and put it away - don't forget to turn off the water! 	6. Close toothpaste 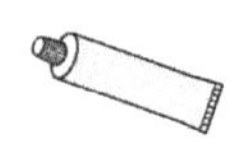

In this example, Ava is exercising the CT concept of *algorithms,* a sequential step-by-step process that can help solve problems or complete tasks (Bers, 2020; ISTE & CSTA, 2011; CSunplugged.org). In a typical coding task, children might explore algorithms by creating a sequence of coded instructions for a robot to carry out, or by using directional steps (e.g. turn left, one step forward) to program an on-screen character to navigate a maze. Here, Ava applies the same instructional logic to a non-technological activity, brushing her teeth.

By following a sequence of steps, the activity reinforces the idea that order matters to assure a process works as intended. Ava also worked with a parent to create the chart herself, meaning she was engaged in constructing the algorithm, not just following it. This way, she comes to know that there is meaning to her daily ritual, and that following the order of her algorithm is important to ensure that her teeth will be clean and healthy. To help children practice algorithmic thinking, you can work to build a chart, song, or other mnemonic to connect children's developing CT skills to their other daily practices, such as personal hygiene habits.

Debugging and Mismatched Socks

It's Ava's first day of Kindergarten and she's excited to dress up for the occasion. Ava picks out her favorite outfit for school and takes a look in the mirror. "Oh no, my socks don't match!" When she opens her sock drawer she can't seem to find a pair that are the same. Every pair of socks in the drawer is mismatched! She takes all of the socks out of the drawer and places them on the bed and starts to sort them. First she puts all the white socks together, then all the red, blue, etc. She then notices that they are still not matched. She looks at the white socks and sees some have patterns and some are plain white, so she separates those. She does the same for the other colors. She pairs up all of the socks by color and pattern only to find that two of the socks don't match each other or any others. She calls her father over to ask what to do. Her father suggests going downstairs to check in the laundry. She goes downstairs, finds the socks and pairs them up. The problem is solved, and she remembers to change into a matched pair before she goes to school! They also agree to keep the socks balled together in the future to avoid this problem.

Debugging is the two-part act of (1) exploring a system for an issue or "bug" that is causing the smooth functioning of the system to break down, and (2) working to iterate through solutions until the bug is resolved (Bers, 2020). Although the term debugging was coined to refer to issues in computational machinery, it also applies to any system with multiple parts working together for a single function. In this story, Ava engages in a form of "debugging" in order to solve the problem of her mismatched socks. The "system" in this case is the flow of her laundry from dirty to clean to sorted. Ava recognized that something was wrong with the process of matching socks in her drawer and went through multiple iterations in order to fix the problem. She completes the debugging process by restoring order to her drawer, correcting her mismatched socks, and coming up with a plan (with the help of her father) to prevent a similar problem in the future.

Modularity at the Breakfast Table

After finally picking her outfit, Ava runs to the kitchen - it smells delicious! While her father finishes cooking and mother is busy helping her brother with homework, Ava knows it's her job to set the table. She is so excited to eat that she sets her own place first - getting a plate, then a fork and knife, and finally a glass and putting them at her seat. She looks up and realizes that she has walked from the cupboard to the table three times, and has only set one place - there must be an easier way! She moves back to the cupboard and this time she stops to think. Her father, mother, and her brother will all need place settings. She counts out family members on her hands - 1, 2, 3. Then, she takes three plates and carefully carries them to the table. She checks the example place setting she left at her own chair, and comes back with three forks and knives, and then again returns with three cups. Just as she is separating out the dishes into three matched place settings, everyone is ready to sit down to her favorite breakfast - waffles!

When Ava sets the table, she is engaging in an idea similar to algorithms, but a bit different. *Modularity* is the process of breaking down tasks or procedures into simpler units. Modularity makes complex processes easier to manage by decomposing many steps of a problem. Examples of modularity in technology-mediated activities include separately coding two units (for example, characters in an animation) to interact with each other, or breaking down a longer coding process (e.g. code a robot to navigate across a room) into shorter steps (e.g. code a robot to navigate to the door, then to the table, etc.). Ava made "modules", or units, out of each place setting composed of one plate, one cup, and one set of silverware.

There are many ways you can explore modularity without using screens or technologies. Activities that involve planning are usually great opportunities to break down complex tasks. You can help children get excited to plan their birthday parties - complete with a guest list, invitations, favors, and games, all of which require children to break down the tasks into smaller steps. You can even explore modularity by breaking apart steps to a favorite song or dance and thinking about the repeating "modules" and the order they are sequenced.

Control Structures for Walking to School

Ava leaves with her mother to go to school. She skips, jumps, and runs along the sidewalk ahead of her mother, but she knows to always stop and wait when she gets to a crosswalk. At the corner in front of her school, she holds her mother's hand and closely watches the crossing guard, Ms. Doyle. She remembers what she learned last year, when her preschool teacher helped her whole class practice the special rules for crossing the street: If Ms. Doyle is holding up the red sign, that means Ava should STOP and wait. If Ms. Doyle holds up the green circle sign, that means it's safe to walk. Ava waits patiently while Ms. Doyle shows her red sign and lets some cars pass. She and her mother do not begin walking until they see Ms. Doyle holding up her green circle sign.

Control structures, sometimes called event-based codes, determine the order and timing of events as well as the sequence in which instructions are followed. Machines that use sensors, such as robotic vacuum cleaners that use distance sensors to navigate a room, or automatic doors that open when a person activates the infrared sensor, are real-world examples of technologies that use control structures. Use of phrases with words such as *when*, *while*, *until*, and *if* are clues that children are exploring Control Structure reasoning. In this unplugged example, Ava shows a good understanding that traffic signs affect the timing of safe street crossing, which is an exercise in control processes. In her understanding of the meaning of the signs Ava is also exercising the CT process of representation which we explore next.

Representation

After hugging her mother goodbye, Ava runs into her new classroom. She gets excited when she finds her friend Liam from her preschool is in her new class! Mr. Oladeji tells everyone to find their nametag at a table at the front of the room. Ava and Liam look at all the nametags - there are so many! Ava looks carefully for any names that start with the letter "L". "Liam! I found yours!" Ava waves Liam's name tag at her friend, and then keeps looking until she finds the name tag that spells A-V-A, but is confused when she finds three tags with her name on them. When she asks her teacher, he tells her there are two other students named Ava in their classroom, so she and the other Avas should pick a colored sticker so they can tell their name tags apart. She picks a blue star since blue is her favorite color! That will make it easy to find every morning. She walks to her seat, ready for the day.

Representation is the notion of an equivalence of symbols with concepts, objects, shapes, and other things. In computer-based coding, representation might come out in understanding the specific symbols of a block-based coding language, such as the color- and shape-based language system used by the Cubetto robot (https://www.primotoys.com/). It might also come out in text-based coding languages, for example, in a function or subroutine where typing the command "time" represents a long series of coding steps in which the computer accesses, sorts, and displays information about the current time of day. In preschool and Kindergarten, children learn that letters can be used to represent sounds, numbers represent quantities, and that other types of symbols (such as stop signs and gendered pictures on bathrooms) have meanings. In the example here, Ava learns that the symbol of a blue star can also be a useful representation for herself when letters and words may not be sufficient.

Hardware/Software

During her first day of kindergarten Ava's teacher asks, "who likes music?" Ava raises her hand and says, "I love to listen to music when I travel in the car". The teacher asks, "can music only come from a radio?" One of Ava's classmates says she listens to music on a computer and another hears music on her parents' cell phone. The teacher asks if anyone plays an instrument. Ava says she's learning to play piano. The teacher asks, "what's the difference between music on the radio and music on the piano?" Ava says, "when I turn on the radio the music just comes out, but I have to remember which keys to press for the piano to make music." The teacher smiles and asks, "Is it the piano that makes the music or is it you?" Ava says, "I guess it is both". The piano makes the sounds, but I have to tell it which notes to play. The teacher says "that is right! The radio, the computer, the cellphone, and the piano are all types of machines that can make sounds, but we need to tell the machines what to do in order for it to come out sounding like music."

Ava's teacher is helping her to understand an analogous concept to hardware and software as it relates to music and musical instruments. Children of this age are learning that technological objects are not magical or alive but in fact need to be constructed and instructed to perform the way they do. Computers use hardware and software to run. Hardware is physical devices (e.g., keyboard, computer screen, motherboard) and software is codes that tell the computer how to operate. In this example, Ava's teacher points out that both digital and analog technologies can function as "hardware". Coded software or sometimes our physical actions serve as the inputs needed to control the machine.

Of all the CT concepts explored in this chapter, the relationship between hardware and software is one of the most challenging for making unplugged learning connections. This is because computers, with their internal lightning-fast processors, simplify the hardware/software. This metaphor is made more complex in the non-digital world by the intervention of our natural processors - the brain. In Ava's example, the piano and radio are easily identifiable as the hardware that "enacts" the musical performance. In the case of the radio, the software is also easily named - the computer file that stores the music recording. Naming the "software" guiding a piano performance is more nebulous. Is it Ava, the player striking the keys? Or, the written music that reminds Ava of the structure and sequence of notes? Or is it Chopin, the original designer and author of the sheet music sitting on Ava's piano? Any of these answers could be argued, although the most precise analogy to computer code probably points to the sheet music as an analogue for software, and Ava (the player) is the "processor" translating the written instructions to sounds on the piano hardware.

Explaining this complex system to children, without the ability to show them a processing unit inside of a computer, leads to some necessary challenges for the educator, and probably some confusion in the learner. One way around this is to use a simplified metaphor, such as comparing it to music and musical instruments. Although this is an imperfect analogy, it has the benefit of leveraging children's natural inclination toward concrete thinking (for a deeper discussion of concrete thinking in the context of computer science education, see Papert, 2005). By using a basic metaphor, children can develop a heuristic that can serve them as a foundation for later, more advanced exploration of hardware and software.

Design Process

When her mother picks Ava up at school, she can hardly wait to tell her about her first homework assignment - she gets to build a structure out of marshmallows and toothpicks! When they get home Ava starts planning the marshmallow structure. Ava made lots of marshmallow towers at STEM camp this summer, so decides to challenge herself and build a marshmallow bridge. The only problem is, she can't figure out how to make it stay upright without falling down in the middle. She asks her brother for help. He looks at the table and asks "Where's your design?" Ava looks up, curious. "You know," her brother says, "your plan. Your blueprint. The picture that you're using to help you build the whole thing." Ava's face lights up. "You're right! I forgot, we used to make blueprints at camp!" She runs to get paper and her pencil box, and begins furiously sketching out her bridge. Her mother smiles when Ava adds her final touches to the drawing, and offers to hold up her design while Ava begins to build it. Ava checks the plan, and this time adds a few supporting marshmallows in the center. Now it looks sturdy! She asks her brother to test it for her. He picks up the box of leftover toothpicks and places it on the bridge - but it crashes down! "Oh no! I need to add more middle parts," Ava says and she picks up the broken bridge carefully. After two more tests and redesigns, her bridge finally holds up to the toothpick box test! Her brother offers to help her carry it into school tomorrow to make sure it doesn't break before she can share it with her class.

In this example, Ava goes through the various steps of the design process. She first *imagines* what she will make and *asks* a question about how to make a sturdy structure. She *plans* using a blueprint, she *creates* the structure, and *tests* its sturdiness. She also *improves* the structure and *shares* with her family and classmates. The design process is an open-ended and iterative process used to create new things. There's no official starting or ending point to the design process. Children can begin at any step, move in any direction and repeat the cycle as many times as they see fit.

In contrast to the CT concept of hardware and software, introducing the design process is one of the easiest computational thinking skills to integrate into unplugged everyday life. Design is a human-driven activity, and can take place in almost any modality and in the context of any topic. Scientists engage in a form of the design process (they prefer the term "scientific method") when planning and carrying out experiments. Writers use the functionally identical "writing process" to brainstorm, draft, and revise their stories. Computer Scientists learn to engage in the iterative process of writing, compiling, running, and debugging their code until it functions as intended. When children explore robotics for the first time they can design the hardware that executes their program, the program itself, or a combination of both. Fortunately for unplugged CT learning, children can explore the steps of the design process using familiar materials and ideas. Crayons, paper, and chalk can help with planning and blueprinting, and traditional blocks, craft supplies, or recycled and found materials are great for modeling their designs. Any activity that engages children in identifying a problem, formulating a solution, and working to build it into a reality is a fantastic opportunity to engage in the steps of the design process.

RESOURCES

In this section we list some of the resources that can aid in the implementation of unplugged learning for early childhood education and allow creation of new activities.

Websites such as www.csunplugged.org, www.code.org, https://www.barefootcomputing.org/, www.kodable.com, and www.csinsf.org, are excellent resources for finding unplugged CT as well as unplugged coding activities and curricula for young children. These websites include unplugged CT activities that engage children in various subdomains including algorithms, debugging, and decomposition. Unplugged coding cards for various early childhood technologies can be found on sites such as https://www.scratchjr.org/ that allows download and printing of free large ScratchJr unplugged cards. Similar cards for the KIBO robotics platform can be purchased at https://shop.kinderlabrobotics.com. These coding cards are an excellent way to introduce children to programming languages for the first time and may be more accessible to some children than manipulating smaller blocks. You can also purchase ScratchJr coding cards which includes both interface and coding as well as unplugged activity ideas for educators.

The DevTech Research Group has created the Coding as Another Language (CAL) ScratchJr and KIBO curricula for PreK-2nd grade that integrate both traditional coding, unplugged coding, and literacy activities. You can learn more and download the CAL curricula for free here: sites.tufts.edu/codingasanotherlanguage.

We and other researchers believe that it is vital that unplugged CS education connects to children's everyday lives. Therefore, we encourage educators and researchers to create their own activities that are personally meaningful to them and their students.

If you are interested in creating your own activities here are guiding questions to inspire children's learning:

- Introduce Design: Can we make a plan or design for this idea? Can we iterate through multiple revisions?
- Practice Sequencing: How does order matter? what would happen if we changed the order of steps?
- Solve Problems by Debugging: Is there a step-by-step approach we could take to identify the problem? How can we try to fix it?

Many educators find it helpful to start by teaching young children unplugged coding/CT and then graduate to traditional lessons of coding. Here are some concepts to keep in mind during that process:

- Consider the age range the tool was designed for. Will it be too simple or too complex for your learner?
    - Start with tangible coding languages and tools for younger learners before moving to screen-based ones
    - For pre-readers and beginning-readers, look for coding tools that don't rely on text-based language. Draw or print out images of the coding symbols to use in off-screen games and activities.
    - For children who are ready for a challenge, look for tools that support open-ended coding experiences, rather than puzzle-style or step-by-step coding games
- Start small.
    - Introduce a robot without its power source or batteries the first time you share it with a child.
    - When beginning to learn a coding language, focus on one or two basic commands to master before moving into more complex instructions.
- Make connections to real-world technology - You can find sensors, cameras, and computers all around!
    - Point out when doors and sinks motion-activated, or when interactive vending machines use robotic arms
    - Have discussions about whether certain technologies are robots, with moving parts that can be programmed (e.g. self-driving cars, motion-activated doors) and which are non-robotic machines (e.g. typewriters, televisions, flashlights).
- Finally, consider a mix of both plugged and unplugged experiences to engage children in all the kinds of learning they might encounter in the real world! Many libraries, makerspaces, and museums will have tools and technologies available for children to explore. This can be a great first introduction to the world of plugged-in CT and CS exploration.

REFLECTING ON THE IMPORTANCE OF UNPLUGGED LEARNING

Throughout this chapter we have argued that unplugged CT/CS learning can support children's later academic success. As we outlined in the beginning of this chapter, preliminary evidence from research supports the perspective that unplugged learning can be an important foundation for more formal CS education (Hermans & Aivaloglou, 2017; Metin, 2020; Wohl et al., 2015). In addition, we propose two other arguments supporting unplugged learning in early childhood.

First, we can look to other, more established learning domains for examples of how to introduce young children to life-long concepts. In the field of language and literacy, educators have long known about the importance of "print awareness," or highlighting text and letter symbols through child-accessible signage and books. If we accept the premise that coding and CS comprise a computational literacy (e.g., Bers, 2020), then we might interpret the unplugged activities presented in this chapter as suggestions for promoting "computational awareness" in learners before being formally introduced to CS education.

Second, regardless of the impact of unplugged CT learning on children's later academic success, research is conclusive about the impacts of early exposure to novel domains on children's developing identity awareness (Kuhl et al. 2019; Sullivan, 2019). Researchers have even demonstrated that early exposure to STEM domains through playful and hand-on activities has contributed to disrupting the formation of stereotypes that may inhibit children (particularly those from resource-stretched backgrounds who are already facing a disadvantage in CS pathways) from engaging in those professions later in life (Karpiński et al. 2021; Sullivan, 2019). In the face of disheartening gaps, it is critical for educators and families to explore the use of unplugged CT activities to achieve more equitable CS education. Unplugged learning might mitigate SES-based discrepancies in student performance by reinforcing critical CT skills without the need for prohibitively expensive or rare technologies. Although accessibility issues remain, we agree with others that a model of unplugged learning outside the formal classroom may still be a useful starting point to engage learners who have no other recourse to explore CT due to inaccessible technologies (Manabe et al. 2011). Since unplugged lessons also tend to be a thoughtful reframing of activities that children and families already do or know how to do, they can also offer adults a chance to demystify CS concepts for themselves, leading to more success teaching and modeling this kind of thinking for children (e.g., Curzon et al. 2014).

Taking these arguments together, we propose that implementing unplugged activities in early childhood may have lasting benefits in terms of children's developing interests, identities, and academic readiness to engage in later CS learning. Unplugged learning may even support foundational learning in other domains identified in research as connected to CT. Our advocacy of unplugged learning does not detract from our enthusiasm for approaches such as learning to code which can have additive or synergistic benefits for children's cognitive development. There is uncertainty about whether unplugged learning alone achieves the same outcomes as coding activities in terms of mastery of CT concepts. More research is needed to determine how to best integrate unplugged and coding approaches into a unified method of CS education for young children. In addition, further refinement of unplugged curricular activities is needed so that best educational practices can be identified.

REFERENCES

Barr, V., & Stephenson, C. (2011). Bringing computational thinking to K-12: What is Involved and what is the role of the computer science education community? *ACM Inroads*, *2*(1), 48–54. doi:10.1145/1929887.1929905

Basu, S., Mustafaraj, E., & Rich, K. (2016). CIRCL Primer: Computational Thinking. In *CIRCL Primer Series*. Retrieved from https://circlcenter.org/computational-thinking

Bell, T., & Lodi, M. (2019). Constructing Computational Thinking Without Using Computers. *Constructivist foundations*, Vrije Universiteit Brussel. *Constructionism and Computational Thinking*, *14*(3), 342–351.

Bell, T., & Vahrenhold, J. (2018). CS unplugged—How is it used, and does it work? In H.-J. Böckenhauer, D. Komm, & W. Unger (Eds.), Adventures between lower bounds and higher altitudes: Essays dedicated to Juraj Hromkovič on the occasion of his 60th birthday. Springer International Publishing.

Bers, M. U. (2018). *Coding as a playground: programming and computational thinking in the early childhood classroom*. Routledge., doi:10.4324/9781315398945

Bers, M. U. (2020). *Coding as a Playground: Programming and Computational Thinking in the Early Childhood Classroom* (2nd ed.). Routledge Press. doi:10.4324/9781003022602

Black, J., Brodie, J., Curzon, P., Myketiak, C., McOwan, P. W., & Meagher, L. R. (2013). Making computing interesting to school students: Teachers' perspectives. In *Proceedings of the 18th ACM Conference on Innovation and Technology in Computer Science Education*. Association for Computing Machinery. 10.1145/2462476.2466519

Brackmann, C. P., Román-González, M., Robles, G., Moreno-León, J., Casali, A., & Barone, D. (2017). Development of computational thinking skills through unplugged activities in primary school. *Proceedings of the 12th Workshop on Primary and Secondary Computing Education*, 65-72. 10.1145/3137065.3137069

Caldwell, H., & Smith, N. (2016). *Teaching computing unplugged in primary schools: Exploring primary computing through practical activities away from the computer*. Learning Matters.

Curzon, P., McOwan, P. W., Plant, N., & Meagher, L. R. (2014). Introducing teachers to computational thinking using unplugged storytelling. *Proceedings of the 9th workshop in primary and secondary computing education*, 89-92. 10.1145/2670757.2670767

del Olmo-Muñoz, J., Cózar-Gutiérrez, R., & González-Calero, J. A. (2020). Computational thinking through unplugged activities in early years of Primary Education. *Computers & Education*, *150*, 103832. doi:10.1016/j.compedu.2020.103832

Fraillon, J., Ainley, J., Schulz, W., Friedman, T., & Duckworth, D. (2020). *Preparing for life in a digital world: IEA International computer and information literacy study 2018 international report* (Vol. 297). Springer Nature. doi:10.1007/978-3-030-38781-5

Hermans, F., & Aivaloglou, E. (2017). To scratch or not to scratch?: A controlled experiment comparing plugged first and unplugged first programming lessons. In *Proceedings of the 12th Workshop on Primary and Secondary Computing Education*. Association for Computing Machinery. 10.1145/3137065.3137072

Huang, W., & Looi, C. K. (2020). A critical review of literature on "unplugged" pedagogies in K-12 computer science and computational thinking education. *Computer Science Education*, 1–29.

International Society for Technology in Education (ISTE) & The Computer Science Teachers Association (CSTA). (2011). *CT leadership toolkit.* Retrieved from https://cdn.iste.org/www-root/2020-10/ISTE_CT_Leadership_Toolkit_booklet.pdf?_ga=2.15251892.309077916.1613247518-1278422219.1611941118

Karpiński, Z., Di Pietro, G., & Biagi, F. (2021). *Computational thinking, socioeconomic gaps, and policy implications.* IEA Compass: Briefs in Education Series (12). Retrieved from: https://www.iea.nl/publications/series-journals/iea-compass-briefs-education-series/january-2021-computational

Kuhl, P. K., Lim, S. S., Guerriero, S., & van Damme, D. (2019). How stereotypes shape children's STEM identity and learning. In *Developing Minds in the Digital Age: Towards a Science of Learning for 21st Century Education*. OECD Publishing. doi:10.1787/43e5bb4c-en

Manabe, H., Kanemune, S., Namiki, M., & Nakano, Y. (2011). CS unplugged assisted by digital materials for handicapped people at schools. In *Proceedings of the 5th International Conference on Informatics in Schools: Situation, Evolution and Perspectives*. Springer-Verlag. 10.1007/978-3-642-24722-4_8

Margolis, J., Estrella, R., Goode, J., Holme, J. J., & Nao, K. (2017). *Stuck in the shallow end: Education, race, and computing*. MIT Press.

McLennan, D. P. (2017). Creating coding stories and games. *Teaching Young Children, 10*(3). 18-21. Retrieved October 02, 2019 from https://www.naeyc.org/resources/pubs/tyc/feb2017/creating-coding-stories-and-games

Metin, S. (2020). Activity-based unplugged coding during the preschool period. *International Journal of Technology and Design Education*, 1–17.

Papert, S. (1980). *Mindstorms: Children, Computers, and Powerful Ideas*. Basic Books.

Papert, S. (2005). You can't think about thinking without thinking about thinking about something. *Contemporary Issues in Technology & Teacher Education*, *5*(3), 366–367.

Przybylski, A. K., & Weinstein, N. (2019). Digital Screen Time Limits and Young Children's Psychological Well-Being: Evidence From a Population-Based Study. *Child Development*, *90*(1), e56–e65. doi:10.1111/cdev.13007 PMID:29235663

Relkin, E., de Ruiter, L. E., & Bers, M. U. (2021). Learning to Code and the Acquisition of Computational Thinking by Young Children. *Computers & Education*, *169*, 104222. Advance online publication. doi:10.1016/j.compedu.2021.104222

Resnick, M., & Silverman, B. (2005). Some reflections on designing construction kits for kids. *Proceeding of the 2005 Conference on Interaction Design and Children - IDC '05*, 117–122. 10.1145/1109540.1109556

Saxena, A., Lo, C. K., Hew, K. F., & Wong, G. K. W. (2020). Designing Unplugged and Plugged Activities to Cultivate Computational Thinking: An Exploratory Study in Early Childhood Education. *The Asia-Pacific Education Researcher*, *29*(1), 55–66. doi:10.100740299-019-00478-w

Strawhacker, A. L., Lee, M. S. C., & Bers, M. U. (2017). Teaching tools, teachers' rules: Exploring the impact of teaching styles on young children's programming knowledge in ScratchJr. *International Journal of Technology and Design Education*. Advance online publication. doi:10.100710798-017-9400-9

Sullivan, A., Elkin, M., & Bers, M. U. (2015). KIBO Robot Demo: Engaging young children in programming and engineering. *Proceedings of the 14th International Conference on Interaction Design and Children (IDC '15)*. 10.1145/2771839.2771868

Sullivan, A. A. (2019). *Breaking the STEM stereotype: Reaching girls in early childhood*. Rowman & Littlefield Publishers.

The Computer Science Teachers Association (CSTA). (2021). *K-12 CS Education Glossary*. https://www.csteachers.org/page/glossary

Thies, R., & Vahrenhold, J. (2012). Reflections on Outreach Programs in CS Classes: Learning Objectives for" Unplugged" Activities. *Proceedings of the 43rd ACM technical symposium on Computer Science Education*, 487-492. 10.1145/2157136.2157281

Thies, R., & Vahrenhold, J. (2013). *On plugging unplugged into CS classes*. doi:10.1145/2445196.2445303

Tucker, A., McCowan, D., Deek, F., Stephenson, C., Jones, J., & Verno, A. (2006). *A model curriculum for K–12 computer science: Report of the ACM K–12 task force curriculum committee* (2nd ed.). Association for Computing Machinery.

Upadhyaya, B., McGill, M. M., & Decker, A. (2020). A Longitudinal Analysis of K-12 Computing Education Research in the United States: Implications and Recommendations for Change. *Proceedings of the 51st ACM Technical Symposium on Computer Science Education*, 605-611. 10.1145/3328778.3366809

Wang, J., & Hejazi Moghadam, S. (2017, March). Diversity barriers in K-12 computer science education: structural and social. *Proceedings of the 2017 ACM SIGCSE Technical Symposium on Computer Science Education*, 615-620. 10.1145/3017680.3017734

Wing, J. (2011). Research notebook: computational thinking—What and why? *The Link Magazine*. Retrieved from: https://www.cs.cmu.edu/link/research-notebookcomputational-thinking-what-and-why

Wing, J. M. (2006). Computational thinking. *CACM Viewpoint*, *49*(3), 33–35. doi:10.1145/1118178.1118215

Wohl, B., Porter, B., & Clinch, S. (2015). Teaching computer science to 5–7 year-olds: An initial study with scratch, cubelets and unplugged computing. *Proceedings of the Workshop in Primary and Secondary Computing Education*, 55–60. 10.1145/2818314.2818340

ADDITIONAL READING

Bell, T., Alexander, J., Freeman, I., & Grimley, M. (2009). Computer science unplugged: School students doing real computing without computers. *The New Zealand Journal of Applied Computing and Information Technology, 13*(1), 20–29.

Bell, T., Duncan, C., & Atlas, J. (2016). Teacher feedback on delivering computational thinking in primary school. *Proceedings of the 11th Workshop in Primary and Secondary Computing Education*, 100-101. 10.1145/2978249.2978266

Bell, T., Witten, I., & Fellows, M. (2002). *Computer science unplugged.* Department of Computer Science, University of Canterbury.

Curzon, P. (2013). cs4fn and computational thinking unplugged. *Proceedings of the 8th workshop in primary and secondary computing education*, 47-50. 10.1145/2532748.2611263

Faber, H. H., Wierdsma, M. D. M., Doornbos, R. P., van der Ven, J. S., & de Vette, K. (2017). Teaching computational thinking to primary school students via unplugged programming lessons. *Journal of the European Teacher Education Network, 12*, 13–24.

Looi, C. K., How, M. L., Longkai, W., Seow, P., & Liu, L. (2018). Analysis of linkages between an unplugged activity and the development of computational thinking. *Computer Science Education, 28*(3), 255–279. doi:10.1080/08993408.2018.1533297

Relkin, E., de Ruiter, L., & Bers, M. U. (2020). TechCheck: Development and validation of an unplugged assessment of computational thinking in early childhood education. *Journal of Science Education and Technology, 29*(4), 482–498. doi:10.100710956-020-09831-x

Thies, R., & Vahrenhold, J. (2016). Back to school: Computer Science unplugged in the wild. In *Proceedings of the 2016 ACM Conference on Innovation and Technology in Computer Science Education*, 118-123. https://dl.acm.org/doi/pdf/10.1145/2899415.2899442

Zapata-Ros, M. (2019). Computational thinking unplugged. *Education in the Knowledge Society*, *20*, 1–29.

KEY TERMS AND DEFINITIONS

Coding: Also called programming, coding is the process of designing and building an executable computer program to accomplish a specific computing result or to perform a specific task.

Computer Science: The study of computers and computing as well as their theoretical and practical applications.

Early Childhood Education: Education of children from birth through age eight.

KIBO: A screen-free programmable robotics kit for young children with blocks, sensors, modules, and art platforms.

Programming: Also called coding, computer programming is the process of designing and building an executable computer program to accomplish a specific computing result or to perform a specific task.

ScratchJr: A free block-based programming application for young children

STEM: An integrated educational approach involving disciplines of Science, Technology, Engineering, and Mathematics.

Unplugged: Describes activities such as games and puzzles that aid the teaching and learning of computer science but without requiring the use of technology.

Section 2
Connections

Chapter 4
The Role of Executive Function and Self-Regulation in the Development of Computational Thinking

Elizabeth Kazakoff Myers
WGBH Educational Foundation, USA

ABSTRACT

This chapter summarizes theoretical connections between computational thinking through learning to code, self-regulation, and executive function and discusses why it is important to continue exploring the intersection of executive function, self-regulation, and computational thinking, including the need to revisit the socio-cultural underpinnings of foundational self-regulation, executive function, and school readiness research. As an example, findings from a 2014 study that explored the relationship between self-regulation and computational thinking when learning to code are shared. Research supports the idea of teaching computational thinking skills within an integrated early childhood curriculum to support the development of well-prepared citizens for the 21st century by drawing on the connections between executive function, self-regulation, and computational thinking.

INTRODUCTION

As other chapters in this book will attest, coding is becoming an increasingly essential skill. As Bers references in Chapter 1, "there are an estimated 500,000 openings for computing jobs nationwide, and a lack of adequately trained people to fill them

DOI: 10.4018/978-1-7998-7308-2.ch004

(Code.org, 2018; Fayer, Lacey, & Watson, 2017)." While job training and 21st century skill development is one reason to advocate for coding education, I argue adoption of a computational thinking curriculum in early childhood by educators, parents, and children is much more likely when coding is connected to other aspects of the early childhood curriculum. More specifically, connecting computational thinking to core academic domains like math and literacy as well as the development of cognitive skills, such as executive function and self-regulation.

When integrating computational thinking into classroom through educational technologies, it is important to note that technological tools are artifacts mediated by social groups and cultural mores (Moll, 2014). Younger children do not provide themselves with the digital technologies in their lives; parents, families, and schools are the ones to make the purchases or hand the child the devices (Gutnick, A. L., Robb, M., Takeuchi, L., & Kotler, J., 2010). The technological tools the child is exposed to are influenced by societal and cultural factors. Furthermore, socioeconomic status of families not only underpins digital divide, but is also a predictor of all domains of executive function (Mulker Greenfader, 2019). Schools can play a key role in narrowing digital gaps, providing more equal opportunities for technology devices, exposure to computational thinking, and the development of executive function skills.

My early research focused on connections between coding and literacy (Kazakoff & Bers, 2012; Kazakoff, Sullivan, & Bers, 2013; Kazakoff & Bers, 2015). My 2014 dissertation, *Cats in Space, Pigs Who Race: Does self-regulation play a role when kindergartners learn to code?* was one of the first studies to examine executive function within the context of the development of computational thinking and coding skills in early childhood. Self-regulation (controlling ones behaviors) and executive function (directing ones thoughts and behaviors towards a problem solving goal) are also skills seen by educators and parents as desirable for traditional ideas of kindergarten readiness (Center for the Developing Child, n.d.; Finders, et. al., 2021).

My work specifically focused on the consideration of the role self-regulation plays when learning to code. This work was inspired by several years of interactions with young children and digital tools, where it became clear to me there were many factors that determined how well children learned novel technologies and coding languages. As an illustrative example, take Danielle and Jennifer, whom I met early in my research.

Danielle and Jennifer were friends and neighbors in the same first grade classroom and described by their parents as excellent, curious students. Their parents had signed them up for a research study on a new coding language for young children where they were able to program a robot's movements with wooden blocks affixed with scannable barcodes that translated an action on the block to a movement for the robot. One Saturday morning, Danielle and Jennifer arrived at the DevTech Lab research lab presenting with equivalent levels of excitement to learn about robots.

When introduced to the robotics lesson, Danielle and Jennifer had similar ideas about decorating their robots to be ballerina robots and programming them to move forward three steps, do three spins, and move backwards three steps. They got to work on their ideas and decorated their robots with pipe cleaners, construction paper, and glitter, transforming the robots into whimsical ballerinas. However, when it came time to code, Jennifer meticulously built and tested each block of code on her ballerina robot, whereas Danielle skipped coding altogether and insisted on dancing around the room with her robot in hand -- making her robot dance through manual effort and not the coding blocks. Danielle expressed interest and excitement in Jennifer's robot dancing on its own "like magic" but seeing her friend code was not enough for Danielle to overcome the instant gratification of dancing with her robot in the moment instead of programming it to dance on its own.

While every child is different, these two general reactions to learning to code presented regularly in my early years of building coding languages for children. When coding with existing programming languages of the time, children of kindergarten and first grade ages observed in studies varied widely in their abilities to overcome "instant gratification" activities and interface components (i.e., buttons that would randomly insert characters or make a character grow big and shrink small) in favor of systematically programming their great ideas. I hypothesized that self-regulation was a key developmental area that appeared to be a particular barrier for children five to seven-year-olds, particularly when using the original version of the Scratch software (for children ages 8 and older).

As such, my research within the development of ScratchJr explored how differences in self-regulation impacted computational thinking and learning to code by examining the possible bi-directional relationship between self-regulation level and coding ability when using ScratchJr. The hope was that coding languages could be developed in more inclusive ways to provide access to all young students in all classrooms, rather than becoming tools only for out-of-school-time enrichment for students who were developmentally ready for existent computational activities. In the nearly eight years since I last focused on this work, the importance of coding and computational thinking for young children has exploded in popularity, as has the call for more studies to demonstrate causal impact of executive function focused interventions in classrooms (Jacob & Parkinson, 2015). It is also important to note that self-regulation and delay of gratification research has grown heavily out of Mischel, et al's "Marshmallow Studies" (Mischel, Shoda, & Rodriguez, 1989) which have recently been revisited and critiqued for their selection bias (Watts, Duncan, & Quan, 2018).

The purpose of this chapter is to (1) summarize theoretical connections between computational thinking through learning to code, self-regulation, and executive function, (2) to summarize findings of a 2014 study which explored the relationship between self-regulation and computational thinking when learning to code, and (3) to highlight why it is still important to continue exploring the intersection of executive function, self-regulation, and computational thinking, including the need to revisit the socio-cultural underpinnings of foundational self-regulation, executive function, and school readiness research (which is beyond the scope of this chapter).

COMPUTATIONAL THINKING

ISTE and the Computer Science Teachers Association (CSTA) define computational thinking as a problem solving process (ISTE & CSTA, 2011) and problem solving will become a common theme throughout this chapter. While this definition (one of many) emerged from the computer science education field, computational thinking is applicable across many disciplines and domains (Digital Promise, 2017). As Bers states in Chapter 1, "although computational thinking has received considerable attention over the past several years, there is little agreement on what a definition for computational thinking might encompass (Allan et al., 2010, Barr & Stephenson, 2011, Grover & Pea, 2013, National Academies of Science, 2010, Relkin, 2018, Relkin & Bers, 2019, Shute, Sun, & Asbell-Clarke, 2017, Grover & Pea, 2013, & Guzdial, 2008)."

Brennan and Resnick (2012) defined computational thinking along three specific dimensions that apply to Scratch (and, in turn, ScratchJr). The dimensions are computational concepts, computational practices, and computational perspectives. Computational concepts refer to areas such as: sequencing of programming instructions; parallel programming by either giving multiple programs to one character or giving two or more characters programs to act out together; and programming events, such as "start on green flag" (Brennan & Resnick, 2012). Computational practices include iterative design, testing and debugging, and abstracting and modularizing. Computational perspectives refer to the connections children make between the technological tool and the real world, by asking questions and making connections with others (Brennan & Resnick, 2012).

This definition of computational thinking, specifically related to Scratch/ScratchJr, is mentioned to highlight the connections between computational thinking, math, literacy, and problem solving. Math, literacy, and problem solving have all been correlated with self-regulation in prior studies (Espy, et. al., 2004; McClelland, et.

al., 2007; Zelazo, Carter, Reznick, & Frye, 1997) and it is possible that activities that utilize and build computational thinking, such as learning to code with ScratchJr, may also contribute to the development of cross-domain skills in math, literacy, problem solving, and self-regulation. Of course, there is still much work to be done in these areas to demonstrate concrete, measurable connections, but these theoretical connections served as the basis for initial work on connecting self-regulation and computational thinking through learning to code. Demonstrating that young children can grasp computational thinking concepts such as sequencing, patterns, modularity, cause and effect, and problem solving when presented with them in a developmentally appropriate way has been confirmed by extensive research (Bers, 2018).

EXECUTIVE FUNCTION AND SELF-REGULATION

Computational thinking, as a means of problem solving, taps into similar and overlapping cognitive functions -- many of which are considered under the umbrella of executive function, and, by extension, self-regulation. An emphasis can be made on the connection between self- regulation and learning to code as coding has previously been defined as problem solving (Brennan & Resnick, 2012; Yelland, 2005) and problem solving is a self-regulative process in itself (Zelazo, Carter, Reznick, & Frye, 1997). In particular, executive function plays a role in a child's capacity to iteratively revise a hypothesis. Testing and revising hypotheses is a key component of not only the scientific method but also testing and debugging and computational thinking while learning to code. Through hypothesis testing, children engage in meaningful science and engineering exploration in early childhood while working to build executive function and computational thinking skills through scaffolded scientific inquiry (Gropen, Clark-Chiarelli, Hoisington, & Ehrlich, 2011).

Broadly, self-regulation refers to an integrative process that governs emotional, cognitive, and behavioral functioning (Baumeister & Vohs, 2004; Gestdotteir & Lerner, 2008; McClelland, Ponitz, Messersmith, & Tominey, 2010). Notably, the Handbook of Self-Regulation (Baumeister & Vohs, 2004) states that all contributing authors had different definitions of self-regulation but agreed on a common theme: "self-regulation refers to the exercise of control over oneself, especially with regard to bringing the self into line with preferred...standards (p. 2)" through controlling one's attention, thoughts, emotions, and actions (McClelland & Cameron, 2012).

Examining self-regulation from an education perspective primarily focused on behaviors, thoughts, and feelings associated with school success (McClelland, Pontiz, Messersmith, & Tominey, 2010) and is defined by the behavioral aspects of self-regulation: flexible attention, working memory, and inhibitory control (McClelland & Cameron, 2011). The separate yet integrated tasks of flexible attention, working memory, and inhibitory control appear to be most relevant for learning in school, particularly when following directions or persisting on difficult tasks (McClelland & Cameron, 2011).

Executive function and self-regulation are related, in that executive function serves as the construct that unites working memory, attention, and inhibitory control for the purposes of planning, problem solving and goal-directed activity (Blair 2002; Blair & Razza, 2007). These three aspects of self-regulation specifically and executive function broadly are critical for success in the early childhood classroom and predictive of academic success (Diamond, 2002; Sameroff & Hait 1996). Self-regulation is critical to school success. It predicts school readiness over and above other factors, such as general cognitive skills and family background (Blair & Razza, 2007) and growth in behavioral regulation predicts growth in math and literacy skills (McClelland, et. al., 2007). Furthermore, through their early work on a large-scale study of children using the LOGO programming language, Clements, Battista, & Sarama (2001) demonstrated that children in Grades K – 6 scored significantly higher on tests of mathematics, reasoning, and problem solving after learning to code with LOGO. The researchers theorize that when children engage in coding activities (through a process which would now be called computational thinking): creating sequences of commands for the computer to read, the children externalize their inner thought process. This externalization of inner thoughts may make a child's thought process more readily available for reflection and understanding (Clements, Battista, & Sarama, 2001). Reflection and understanding of one's thought process is also known as metacognition, which is a component of executive function (Garon, Bryson, & Smith, 2008). Perhaps then, learning to code promotes metacognition, which in turn, could contribute to children's development of executive function and self-regulation.

More recently, a 2020 study looked specifically at a correlation between the BRIEF2 assessment of executive function and debugging in Scratch for 11-year-old students. The study found that BRIEF2 scores that demonstrated executive function maturity were strongly correlated with students' Scratch debugging scores (Robertson, Gray, Martin, & Booth, 2020). The BRIEF2 (and earlier version, BRIEF) is a comprehensive, clinical measure of multiple facets of executive function including working memory, inhibitory control, and (shifting) attention plus emotional control, initiation, planning, organization of materials, and monitoring (Gioia, Isquith, Guy, Kenworthy, 2000). The authors conclude that study was another piece of the puzzle connecting computational thinking and cognitive variables. The authors also emphasize the need to scaffold students based on executive function levels and join the literature suggesting that executive function skills could be developed through motivating and engaging computational thinking activities (Robertson, Gray, Martin, & Booth, 2020).

Another small study with first graders in Italy explored the connection between computational thinking and executive function skills primarily focused on response inhibition and planning. Children in the study who learned to code improved planning and inhibition control skills compared to controls. Researchers equated the gains from one month of coding to seven months for control students in planning and inhibition tasks. Although small, this study further supports the theory that computational thinking can boost at least some aspects of executive function in young children (Arfe, Vardangega, Monturori, & Lavanga, 2019).

A CASE STUDY WITH SCRATCHJR

This section describes the results of a previously unpublished case study. In two kindergarten classrooms, the relationship between self-regulation and children's experiences in learning to code using ScratchJr was explored. Specifically, this work examined whether initial levels of self-regulation made a difference in ScratchJr performance during the first eight ScratchJr lessons, and if varied lengths of exposure to ScratchJr correlated with changes in self-regulation scores. The central research question was: *To what extent does self-regulation have a role in learning to code with a novel computer programming software in kindergarten classrooms, and does length of exposure to the programming software correlate with the post-test self-regulation scores of these kindergarten students?*

In learning to code with ScratchJr, children exercise many intersecting concepts across computational thinking, executive function, and self-regulation described above. For example, a child may need to: develop a concept (*creativity*); think about how to compensate when the exact character they want or need does not exist (*out of the box thinking; working memory*); debug code that is not working correctly (*problem solving; staying focused on a task; attention);* and continue working on code despite temptations to explore other areas of a computing device (*stay on task; sustain focus; seek long term reward; inhibition control*)..

ScratchJr, was developed with early childhood developmental theory in mind, paying particular attention to young children's self-regulation, early math ability, and early literacy skills. For example, the number of coding blocks children are presented with in the ScratchJr programming language is limited due to children's developing attention and working memory. There are very few words in ScratchJr in consideration of young children's developing literacy skills, although children can add text to practice writing in addition to sequencing their words within "phrases" or "sentences" of code.

Almost no "instant gratification" buttons (blocks you can click for an immediate action or reward like a plus sign that make your character grow huge) exist in

ScratchJr, in consideration of young children's developing inhibitory control. The lack of such buttons encourages programming over simply clicking for cause and effect. The number range children can work with in ScratchJr was limited in this study to 20 horizontally and 15 vertically based on young children's number sense and inhibition control (i.e., a limit on how large a number can be input into the software).

Of note, the ScratchJr curriculum used in this study was a pilot version and a somewhat hybrid approach to curriculum design. Part of the ScratchJr project as a whole included testing various curriculum with different levels of scaffolding. The children who took part in this version of the study used the most open-ended, least-scaffolded version of the curriculum, which included one day of open-ended exploration, five structured lessons, and two days of semi-structured projects. At the conclusion of the phase of the ScratchJr project described here, a new and significantly more structured curriculum was introduced for future ScratchJr studies.

The focus of this study was to understand how to develop novel coding tools usable by more than the classroom's top performing, most attentive, or most engaged kindergarten students. As such, this study focused on how differences in initial levels of self-regulation and the development of computational thinking skills through learning to code with ScratchJr might be related and, therefore, how do developers ensure coding tools for children are developmentally appropriate for all classrooms?

Measures

Self-regulation was measured through the Head Toes Knees Shoulders (HTKS) assessment (McClelland, et. al., 2007, Cameron, et. al., 2008, Ponitz, McClelland, Matthews, & Morrison, 2009). The HTKS validly and reliably measures working memory, inhibition control, and attention through an assessment directly with children as demonstrated in several prior studies utilizing the measure (McClelland & Cameron, 2012; Ponitz, McClelland, Matthews, & Morrison 2009; McClelland, et al., 2007) and scores range from 0-60. Working memory, inhibition control, and attention are exercised when using ScratchJr in the following ways:

Working Memory. Children must remember their coding project goal, the coding blocks that correspond to the actions they would like for their characters, and the series of instructions that they place on each of their characters.

Attentional Flexibility. Children must switch between characters, backgrounds, and codes for their different characters; switch between categories of coding blocks; and switch between multiple pages within each coding project.

Inhibition Control. Children need to inhibit a dominant response in favor of a more productive one, for example, resisting an urge to spend a majority of their time in the ScratchJr paint tool rather than programming their character.

Computational thinking and coding skills were measured through several researcher-developed measures created through video coding of students and their ScratchJr programs: Programming Score, Goal Completion Score, and Time on Task Score. Participants received a Programming, Goal Completion, and Time on Task Score for each ScratchJr coding lesson.

Programming score. Programming score was calculated from a ScratchJr interface checklist, based on watching the videos of children working with ScratchJr. Programming score was intended to be a measure of what targeted blocks and interface elements of the ScratchJr Lesson the child attempted and understood. Coders watched videos of the participants coding with ScratchJr and completed checklists of all possible coding blocks and interface elements used and understood or attempted. Programming scores ranged from 0-5 based on the percent of targeted blocks and interface elements attempted and understood.

Goal Completion Score. Goal completion scores assessed if the child worked toward and completed the day's assigned ScratchJr lesson. The child received a score of 0 if they did not attempt the lesson, a score of 1 if the child attempted the lesson, and a score of 2 if the child completed the lesson.

Time on Task Score. Time on Task Score was a calculation (0-100%) of how frequently a child was focused on the ScratchJr Lesson, another area of the ScratchJr interface (e.g., the paint editor), or away from the computer. Time on Task was included as a variable since attention span and staying on task towards goals is predictive of future success and is related to self-regulation (McClelland, Acock, Piccinin, Rhea, & Stallings, 2013).

Sample

Participants were members of two kindergarten classrooms located in a densely populated suburb of Boston, Massachusetts with a range of economic and racial-ethnic diversity. Participating teachers and parents/guardians of students provided signed informed consent prior to participation. The initial sample size for this study was 38 students: 19 in Classroom Pre16 (labeled as such because the students took a self-regulation pre-test followed by 16 ScratchJr lessons) and 19 students in Classroom Pre8 (labeled as such because they took a self-regulation pre-test followed by 8 ScratchJr lessons). The average age at the start of the study was 5.51 years old (SD = 0.30). For the study year, the participating school reported in this study was composed of students identified as 40.1% Latinx/Hispanic, 37.1% White, 14.6% African American, 6.7% Asian, and 1.5% Multi-Race/Non-Hispanic students. About 75% of the school was classified as "high needs," including students with a first language other than English (41.2% of the total student population), low-income students (68.4% of students), students receiving free or reduced lunch (68.2% of students), and students classified as Special Education (24.7% of students). Consent forms were completed in English, Spanish, and Portuguese. The classrooms did not vary significantly in terms of age, sex, or pre-test score on the HTKS.

ScratchJr Coding Lesson Descriptions

Introduction/Free Explore. After a brief overview of the interface of the ScratchJr software, children were encouraged to explore the tool on their own and discuss what they had discovered with the researchers, their teachers, and each other.

Airplane Fly Across USA ("Airplane Lesson"). This lesson introduced the "start on flag" block, motion blocks, and the method for changing number parameters. Children programmed an airplane to fly across a map of the USA.

Character Race ("Race Lesson"). In this lesson, students programmed three characters to race against each other at varying speeds. This lesson introduced the set speed block and reinforced the motion and "start on flag" blocks.

Figure 1. Lesson Three. A screenshot of ScratchJr, Lesson 3, where pig, caterpillar, and chicken are programmed to race. The "Start on Flag," "Set Speed," "Move Forward," and "End" blocks are shown.

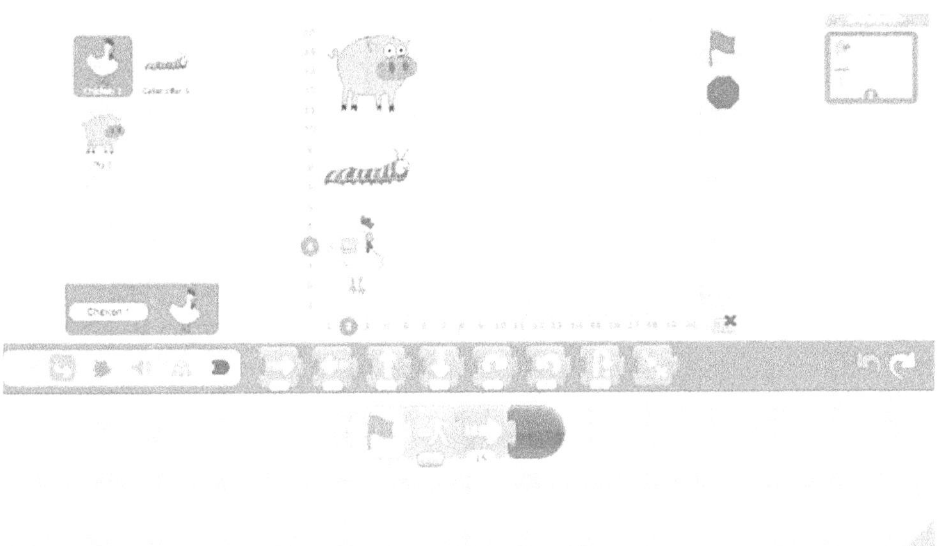

Characters Dance ("Dance Lesson"). The dance lesson introduced the concepts of multiple scripts on one character and sound blocks. Children programmed one character to provide both the music and dance steps, while they programmed a second character to provide just dance steps. This lesson also introduced the "start on bump" block, which initiates a program when two characters bump into each other and the "repeat forever" block, which plays a programming script continuously.

Sunset ("Sunset Lesson"). Children programmed a sunset in this lesson. The show, hide, and go home (reset) blocks were introduced. Start blocks and motion blocks were reviewed. Children that finished the sunset lesson early had the opportunity to attempt a "moon rise" lesson.

Characters Greet Each Other ("Greet Lesson"). In this lesson, children learned how to program characters to have a conversation when they bump into each other. This lesson reinforced the "start on bump" start block and introduced speech bubble blocks, "send a message", and "start on message received" blocks.

"About Me" Project (Project Lesson). Children were instructed to make four ScratchJr pages about themselves, first brainstorming with paper and crayons. The four pages were (1) a picture of themselves with their name, (2) a page about school, (3) a page about home, and (4) a page about what the student wanted to be when they grew up. This task proved too intense for a two-day project. Final projects were considered successful if two of these four pages were created and something was programmed. Children received a lesson on how to use the "add a page" feature before beginning their projects.

Findings

Children in this study were able to use and understand the ScratchJr interface and coding blocks (Programming Score) with not significant differences based on self-regulation pre-test scores. Meaning, ScratchJr appears to be developmentally appropriate to code in for the wide variety of self-regulation levels found at baseline in kindergarten classrooms. Working to ensure the software is developmentally appropriate is working to ensure the software tool itself does not hinder use by students from variable backgrounds. Baseline levels of self-regulation, as measured by the HTKS assessment, did not make a statistical difference when the children were attempting to code with ScratchJr.

Baseline self-regulation levels did matter outside the scope of the ScratchJr interface when the children progressed to projects that were more open-ended, self-directed, and goal-oriented (measured by Goal Completion Score). This variable that heavily relied upon attention and inhibitory control when considering baseline level of self-regulation. Perhaps it is not necessarily that the children with lower levels of self- regulation had more difficulty reaching goals, but instead, that the curricular goals themselves were not motivating enough for students. Substantive anecdotes where children were more successful at lessons about which they were more excited and enthusiastic (repeatedly revisiting Race programs or finding Greet program to be very funny) point to an argument for exploring the importance of engagement and motivation in the context of computational thinking and the development of self-regulation and executive function skills. Time on Task data points to the idea that children can attend more to lessons they find engaging, relevant, and structured, like the Race lesson, but have more difficulty focusing when a goal is too open-ended and the lesson plan did not provide enough scaffolding, as in the students' About Me projects which were personally meaningful but lacked scaffolding.

IMPACT

Although there was not enough power for significance, the differences in self-regulation scores between pre-test and post-test for the Pre8 (eight coding lessons) and Pre16 (sixteen coding lessons) groups are of interest. While self-regulation scores increased for both groups between pre-test and post-test, and the gains in self-regulation scores for the Pre16 group were larger than those for the Pre8 group, the differences between the groups were not significant. Perhaps there is a relationship where eight lessons were enough to gain some familiarity with coding concepts and understanding of the blocks in ScratchJr, but a child may need somewhere between eight and sixteen lessons to begin to work on the computational thinking, problem solving, and debugging that could lead to marked improvements in self-regulation and executive function. However, this study was statistically underpowered to see significant differences in pre-post HTKS self-regulation scores. It is also possible that each teacher could have had a unique impact on self-regulation gains through other aspects of their curriculum. Future studies exploring the relationship between computational thinking and executive function must have numerous treatment and control groups to account for the likely influence of teachers on development of both computational thinking and executive function skills.

IMPLICATIONS AND CONCLUDING THOUGHTS

As noted throughout this book, coding is becoming an increasingly essential skill. A 2017 National Association for the Education of Young Children (NAEYC) blog post highlighted similarities between computational thinking (CT), higher order thinking (HOT), and executive functioning (EF) skills because at their core, CT skills support reasoning, critical thinking, and problem solving (Kaldor, 2017). However, the American Educational Research Association promoted a 25-year meta-analysis in 2015 that found no conclusive evidence that interventions designed to develop students' executive function skills are causally connected to improved academic performance and concluded money should not be spent on these interventions (AERA, 2015; Jacob & Parkinson, 2015). The meta-analysis did conclude, however, that there is a strong correlation between a children's executive function skills and their achievement level (Jacob & Parkinson, 2015) as was explored in the ScratchJr study reported herein.

In conclusion, even if there is no conclusive causal relationship between interventions designed to develop executive function skills on the whole and resulting gains in achievement, it is still worth teaching students interdisciplinary skills that utilize executive function. Given that executive function is strongly correlated with academic and life success (albeit a western ideal of success – a topic beyond the scope of this chapter, see Jaramillo, et. al, 2017 for a discussion), educators should integrate executive function exercises and executive function supportive skills into the curriculum. Furthermore, greater focus should be placed on determining causal relationships of classroom interventions that develop skills within the subdomains of executive function (such as inhibition control, working memory, or attention) and outcomes on academic achievement.

Jacob & Parkinson state "Although investing in executive function interventions has strong intuitive appeal, we should be wary of investing in these often expensive programs before we have a strong research base behind them" (AERA, 2015). This seems to ignore the robust and growing early childhood computational thinking curricula: ScratchJr is free on PBS Kids, for instance, WGBH/PBS are in the process of developing a computational thinking TV show, with digital tools, and hands-on materials for young children, and activities that practice story sequencing on paper or digitally are also low to no cost to families.

Young children grow up surrounded by digital devices and yet know very little about how these tools work (Bers, 2008). This situation is especially true for children from families living in poverty or dealing with systemic racism or otherwise facing education opportunity gaps -- all children whose families have less exposure to new digital devices on a daily basis (Gee, 2013a; 2013b, Gutnick, Robb, Takeuchi, & Kotler, 2010). Understanding how different levels of self-regulation may influence a child's ability to navigate a digital tool and learn to code can inform the design and evaluation of digital tools for young children in order to ensure the most universal access and most successful use of digital devices possible for all children. Given the preliminary findings about the potential bi-directional relationship between coding tools, digital devices, computational thinking and executive function, I argue it is very much worth it to invest in interventions that support multiple facets of a child's education in the 21st century.

Perhaps learning to code with tools like ScratchJr is "only" utilizing and building computational thinking skills, and that, in itself, would be enough. However, students are likely also building their executive function skills. While more research needs to be conducted, it would seem to be a disservice to children to advocate for limiting interventions, especially ones that are multifaceted and cross disciplinary. Teaching computational thinking skills with an integrated curriculum, rather than one that separates coding from social and emotional and cognitive processing skills could potentially create more well-balanced and well-prepared citizens for the 21st century, which will likely need collaborative, creative, problem-solvers both in school and in future careers.

REFERENCES

AERA. (2015). (2015, March 5). *Study: Little Evidence That Executive Function Interventions Boost Student Achievement* [Press release]. Retrieved from https://www.aera.net/Newsroom/News-Releases-and-Statements/Study-Little-EvidenceThat-Executive-Function-Interventions-Boost-Student-Achievement

Allan, W., Coulter, B., Denner, J., Erickson, J., Lee, I., Malyn-Smith, J., & Martin, F. (2010). *Computational thinking for youth.* White Paper for the ITEST Small Working Group on Computational Thinking (CT).

Arfe, B., Vardangega, T., Monturori, C., & Lavanga, M. (2019). Coding in primary grades boosts children's executive functions. *Frontiers in Psychology*, *10*, 2713. doi:10.3389/fpsyg.2019.02713 PMID:31920786

Barr, V., & Stephenson, C. (2011). Bringing computational thinking to K-12: What is Involved and what is the role of the computer science education community? *ACM Inroads*, *2*(1), 48–54. doi:10.1145/1929887.1929905

Baumeister, R. F., & Vohs, K. D. (2004). *Handbook of self-regulation: Research, theory, and applications*. Guilford Press.

Bers, M. U. (2018). *Coding as a playground: Programming and computational thinking in the early childhood classroom*. Routledge.

Blair, C. (2002). School readiness: Integrating cognition and emotion in a neurobiological conceptualization of child functioning at school entry. *The American Psychologist*, *57*(2), 111–127. doi:10.1037/0003-066X.57.2.111 PMID:11899554

Blair, C., & Razza, R. P. (2007). Relating effortful control, executive function, and false belief understanding to emerging math and literacy ability in kindergarten. *Child Development*, *78*(2), 647–663. doi:10.1111/j.1467-8624.2007.01019.x PMID:17381795

Brennan, K., & Resnick, M. (2012, April). New frameworks for studying and assessing the development of computational thinking. In *Proceedings of the 2012 annual meeting of the American educational research association* (*Vol. 1*, p. 25). Academic Press.

Bull, R., & Scerif, G. (2001). Executive functioning as a predictor of children's mathematics ability: Inhibition, switching, and working memory. *Developmental Psychology*, *19*(3), 273–293. PMID:11758669

Cameron, C., McClelland, M. M., Jewkes, A., Connor, C., Farris, C., & Morrison, F. (2008). Touch your toes! Developing a direct measure of behavioral regulation in early childhood. *Early Childhood Research Quarterly*, *23*(2), 141–158. doi:10.1016/j.ecresq.2007.01.004

Center for the Developing Child. (n.d.). *A Guide to Executive Function.* Retrieved from https://developingchild.harvard.edu/guide/a-guide-to-executive-function/

Clements, D. H., Battista, M. T., & Sarama, J. (2001). LOGO and Geometry. Journal for Research in Mathematics Education Monograph Series, 10.

Clements, D.H., Sarama, J., Unlu, F., Layzer, C. (2012, March). *The efficacy of an intervention synthesizing scaffolding designed to promote self- regulation with an early mathematics curriculum: Effects on executive function.* Presentation at Society for Research on Educational Effectiveness (SREE), Washington, DC.

Copple, C., & Bredekamp, S. (2009). *Developmentally appropriate practice in early childhood programs serving children from birth through age 8.* National Association for the Education of Young Children.

CSTA & ISTE. (2011). *Operational Definition of Computational Thinking for K-12 Education.* http://www.iste.org/docs/pdfs/Operational-Definition-of-Computational-Thinking.pdf

Diamond, A. (2002). Normal development of prefrontal cortex from birth to young adulthood: Cognitive functions, anatomy, and biochemistry. In D. T. Stuss & R. T. Knight (Eds.), *Principles of frontal lobe function* (pp. 466–503). Oxford University Press. doi:10.1093/acprof:oso/9780195134971.003.0029

Espy, K. A., McDiarmid, M. M., Cwik, M. F., Stalets, M. M., Hamby, A., & Senn, T. E. (2004). The contribution of executive functions to emergent mathematic skills in preschool children. *Developmental Neuropsychology*, *26*(1), 465–486. doi:10.120715326942dn2601_6 PMID:15276905

Finders, J. K., McClelland, M. M., Geldhof, G. J., Rothwell, D. W., & Hatfield, B. E. (2021). Explaining achievement gaps in kindergarten and third grade: The role of self-regulation and executive function skills. *Early Childhood Research Quarterly*, *54*, 72–85. doi:10.1016/j.ecresq.2020.07.008

Garon, N., Bryson, S. E., & Smith, I. M. (2008). Executive Function in Preschoolers: A Review Using an Integrative Framework. *Psychological Bulletin*, *134*(1), 31–60. doi:10.1037/0033-2909.134.1.31 PMID:18193994

Gestsdottir, S., & Lerner, R. M. (2008). Positive development in adolescence: The development and role of intentional self-regulation. *Human Development*, *51*(3), 202–224. doi:10.1159/000135757

Gioia, I., & Guy, K. (2000). *Behavior Rating Inventory of Executive Function*. Psychological Assessment Resources.

Gropen, J., Clark-Chiarelli, N., Hoisington, C., & Ehrlich, S. B. (2011). The importance of executive function in early science education. *Child Development Perspectives*, *5*(4), 298–304. doi:10.1111/j.1750-8606.2011.00201.x

Grover, S., & Pea, R. (2013). Computational thinking in K–12: A review of the state of the field. *Educational Researcher*, *42*(1), 38–43. doi:10.3102/0013189X12463051

Guzdial, M. (2008). Education Paving the way for computational thinking. *Communications of the ACM*, *51*(8), 25–27. doi:10.1145/1378704.1378713

Jaramillo, J. M., Rendón, M. I., Muñoz, L., Weis, M., & Trommsdorff, G. (2017). Children's self-regulation in cultural contexts: The role of parental socialization theories, goals, and practices. *Frontiers in Psychology*, *8*, 923. doi:10.3389/fpsyg.2017.00923 PMID:28634460

Kaldor, T. (2017). *The T in STEM: Creating Play-Based Experiences That Support Children's Learning of Coding and Higher Order Thinking*. Retrieved from https://www.naeyc.org/resources/blog/creating-play-based-experiences

Kamps, D., Abbott, M., Greenwood, C., Wills, H., Veerkamp, M., & Kaufman, J. (2008). Effects of small-group reading instruction and curriculum differences for students most at risk in kindergarten: Two-year results for secondary- and tertiary-level interventions. *Journal of Learning Disabilities*, *41*(2), 101–114. doi:10.1177/0022219407313412 PMID:18354931

Kazakoff, E. R., & Bers, M. (2012). Programming in a robotics context in the kindergarten classroom: The impact on sequencing skills. *Journal of Educational Multimedia and Hypermedia*, *21*(4), 371–391.

Kazakoff, E. R., & Bers, M. U. (2014). Put your robot in, Put your robot out: Sequencing through programming robots in early childhood. *Journal of Educational Computing Research*, *50*(4), 553–573. doi:10.2190/EC.50.4.f

Kazakoff, E. R., Sullivan, A., & Bers, M. (2013). The effect of a classroom-based intensive robotics and programming workshop on sequencing ability in early childhood. *Early Childhood Education Journal*, *41*(4), 245–255. doi:10.100710643-012-0554-5

Lewin-Bizan, S. G., & Urban, J. B. (Eds.). Thriving in childhood and adolescence: The role of self-regulation processes. New Directions for Child and Adolescent Development, 133, 29–44.

McClelland, M. M., & Cameron, C. E. (2011). Self-regulation and academic achievement in elementary school children. *New Directions for Child and Adolescent Development*, *2011*(133), 29–44. doi:10.1002/cd.302 PMID:21898897

McClelland, M. M., & Cameron, C. E. (2012). Self-regulation in early childhood: Improving conceptual clarity and developing ecologically valid measures. *Child Development Perspectives*, *6*(2), 136–142. doi:10.1111/j.1750-8606.2011.00191.x

McClelland, M. M., Cameron, C. E., Connor, C. M., Farris, C. L., Jewkes, A. M., & Morrison, F. J. (2007). Links between behavioral regulation and preschoolers' literacy, vocabulary and math skills. *Developmental Psychology*, *43*(4), 947–959. doi:10.1037/0012-1649.43.4.947 PMID:17605527

McClelland, M. M., Ponitz, C. C., Messersmith, E., & Tominey, S. (2010). Self-regulation: The integration of cognition and emotion. In The Handbook of Life-Span Development. Vol. 1: Cognition, Neuroscience, Methods (pp. 509–553). Hoboken, NJ: Wiley.

Mischel, W., Shoda, Y., & Rodriguez, M. L. (1989). Delay of gratification in children. *Science*, *244*(4907), 933–938. doi:10.1126cience.2658056 PMID:2658056

Morrison, F. J. (2008). Touch your toes! Developing a direct measure of behavioral regulation in early childhood. *Early Childhood Research Quarterly*, *23*(2), 141–158. doi:10.1016/j.ecresq.2007.01.004

Mulker Greenfader, C. (2019). What is the role of executive function in the school readiness of Latino students? *Early Childhood Research Quarterly*, *49*(4), 93–108. doi:10.1016/j.ecresq.2019.02.011

Paris, A. H., & Paris, S. G. (2003). Assessing narrative comprehension in young children. *Reading Research Quarterly*, *38*(1), 36–76. doi:10.1598/RRQ.38.1.3

Ponitz, C. C., McClelland, M. M., Matthews, J. S., & Morrison, F. J. (2009). A structured observation of behavioral self-regulation and its contribution to kindergarten outcomes. *Developmental Psychology*, *45*(3), 605–619. doi:10.1037/a0015365 PMID:19413419

Promise, D. (2017). *Computational Thinking for a Computational World.* Retrieved from https://digitalpromise.org/wp-content/uploads/2017/12/dp-comp-thinking-v1r5.pdf

Relkin, E. (2018). *Assessing Young Children's Computational Thinking Abilities* (Masters Thesis). Tufts University.

Relkin, E., & Bers, M. U. (2019). *Designing an assessment of computational thinking abilities for young children. In STEM for Early Childhood Learners: How Science, Technology, Engineering and Mathematics Strengthen Learning*. Routledge.

Robertson, J., Gray, S., Toye, M., & Booth, J. N. (2020). The relationship between executive functions and computational thinking. *International Journal of Computer Science Education in Schools*, *3*(4), 35–49. doi:10.21585/ijcses.v3i4.76

Rothbart, M. K. (2007). Temperament, development, and personality. *Current Directions in Psychological Science*, *16*(4), 207–212. doi:10.1111/j.1467-8721.2007.00505.x

Rothbart, M. K., Sheese, B. E., & Posner, M. I. (2007). Executive attention and effortful control: Linking temperament, brain networks, and genes. *Child Development Perspectives*, *1*(1), 2–7. doi:10.1111/j.1750-8606.2007.00002.x

Sameroff, A. J., & Haith, M. M. (1996). *The Five to seven year shift: The age of reason and responsibility*. University of Chicago Press.

Schunk, D. H., & Zimmerman, B. J. (1997). Social origins of self-regulatory competence. *Educational Psychologist*, *32*(4), 195–208. doi:10.120715326985ep3204_1

Shute, V. J., Sun, C., & Asbell-Clarke, J. (2017). Demystifying computational thinking. *Educational Research Review*, *22*, 142–158. doi:10.1016/j.edurev.2017.09.003

Vohs, K. D., & Baumeister, R. F. (2004). Understanding self-regulation: An Introduction. In R. F. Baumeister & K. D. Vohs (Eds.), *Handbook of Self-Regulation: Research, theory, and applications* (pp. 1–9). Guilford Press.

Yelland, N. (2005). Mindstorms or a storm in a teacup? A review of research with Logo. *International Journal of Mathematical Education in Science and Technology*, *26*(6), 853–869. doi:10.1080/0020739950260607

Zelazo, P. D., Carter, A., Reznick, J. S., & Frye, D. (1997). Early development of executive function: A problem solving framework. *Review of General Psychology*, *1*(2), 198–226. doi:10.1037/1089-2680.1.2.198

ADDITIONAL READING

Arfe, B., Vardangega, T., Monturori, C., & Lavanga, M. (2019). Coding in primary grades boosts children's executive functions. *Frontiers in Psychology*, *10*, 2713. doi:10.3389/fpsyg.2019.02713 PMID:31920786

Baumeister, R. F., & Vohs, K. D. (2004). *Handbook of self-regulation: Research, theory, and applications*. Guilford Press.

Cameron, C., McClelland, M. M., Jewkes, A., Connor, C., Farris, C., & Morrison, F. (2008). Touch your toes! Developing a direct measure of behavioral regulation in early childhood. *Early Childhood Research Quarterly*, *23*(2), 141–158. doi:10.1016/j.ecresq.2007.01.004

Center for the Developing Child. (n.d.). *A Guide to Executive Function*. Retrieved from https://developingchild.harvard.edu/guide/a-guide-to-executive-function/

CSTA & ISTE. (2011). *Operational Definition of Computational Thinking for K-12 Education*. http://www.iste.org/docs/pdfs/Operational-Definition-of-Computational-Thinking.pdf

McClelland, M. M., & Cameron, C. E. (2012). Self-regulation in early childhood: Improving conceptual clarity and developing ecologically valid measures. *Child Development Perspectives*, *6*(2), 136–142. doi:10.1111/j.1750-8606.2011.00191.x

Mischel, W., Shoda, Y., & Rodriguez, M. L. (1989). Delay of gratification in children. *Science*, *244*(4907), 933–938. doi:10.1126cience.2658056 PMID:2658056

Ponitz, C. C., McClelland, M. M., Matthews, J. S., & Morrison, F. J. (2009). A structured observation of behavioral self-regulation and its contribution to kindergarten outcomes. *Developmental Psychology*, *45*(3), 605–619. doi:10.1037/a0015365 PMID:19413419

Robertson, J., Gray, S., Toye, M., & Booth, J. N. (2020). The relationship between executive functions and computational thinking. *International Journal of Computer Science Education in Schools*, *3*(4), 35–49. doi:10.21585/ijcses.v3i4.76

KEY TERMS AND DEFINITIONS

Attentional Flexibility: Ability to shift ones focus and attention.

Computational Thinking: Thinking or problem solving systematically like a computer.

Executive Function: Cognitive construct that unites working memory, attention, and inhibitory control for the purposes of planning, problem solving and goal-directed activity.

HTKS: Heads-toes-knees-shoulders assessment of self-regulation for young children developed by McClelland, et. al. measuring a child's working memory, attention, and inhibition control.

Inhibition Control: Ability to prioritize one's actions or behaviors and resist impulses.

Instant Gratification: Immediate availability or action, usually accompanied by a lack of inhibition control in pursuit of a more substantial or significant reward.

ScratchJr: A digital block-based coding language for young children.

Self-Regulation: The behavioral aspects of executive function, including working memory, attention, and inhibition control.

Working Memory: Ability to hold and process thoughts or information.

Chapter 5
Rhyme and Reason:
The Connections Among Coding, Computational Thinking, and Literacy

Madhu Govind
Tufts University, USA

Ziva Reimer Hassenfeld
Brandeis University, USA

Laura de Ruiter
Tufts University, USA

ABSTRACT

The chapter begins with an exploration of computational thinking (CT) and its relationship to computational literacy, followed by a summary of theoretical and empirical work that aims to elucidate the connections among coding, CT, and literacy. The authors argue that these connections thus far have been predominantly one of support (i.e., unidirectional) and motivated by technological and policy advances, as opposed to considering the connections as mutually reinforcing and developmentally coaligned. The authors discuss the coding as another language (CAL) pedagogical approach, a pedagogy that presents learning to program as akin to learning how to use a new language for communicative and expressive functions, emphasizing the bidirectional connections between the two domains. Finally, the authors detail various curricula that use the CAL approach and discuss the implications of CAL for teaching and learning in early childhood.

DOI: 10.4018/978-1-7998-7308-2.ch005

INTRODUCTION

Anyone in the field of computer science education has likely seen or heard the phrase "Coding is the literacy of the twenty-first century," but what meaning does it hold beyond being a lofty metaphor or a catchy marketing slogan? The term "literacy" is often invoked to emphasize the importance of coding in our modern technology-rich world. After all, at the turn of the twentieth century, it was difficult to imagine achieving economic independence or participating in civic society without knowing how to read and write. As such, we might imagine that in the near future, it may be difficult to succeed without some knowledge of coding, or at the very least, a foundational understanding of the computational processes involved in computing.

With the increasing prevalence of technology and the rise of computing jobs (Code.org, 2020; Fayer, Lacey & Watson, 2017), there is no question that coding and computational thinking (CT) have been a growing national and international area of focus. In December 2018, the US White House released a report *Charting a Course for Success: America's Strategy for STEM Education*, in which they named computational literacy as one of the four pathways to success in STEM (Science, Technology, Engineering, and Mathematics) education and the promotion of CT as one of the three objectives for achieving this goal (Committee on STEM Education, 2018). However, there are many unanswered questions about what this pathway looks like, especially in the early years when young children are actively acquiring foundational literacy and language skills. We explore in this chapter the bidirectional and developmentally aligned connections among coding, CT, and literacy. We conclude by suggesting that early childhood coding education need not repeat the mistakes of literacy education. It need not promote inequity and status quo by limiting access or by legitimizing only particular notions of knowledge, truth, and values. By teaching coding through a pedagogy that centers the child and her funds of knowledge, we can show the next generation that coding education is for everyone.

Background

Before CT was popularized as a "universally applicable attitude and skillset everyone, not just computer scientists, would be eager to learn and use" (Wing, 2006, p. 33), there was the notion of computational literacy (diSessa, 2000). Whereas Wing's definition of computational thinking highlighted the universality of the principles behind computer science (CS) that could be used to promote learning in all areas, diSessa's definition of computational literacy extended beyond CS and took into consideration the material, cognitive, and social dimensions of computing.

Many use the terms computational thinking and computational literacy interchangeably, sometimes preferring the former in order to clearly differentiate from digital literacy (Grover & Pea, 2013). However, we argue, as have others (see e.g., Committee on STEM Education, 2018; Li et al., 2020), that the distinction is meaningful. Computational literacy has a broader scope than CT and much greater implications regarding how people think, communicate, and make sense of the world around them. For the purposes of our discussion on the conceptual and pedagogical connections to young children's literacy and language development, we focus specifically on CT.

Given the range of thought on the topic, there is a need for a guiding framework through which we can situate our work. Jacob and Warschauer (2018) propose a three-dimensional framework for exploring the relationship between CT and literacy: 1) understanding the connection between CT and literacy from a cognitive and sociocultural perspective; 2) outlining mechanisms by which existing literacy and language skills can help augment CT; and 3) exploring ways in which CT skills can facilitate the development of traditional and new literacies. To put it succinctly, they see the connection between CT and literacy unfolding in three distinct ways: CT *as* literacy, CT *through* literacy, and literacy *through* CT. This framework proves useful for structuring our discussion in this chapter.

Computational Thinking as Literacy

Understanding the connection between CT and literacy from a cognitive and sociocultural perspective requires a deeper dive into literacy development. There are two great conversations in the study of reading: decoding and comprehension. Most research has focused on decoding – how students turn letters on a page into words. In the last several decades, fierce debates have raged (and continue to rage) about how best to teach decoding. Some say focus on letter-sound correspondence, others point to phonemic awareness, and even others look to whole words. But at a fundamental level, everyone agrees on the rules that govern decoding: the letters "f" "o" and "x" combine to form the word "fox."

Comprehension is the process through which students understand what they are reading. Comprehension is also governed by rules, but there is much less agreement as to the contours of those rules, in part because the process is far less understood (see e.g., Smith, Snow, Serry & Hammond, 2020). Despite the various definitions and conceptions of comprehension, there have been a few consistent perspectives. The cognitive and psychological perspective posits that comprehension involves a complex nonlinear dance of mental processes that transform words on a page into meaning. It includes the application of vocabulary knowledge, but also orthographic, lexical, syntactic and semantic knowledge (Rumelhart, 1994). For example, a child comprehending the sentence "The quick brown dog jumped over the lazy fox" requires the child to convert the code on the page to spoken language, know that "dog" refers to the four-legged barking fluffy animal she sees on leashes in the street, grasp that "over" after "dog" and before "fox" indicates that the quick dog was doing the jumping and not the lazy fox, and understand that it makes sense that a quick dog would jump and a lazy fox would stay put.

This cognitive model of comprehension, however, is critiqued for its detachment from the cultural and social diversity that students and teachers bring to the classroom. Oral and written language, critics argue, is not agnostic like computer code, but always and necessarily a deeply social and cultural phenomenon. Sociocultural researchers start from the position that all students bring their own funds of knowledge to the activity of reading (Handsfield, 2016; Moll et al., 2005; RAND, 2002) and that all reading is always dialogical as a result (Bakhtin, 1981). Thus, the sociocultural view posits that how a child comprehends the sentence "The quick brown dog jumped over the lazy fox" would have as much to do with their cultural associations with dogs, foxes, and the term "lazy" as it would with any syntactic or lexical understanding. For example, you can imagine the adjective "lazy" evoking triggering and hurtful stereotypes for students from traditionally marginalized and discriminated communities. Likewise, students coming from cultures and communities where dogs are not domesticated will draw from a different schema than students from suburban America where many homes keep their pet on a leash.

At the center of the move to sociocultural models of comprehension is the acknowledgment that children grow up in a multimodal world raised in multiliteracies (Serafini & Gee, 2017). The multiliteracies vantagepoint, drawing on semiotic frameworks, argues that people "read" many things that are not words printed on paper, such as road signs, artwork, and even videogames (see Gee, 2007). Another form of text that students are learning to read in the modern moment is computer programs – that which happens *behind* the screen to produce the visual and print texts they see *on* the screen. The reading connection between learning to program and learning to decode and comprehend print texts is without question the most compelling case of CT-related multiliteracies today. And yet the connection is rarely made. Although the field of literacy has embraced technological tools produced by coding (see e.g., International Literacy Association, 2021), the field has yet to consider the bidirectional relationship between coding and literacy except for a few scant articles (e.g., Jacob & Warschauer, 2018; Vee, 2017). We explore this bidirectionality further in the next two sections: CT through literacy and literacy through CT.

Computational Thinking Through Literacy

Efforts to integrate CT into traditional literacy instruction have been largely fueled by technological and policy advances. The rise of coding tools for young children and the increasing adoption of CS standards and frameworks have propelled the pedagogical movement of integration through incorporating coding and CT into traditional literacy instruction. Table 1 illustrates some examples of academic alignment between computer science and English/Language Arts (ELA) standards in the United States. In practice, the integration of coding involves educators finding ways for children to use their existing literacy and language skills and apply them in the context of programming. For example, the practice of writing pseudocode encourages students to use their language abilities to think about a problem and communicate the process—in their natural written and spoken language—for how that problem might be approached (Pane & Myers, 2001). Rather than getting bogged down by the nuances of syntax, children leverage their existing language abilities to think through the logic of the problem, develop their algorithmic plan, and then iteratively revise the plan to match the appropriate computational syntax. The integration of CT in early childhood classrooms may or may not necessarily involve the activity of coding (see Chapter 3 for a longer discussion of unplugged learning). For example, teachers might use CT vocabulary to reinforce classroom routines and practices (e.g., introducing a set of instructions on a handout as an "algorithm" or praising students for their "debugging" skills when they solve a challenging problem). Although the CT terms may not be used *exactly* as defined in a programming context, the practice of reinforcing CT vocabulary in other areas of instruction helps further the goal of cross-curricular integration.

Using programming languages to construct narratives is another method that has been used to infuse CT within literacy practices. For example, Burke and Kafai (2012) explored middle schoolers' storytelling using the Scratch programming language during writer's workshop. Their findings indicated that students' existing knowledge and application of the writing process supported their understanding of CT practices such as designing and debugging. Similar findings were reported with young children. For instance, Portelance and Bers (2015) found that second graders displayed CT concepts such as sequencing and parallel programming while constructing ScratchJr animations and orally sharing their digital artifacts with peers.

Table 1. Standards alignment between computer science/computational thinking (CT) and English/Language Arts (ELA)

CT Concept (Analogous Literacy/Language Connections)	Related Computer Science Teachers Association (CSTA) Standards For K-2	Related Common Core ELA Standards for First Grade
Algorithms (Sequencing)	1A-CS-08: Model daily processes by creating and following algorithms (sets of step-by-step instructions) to complete tasks. 1A-CS-10: Develop programs with sequences and simple loops, to express ideas or address a problem.	CCSS.ELA-LITERACY.RL.1.2: Retell stories, including key details, and demonstrate understanding of their central message or lesson. CCSS.ELA-LITERACY.W.1.3: Write narratives in which they recount two or more appropriately sequenced events, include some details regarding what happened, use temporal words to signal event order, and provide some sense of closure. CCSS.ELA-LITERACY.W.1.7: Participate in shared research and writing projects (e.g., explore a number of "how to" books on a given topic and use them to write a sequence of instructions).
Modularity (Phonological Awareness and Decoding)	1A-CS-11: Decompose (break down) the steps needed to solve a problem into a precise sequence of instructions.	CCSS.ELA-LITERACY.RF.1.2: Demonstrate understanding of spoken words, syllables, and sounds (phonemes). CCSS.ELA-LITERACY.RF.1.3: Know and apply grade-level phonics and word analysis skills in decoding words.
Representation (Alphabet and Letter-Sound Correspondence)	1A-CS-09: Model the way programs store and manipulate data by using numbers or other symbols to represent information.	CCSS.ELA-LITERACY.RF.1.1: Demonstrate understanding of the organization and basic features of print. CCSS.ELA-LITERACY.RF.1.3: Know and apply grade-level phonics and word analysis skills in decoding words.
Hardware/ Software (Tools of Communication and Language)	1A-CS-02: Use appropriate terminology in identifying and describing the function of common physical components of computing systems (hardware). 1A-CS-03: Describe basic hardware and software problems using accurate terminology. 1A-CS-16: Compare how people live and work before and after the implementation or adoption of new computing technology.	CCSS.ELA-LITERACY.W.1.6: With guidance and support from adults, use a variety of digital tools to produce and publish writing, including in collaboration with peers.
Design Process (Writing Process)	1A-CS-12: Develop plans that describe a program's sequence of events, goals, and expected outcomes. 1A-CS-15: Using correct terminology, describe steps taken and choices made during the iterative process of program development.	CCSS.ELA-LITERACY.W.1.5: With guidance and support from adults, focus on a topic, respond to questions and suggestions from peers, and add details to strengthen writing as needed. CCSS.ELA-LITERACY.SL.1.5 Add drawings or other visual displays to descriptions when appropriate to clarify ideas, thoughts, and feelings.
Debugging (Editing and Audience Awareness)	1A-CS-14: Debug (identify and fix) errors in an algorithm or program that includes sequences and simple loops.	CCSS.ELA-LITERACY.SL.1.3: Ask and answer questions about what a speaker says in order to gather additional information or clarify something that is not understood.
Control Structures (Literary Devices)	1A-CS-10: Develop programs with sequences and simple loops, to express ideas or address a problem.	CCSS.ELA-LITERACY.RF.1.4.B: Read grade-level text orally with accuracy, appropriate rate, and expression on successive readings.

Source: (IGI, 2021)

Literacy Through Computational Thinking

The research on how CT can facilitate the development of traditional and new literacies is currently limited in scope, but some work has been done to examine the possible transfer effect of programming on literacy. A meta-analysis of 105 studies that explored the transfer of programming skills to a variety of domains found very little transfer to students' literacy skills (Scherer et al., 2018). The authors concluded that "reading comprehension and writing skills [are] skills that overlap only marginally with programming," (p. 783), although noting that only nine of the 105 studies explored the transfer effect on literacy, and those studies' interventions were not necessarily tailored to foster literacy through programming.

There is some evidence, however, that learning to code can be leveraged to promote young children's language and literacy development. Some studies point to introductory programming as a way to engage students of varying skill levels, highlighting improved student outcomes such as metalinguistic awareness, sequencing and storytelling abilities, and vocabulary and language skills (Burke & Kafai, 2010; Clements, 1999; Fridin, 2014; Movellan, Eckhardt, Virnes, & Rodriguez, 2009; Peppler & Warschauer, 2012). The sequencing and storytelling connection in particular has been explored with early childhood robotics. Kazakoff and colleagues (2013) showed that an intensive robotics and programming intervention as short as one week significantly improved story sequencing abilities among a sample of pre-kindergarten and kindergarten students. Another study by Westlund and Breazeal (2015) indicated that preschool children were able to create stories by engaging in a storytelling game with a social robot. Sullivan and Bers (2015), who conducted a cross-sectional study with 60 pre-kindergarten to second grade students, found that children of all grade levels performed well on sequencing-related programming tasks, with older first and second graders performing slightly better on hard sequencing tasks.

Although these aforementioned studies did not consider literacy as an explicit focus in their curriculum design, these studies indicate, in alignment with Jacob and Warschauer's (2018) framework, that the connections among coding and CT on the one hand, and literacy and language on the other, may be developmentally coaligned. In order to support integration of these two curricular domains in a way that produces meaningful outcomes in both areas, we must first understand the similarities and differences between natural languages and artificial programming languages, which we unpack in the next section.

UNPACKING THE SIMILARITIES AND DIFFERENCES

At their core, both artificial programming languages and natural languages are symbolic, representational systems with a grammar and syntax that can be used to convey meaning and to express ideas that others can interpret (Bers, 2019). Research studies have explored the similarities between programming languages and natural languages, showing how learning to program might be beneficial for learning new languages. For instance, Sara Vogel and colleagues (2020) propose that translanguaging pedagogy—a term used in bilingual education referring to how students use all of their linguistic resources across and beyond their multiple languages to learn—might also be applied to computer science pedagogy to engage children in CT practices alongside supporting their multiple language acquisition.

Some of the recent work at the DevTech Research Group has focused on exploring the relationship between early elementary students' literacy levels and programming skills. Hassenfeld and colleagues (2020), for example, measured 132 second graders' phonological awareness at the beginning of their school year using the Phonological Awareness Literacy Screening (PALS), a diagnostic tool that looks at abilities such as phonemic awareness, alphabet knowledge, letter-sound knowledge and word recognition. Phonological awareness in the early grades is an important predictor of later reading achievement (Hogan, Catts, & Little, 2005). The researchers also assessed students' programming skills in the KIBO programming language and the knowledge of programming concepts using an assessment called KIBO Mastery Challenges (KMCs) at different times over the course of a KIBO robotics curriculum. They found that there was evidence for a weak, positive correlation ($r = 0.3$) between PALS scores and KMC scores. Hassenfeld and colleagues' findings are in line with other studies that looked at the relationship between CT (not programming) and verbal abilities more generally in older children and adolescents (Román-González et al., 2017). These studies indicate there may be an overlap between language and literacy ability in children on the one hand and the ability to learn computational concepts on the other.

In addition to the connections between reading (decoding and comprehension) and coding, there are also interesting connections between writing and coding. Coding and writing are both compositional processes, and they share a subset of activities (Hassenfeld & Bers, 2020). Coding is usually preceded by planning – determining what the purpose of the program is, for example, by creating a flowchart or thinking aloud what you want the computer to do. Writing is also usually preceded by planning and pre-writing, for example, by researching a topic, jotting down notes, or creating a graphic organizer. Then, both programmer and writer create their first program or first draft and evaluate (test) it, becoming observers of how their program performs or how their text reads. This first product is almost never the way the composer had envisioned it. The artifact needs to be debugged (program) or edited and revised (text). There may be mechanical errors (e.g., forgetting an end block in a program; missing a punctuation mark after a sentence) or stylistic errors (e.g., using multiple individual blocks when a repeat loop would be more efficient; using a word that does not have quite the intended meaning). In an iterative process, the composer may review their product, identify these errors, and take steps to correct them until the product matches what they had in mind.

However, editing and revising are often where there are differences between the two processes, at least as observed in young children. When writing, novice writers are often reluctant to edit and revise (see, e.g., Applebee et al., 1986; Fitzgerald & Markham, 1987; Hassenfeld & Bers, 2020). In contrast, novice programmers often dedicate considerable time to getting their program 'right'. Potential reasons lie in the affordances of each medium – with debugging in ScratchJr and KIBO being less cumbersome than erasing penciled text and rewriting – and the role of feedback. Unlike a reader of written text, a computer can provide immediate feedback on a program. The programmer sees right away what is working and what isn't. In other words, the programmer can shift their role between producer and consumer of their computational artifact much more readily (e.g., "Is the character moving as much as it should? Maybe instead of moving 3 steps to the left, the character should move 5 steps"). On the other hand, the process of writing and reading one's own writing for clarity, grammar, and other stylistic aspects requires a qualitatively different level of audience awareness and role navigation between producer and consumer.

Understanding these similarities and differences between coding and writing has important implications for teaching and learning. For example, a teacher who struggles with motivating his students to revise their writing or to compose a story might introduce a coding application as a supportive tool to help students understand audience awareness or to inspire students' story planning (e.g., Delacruz, 2020). In a different vein, a student who is below grade-level in their reading and writing might be encouraged to explore block-based programming to express their creative ideas and produce meaningful computational artifacts (e.g., Peppler & Warschauer, 2012). These examples push us to consider the reframing of CS education in early childhood as one that embraces the connections to literacy and language. This perspective, of course, requires understanding early childhood teachers' dispositions towards both disciplines. We might ask, for example, what are early childhood teachers' views on and priorities regarding literacy education, and what role can coding play to support those priorities? To what extent might the broader discourse on CS education (and the push to fuel the STEM professional pathway) possibly cloud early childhood teachers' perceptions of coding as a sense-making creative activity?

These questions set the stage for our current work developing and implementing a coding and CT curriculum and sustainable professional development model in

K-2. Our work actively engages teachers in viewing coding as another language, exploring cross-disciplinary alignment, and understanding feasible practices for implementation and integration. In the next section, we examine these topics by introducing the pedagogical approach developed by Prof. Marina Bers and members of the DevTech Research Group at Tufts University called "Coding as Another Language" (CAL).

The Coding as Another Language (CAL) Pedagogical Approach

The CAL pedagogical approach is grounded in three theoretical perspectives for how young children learn and develop when engaging with computational tools: 1) constructionism (Papert, 1980), 2) positive technological development (Bers, 2012), and 3) dialogic instruction (Clarke et al., 2015; Littleton & Howe, 2010; Resnick et al., 2018).

1. Constructionist theory, as its root word "construct" implies, is the process-oriented theory that maintains people learn best when they actively build and make things that can be shared with others (as opposed to the product-oriented instructionist approach, in which knowledge is transmitted from instructor to learner).[1] In the context of early childhood coding education, constructionist theory emphasizes programming as an opportunity for young children to construct their own programs and make personally meaningful projects.
2. The Positive Technological Development framework proposed by Bers (2012) identifies six kinds of behaviors that can be fostered in a technology-mediated learning experience: content creation, creativity, communication, collaboration, community building, and choices of conduct. These behaviors are not only supported by the computational tool or activity, but also the context of the learning environment and the ways in which the activity is presented to children by the teacher or facilitator. This latter point is emphasized through the third theoretical perspective: dialogic instruction.
3. In dialogically organized instruction (Nystrand, 1997), the teacher is not positioned as the sole authoritative expert, nor is the child positioned as the novice. Instead, teachers engage their students in authentic explorations of the subject matter and invite open-ended discussions of student ideas and interpretations. As a result, learning is co-constructed collaboratively and through active discourse.

We next describe how these three theoretical perspectives are operationalized into full-length curricula with KIBO robotics and ScratchJr, two block-based programming tools for young children developed by the DevTech Research Group.

Operationalization of the CAL Approach into Early Childhood Curricula

The CAL curricula are organized into grade-level units, all centered around various children's books and an introductory programming language for young children (either the KIBO robotics kit or the screen-based ScratchJr application). Regardless of the programming language used, each curriculum unit follows a similar structure and consists of games, songs, design challenges, free play, expressive explorations, reading and writing activities, and technology circles. Each curriculum unit is aligned with nationally recognized computer science and literacy standards and frameworks, including the International Society for Technology in Education (ISTE) Standards for Students, K-12 Computer Science Framework, Computer Science Teachers Association (CSTA) K-12 Computer Science Standards, and Common Core ELA Standards. We next describe the set of curriculum units split by programming language, first KIBO and then ScratchJr.

The CAL KIBO curriculum uses the KIBO programming language to introduce young children to foundational concepts of coding, CT, and robotics. Formerly known as KIWI in its early research prototype form, KIBO is a screen-free robotics set sold commercially by KinderLab Robotics, Inc. The kit is comprised of a battery-operated robot with an embedded barcode scanner and the following detachable parts: wheels and motors; lightbulb and sound recorder modules that enable the robot to light up and make sounds; sensors that enable the robot to sense light, sound, and distance; tangible wooden programming blocks with barcode stickers; and art platforms enabling children to personalize their robots with arts and crafts. The DevTech Research Group has developed CAL KIBO curricula for Pre-Kindergarten, Kindergarten, First Grade, and Second Grade. Each grade-level curriculum spans between 12 to 30 lessons of approximately 30-60 minutes each. The variability in lesson length and duration takes into account developmental differences of students and the level of programming complexity introduced at each grade level. However, as with any curriculum, the content and pacing can be adjusted to particular learning settings to meet the needs of teachers and students.

The CAL ScratchJr curriculum uses the ScratchJr programming language to teach children to code. The ScratchJr programming language is an introductory, visual programming language for children between five and seven years of age. ScratchJr is a freely available app and, at the time of writing, the most popular free programming language in the world (Bers, 2020). In the app, children can create stories and games by putting together graphical programming blocks that represent different commands, similar to the wooden blocks used with the KIBO robot. All menu options and instructions are represented by symbols and colors, so children at all literacy levels can use it as well. The CAL ScratchJr curriculum consists of

24 individual lessons of approximately 45 minutes each (totaling 18 hours), but the pacing can be adjusted to particular learning settings. Individual curricula have been developed for Kindergarten, First Grade, and Second Grade. The curriculum provides integration between computer science and programming in the context of literacy. Throughout the 24 lessons, students learn to explore two books (different ones for each grade) to write creative, fun programs on ScratchJr. The curriculum culminates with an open-ended project to share with family and friends.

Table 2 illustrates the different types of activities in the CAL KIBO and ScratchJr curricula and highlights the CT concepts and skills that are supported throughout the lessons. The next section describes three example lesson activities from the CAL curricular units with related CT concepts italicized.

Examples of CAL Lesson Activities

How-To Prompts

Functional texts, or texts used for everyday communication that serve a particular purpose (e.g., recipes, manuals, instructions, etc.), are an integral aspect of early elementary literacy education. Children are regularly tasked with following single- and multi-step directions and communicating how to do something on their own. How-To prompts are a low-stress entry point into writing and provide children the opportunity to reflect on the process they undertake to accomplish a task (design process) and to communicate the steps of that process in a sequential and detailed fashion (algorithms). There are several activities in the CAL KIBO and ScratchJr curricular units that engage children in CT through procedural writing or functional texts. For instance, one prompt in the CAL KIBO First Grade curriculum is "What are the steps for making a pizza? What toppings will you put on your pizza? Draw or write these steps in your Design Journal." Another activity is "Program the Hokey-Pokey," in which the class dances to the Hokey-Pokey song and brainstorms a set of programming actions that would correspond to the physical movements for the song. Children then program their KIBO robots or ScratchJr characters to dance the Hokey-Pokey.

Table 2. Summary of CAL KIBO and ScratchJr curricular activity types

Activity	Purpose of Activity	CAL KIBO Example	CAL ScratchJr Example
Warm Up	Playfully introduce or reinforce concepts	Children sing and dance to the "Robot Parts" song that describes how children act as engineers to connect the different parts of the KIBO robot and to program it to move using the blocks (*hardware and software*).	Children are shown a picture of a street and search for symbols (STOP signs, zebra crossings, a shop's sign). The activity gets children started in understanding and expressing that symbols stand for something else (*representation*).
Opening/ Closing Technology Circle	Come together to discuss, share, and reflect on activities and concepts	Children gather in a community circle to share problems they had while creating and scanning KIBO programs and discuss problem-solving strategies (*debugging*).	Children gather in a community circle to talk about the rules and elements of a race (e.g., distance, participants, speed).
Structured Coding Challenge	Engage children in powerful ideas in computer science through learning new coding skills	Children learn about algorithms and how KIBO will perform the actions in the same order that the blocks are assembled and scanned using the robot's embedded barcode scanner (*algorithms*).	Children learn what a parameter is and why parameters are useful (e.g., instead of using six turn blocks, they can use just one turn block and change the number of times it will be used to six).
Expressive Coding Explorations	Practice learned coding skills in an open-ended way	Children engage in an iterative design process to create their own version of the Hokey-Pokey dance using the KIBO blocks. After planning, testing and revising their programs, children share their KIBO dances with their peers (*design process*).	Children apply their knowledge of the speed blocks to program their own ScratchJr race between multiple characters (*control structures*).
Unplugged Time	Promote CT learning, social interaction, and movement without the use of any devices	Children play a game of "Red Light, Green Light" to reinforce the meanings of the green Begin block and the red End block. The activity aims to promote children's understanding of attributes such as color or symbol being used to communicate information (*representation*).	Children play a game of "Programmer Says," similar to the traditional "Simon Says" game, in which children repeat an action (a ScratchJr command) when instructed by the programmer (i.e., the teacher or a child).
Word Time	Engage children in powerful ideas of literacy and language	Children listen to a poem or song and try to identify the repeating words or phrases. This activity is extended into a discussion about repetition as a literary device and how we can use repeat loops in our KIBO programs (*control structure*).	Children learn about the importance of sequencing by planning their own story using a planning sheet with three sections for the beginning, middle, and end of the story with lines for writing and space for drawing (*algorithms*).

Source: (IGI, 2021)

Tools of Communication

People can communicate with one another in a variety of ways, for instance, through oral, gestural, written, and pictorial representations. New technological tools such as telephones, e-mail, video-chatting platforms, and emojis have further expanded the ways in which people can communicate (hardware and software). Each form of communication has its own strengths and limitations and, depending on the context, might be a preferable method of getting the right message across to the recipient. Although natural languages might have some flexibility in interpretation, programming languages do not; without proper syntax and grammar, the computer will not interpret the programmed instructions appropriately. The CAL KIBO and ScratchJr curricular units engage children in exploring different tools of communication and in reflecting on the similarities and differences between natural and artificial programming languages.

For example, one lesson activity involves a game of "Telephone," in which one student thinks of a message and whispers it to the person sitting next to them, who then whispers to the person next to them, and so on until the message gets to the last person. The first and last people deliver these messages out loud, and the class compares and discusses the two messages. Children then play additional modified rounds, such as with a handwritten message or with a printed or typed message. Children discuss the experience of receiving and communicating messages in different forms (representation), and importantly how they would revise their communication if the recipient is confused. This activity of revising is later connected to the importance of troubleshooting errors (debugging) when children are programming with KIBO or ScratchJr.

Creative Writing and Coding Compositions

The final lessons of the CAL KIBO and ScratchJr curricula invite students to compose creative artifacts through writing and programming. Students first engage in a book read-aloud, which serves as inspiration for their final project creations. They ask and imagine an alternative ending for one of the book's characters or (in the case of the Pre-Kindergarten KIBO curriculum and the book *Pete the Cat: Robo-Pete* by James Dean) what their own robot-friend would look like and do. Students compose a written artifact about their project idea or orally share their initial ideas with peers, and then are tasked with programming and designing their final projects. Once students have the first iterations of their creations, they test out their designs, troubleshoot bugs, and then share their projects with peers, families, and community members (design process). Through this process of planning and designing their projects, children engage in computational thinking and making and are able to experience the unfolding of their unique ideas with each medium.

IMPLICATIONS FOR TEACHING AND LEARNING

For a long time, computer science has been viewed as a talent that people either have or don't have, and the computing field has been highly gendered as well (Miller, 2017). However, once the parallels between literacy and coding are appreciated, it becomes clear that CT is a teachable and learnable skill just like reading and writing. Recent research using different methodologies has shown that both girls and boys can improve their CT skills through classroom coding instruction. For example, Pérez-Marín and colleagues (2018) found that 9 to 12-year-old children who were taught computer science concepts using metaphors and the Scratch App (a block-based programming language for children ages eight and up) were able to improve their scores on standardized multiple-choice CT assessments (Román-González et al., 2017). Working with even younger children, Relkin and colleagues examined changes in CT skills in first and second grade students (six- and seven-year-olds) who had been exposed to the CAL KIBO curriculum compared to children who did not. Over the course of the study, children who received CAL KIBO improved on their CT skills as measured by the unplugged CT assessment TechCheck (Relkin et al., 2020), whereas the control group did not. There were no differences between boys and girls. Unlike other studies, this study included a control group, a crucial element for being able to show that the improvements are due to instruction, and not a result of students improving by themselves or through business-as-usual classroom instruction. The study adds to the growing body of evidence (Lye & Koh, 2014) that CT can be taught successfully just like literacy, and that early interventions have strong potential to help dismantle gender stereotypes about the computing field (Sullivan, 2019; see Sullivan's Chapter 11).

CONCLUSION

Just as there is an inextricable, bidirectional link between thought and language (Vygotsky, 2012), there cannot be a complete conversation about coding and CT without a discussion of literacy and language. In this chapter, we presented these two sets of domains as mutually reinforcing and developmentally coaligned. As a literacy, coding engages children in thinking about "powerful ideas" (Papert, 1980) from computer science as well as other domains. Through literacy, coding and CT provide opportunities for children's sense-making. Through coding and CT, children are able to use problem solving as a means towards self-expression and communication, ultimately becoming able to navigate the digital world around them.

While we see great value in appreciating the connections between CT, coding, literacy and language, it is important to highlight that there are also several crucial differences. First and foremost, when children learn to write, they already have spoken language to start from. They need to learn new symbols (letters, punctuation, etc.), but they don't have to learn the grammar or the vocabulary of their language, which they've had years of experience and practice with. In contrast, learning to program means learning a language that has hardly any connection to the language they speak (only via the route of written language, such as "start" written on a start block). There is no spoken language equivalent of the ScratchJr programming language like there is with human languages (Goswami, 2001). In addition, both written and spoken language are very accommodating when it comes to errors. We can process speech effortlessly although most spontaneous utterances are not "grammatical," and we can read texts riddled with spelling errors. A program, however, may not run at all if there is even a single error in it. Language (written and spoken) fulfills many different functions – we can use it to describe, to question, to praise, to scold, to plea, to apologize, to congratulate, and so on. We can modulate our tone and the level of politeness. It's a complex system that has evolved over thousands of years. Programs can do many things, but they do not match the functionality of human language.

These differences notwithstanding, appreciating the parallels can open our eyes to another important point: literacy is and always has been a deeply political issue. Literacy has stood at the heart of access, power and hierarchy in our society. The United States in the twentieth century used literacy as a barrier for people of color, people of low socioeconomic backgrounds, and women. One of the main ways that literacy has been manipulated to support the status quo and disenfranchise is to limit access. At this present moment in time when the country is divided and hateful rhetoric is at a dangerous high, we cannot afford as a nation to repeat the mistakes of print literacy with digital literacy. The future of coding education begins with access. All children deserve to learn how to code and learn from an early age.

However, access is not the only way literacy has been used to disenfranchise. Literacy instruction has also been used to legitimize particular notions of knowledge, truth and values. Assessment of literacy in schools has been used to promote philosophically narrow and biased viewpoints linked to race, class and gender. As Willis and Harris (2000) explain, "Literacy learning and teaching has never been ideologically neutral or culturally unbiased. It has been a series of related political acts of ideological domination and conformity draped under a thick veil of paternalism" (p. 78). The need for more culturally sensitive pedagogies for literacy is a call that goes out against an educational landscape in which demographics suggest that future teachers are "most likely to be white monolingual females from suburban and rural middle-class homes [while] the student population the next century suggest two out of every three students will be children of color" (Willis & Harris, 2000, p. 76).

CAL begins the corollary work of multiliteracies for computational thinking and coding. While every child needs access to the new literacy of coding, they also need instruction in ways that overcome the same narrow, transmissive focused way that literacy has been taught (and is only now beginning to be transformed). CAL is a step in the right direction of pedagogical approaches and curricula that no longer ignore students' lives, contexts and desire to make meaning with the resources of their lived experience, and avoids a pedagogy that continues to exclude and marginalize. Like literacy, coding must start from a pedagogical premise that asks students to bring their questions, interests, and experiences to the task at hand. Only then will the full benefits of CT and literacy come to fruition.

ACKNOWLEDGMENT

This research was supported by the U.S. Department of Defense Education Activity Grant "Breaking the Code for College and Career Readiness" and the U.S. Department of Education [grant number U411C190006]. The authors also wish to thank Dr. Olson Pook for copyediting this chapter.

REFERENCES

Applebee, A. N., Langer, J. A., & Mullis, I. V. S. (1986). The Writing Report Card: Writing Achievement in American Schools. Princeton, NJ: Educational Testing Service; Washington, DC: Office of Educational Research and Improvement.

Bakhtin, M. M. (1981). *The dialogic imagination: Four essays* (M. Holquist, Ed. & Trans.). University of Texas Press.

Bers, M. (2020). *Coding as a Playground: Programming and Computational Thinking in the Early Childhood Classroom* (2nd ed.). Routledge Press. doi:10.4324/9781003022602

Bers, M. U. (2012). Designing Digital Experiences for Positive Youth Development: From Playpen to Playground. Cary, NC: Oxford. doi:10.1093/acprof:oso/9780199757022.001.0001

Bers, M. U. (2019). Coding as another language: A pedagogical approach for teaching computer science in early childhood. *Journal of Computers in Education*, *6*(4), 499–528. doi:10.100740692-019-00147-3

Burke, Q., & Kafai, Y. B. (2010). Programming & storytelling: Opportunities for learning about coding & composition. *Proceedings of the 9th International Conference on Interaction Design and Children*. 10.1145/1810543.1810611

Clarke, S., Resnick, L. B., & Rose, C. P. (2015). *Dialogic instruction: A new frontier.* Academic Press.

Clements, D. (1999). The Future of Educational Computing Research: The Case of Computer Programming. In C. Hoyles & R. Noss (Eds.), Learning mathematics and Logo. Academic Press.

Code.org. (2020). *Leaders and Trendsetters Agree More Students Should Learn Computer Science*. https://code.org/promote

Committee on STEM Education, National Science & Technology Council, the White House. (2018). *Charting a course for success: America's strategy for STEM education*. https://www.whitehouse.gov/wp-content/uploads/2018/12/STEM-Education-Strategic-Plan-2018.pdf

Delacruz, S. (2020). Starting From Scratch (Jr.): Integrating Code Literacy in the Primary Grades. *The Reading Teacher*, *73*(6), 805–811. doi:10.1002/trtr.1909

diSessa, A. A. (2000). *Changing minds: Computers, learning, and literacy*. MIT Press. doi:10.7551/mitpress/1786.001.0001

Fayer, S., Lacey, A., & Watson, A. (2017). *BLS Spotlight on Statistics: STEM Occupations - Past, Present, and Future*. https://hdl.handle.net/1813/79240

Fitzgerald, J., & Markham, L. R. (1987). Teaching children about revision in writing. *Cognition and Instruction*, *4*(1), 3–24. doi:10.12071532690xci0401_1

Fridin, M. (2014). Storytelling by a kindergarten social assistive robot: A tool for constructive learning in preschool education. *Computers & Education*, *70*, 53–64. doi:10.1016/j.compedu.2013.07.043

Gee, J. P. (2007). What Video Games Have to Teach Us About Learning and Literacy. *Cyberpsychology & Behavior*, *12*(1).

Goswami, U. (2001). Early phonological development and the acquisition of literacy. Handbook of Early Literacy Research, 111-125.

Grover, S., & Pea, R. (2013). Computational Thinking in K–12: A Review of the State of the Field. *Educational Researcher*, *42*(1), 38–43. doi:10.3102/0013189X12463051

Handsfield, L. (2016). *Literacy Theory as Practice: Connecting Theory and Instruction in K–12 Classrooms*. Teachers College Press.

Hassenfeld, Z. R., & Bers, M. U. (2020). Debugging the Writing Process: Lessons From a Comparison of Students' Coding and Writing Practices. *The Reading Teacher*, *73*(6), 735–746. doi:10.1002/trtr.1885

Hassenfeld, Z. R., Govind, M., de Ruiter, L. E., & Bers, M. U. (2020). If You Can Program, You Can Write: Learning Introductory Programming Across Literacy Levels. *Journal of Information Technology Education*, *19*, 65–85. doi:10.28945/4509

Hogan, T. P., Catts, H. W., & Little, T. D. (2005). The Relationship between Phonological Awareness and Reading: Implications for the Assessment of Phonological Awareness. *Language, Speech, and Hearing Services in Schools*, *36*(4), 285–293. doi:10.1044/0161-1461(2005/029) PMID:16389701

International Literacy Association. (2021). *Teaching with Tech*. https://www.literacyworldwide.org/blog/digital-literacies/teaching-with-tech

Jacob, S. R., & Warschauer, M. (2018). Computational thinking and literacy. *Journal of Computer Science Integration*, *1*(1). Advance online publication. doi:10.26716/jcsi.2018.01.1.1

Kazakoff, E., Sullivan, A., & Bers, M. U. (2013). The effect of a classroom-based intensive robotics and programming workshop on sequencing ability in early childhood. *Early Childhood Education Journal*, *41*(4), 245–255. doi:10.100710643-012-0554-5

Li, Y., Schoenfeld, A. H., diSessa, A. A., Graesser, A. C., Benson, L. C., English, L. D., & Duschl, R. A. (2020). Computational Thinking Is More about Thinking than Computing. *Journal for STEM Education Research*, *3*(1), 1–18. doi:10.100741979-020-00030-2 PMID:32838129

Littleton, K., & Howe, C. (2010). *Educational Dialogues: Understanding and Promoting Productive Interaction*. Routledge. doi:10.4324/9780203863510

Lye, S. Y., & Koh, J. H. L. (2014). Review on teaching and learning of computational thinking through programming: What is next for K-12? *Computers in Human Behavior*, *41*, 51–61. doi:10.1016/j.chb.2014.09.012

Miller, C. C. (2017). *Tech's Damaging Myth of the Loner Genius Nerd.* https://www.nytimes.com/2017/08/12/upshot/techs-damaging-myth-of-the-loner-genius-nerd.html

Moll, L., Amanti, C., Neff, D., & González, N. (2005). Funds of knowledge for teaching: Using a qualitative approach to connect homes and classrooms. In Funds of Knowledge: Theorizing Practices in Households, Communities, and Classrooms (pp. 71-88). Lawrence Erlbaum Associates.

Movellan, J., Eckhardt, M., Virnes, M., & Rodriguez, A. (2009). Sociable robot improves toddler vocabulary skills. *Proceedings of the 4th ACM/IEEE International Conference on Human Robot Interaction.* 10.1145/1514095.1514189

Nystrand, M. (1997). *Opening Dialogue: Understanding the Dynamics of Language and Learning in the English Classroom.* Teachers College Press.

Pane, J. F., & Myers, B. A. (2001). The impact of human-centered features on the usability of a programming system for children. *Proceedings of CHI EA'02.*

Papert, S. (1980). *Mindstorms: Children, computers, and powerful ideas.* Basic Books.

Peppler, K. A., & Warschauer, M. (2012). Uncovering Literacies, Disrupting Stereotypes: Examining the (Dis)Abilities of a Child Learning to Computer Program and Read. *International Journal of Learning and Media, 3*(3), 15–41. doi:10.1162/IJLM_a_00073

Pérez-Marín, M., Hijón-Neira, R., Bacelo, A., & Pizarro, C. (2018). Can computational thinking be improved by using a methodology based on metaphors and Scratch to teach computer programming to children? *Computers in Human Behavior.*

Portelance, D. J., & Bers, M. U. (2015). Code and Tell: Assessing young children's learning of computational thinking using peer video interviews with ScratchJr. *Proceedings of the 14th International Conference on Interaction Design and Children (IDC '15).* 10.1145/2771839.2771894

RAND Reading Study Group. (2002). *Reading for Understanding, toward an R&D Program in Reading Comprehension.* RAND.

Relkin, E., de Ruiter, L., & Bers, M. U. (2020). TechCheck: Development and Validation of an Unplugged Assessment of Computational Thinking in Early Childhood Education. *Journal of Science Education and Technology, 29*(4), 482–498. doi:10.100710956-020-09831-x

Resnick, L. B., Asterhan, C. S. C., & Clarke, S. (2018). Next Generation Research in Dialogic Learning. In G. E. Hall, L. F. Quinn & D. M. Gollnick (Eds.), Wiley Handbook of Teaching and Learning (pp. 338-323). Wiley-Blackwell.

Resnick, M., Maloney, J., Monroy-Hernandez, A., Rusk, N., Eastmond, E., Brennan, K., Millner, A., Rosenbaum, E., Silver, J., Silverman, B., & Kafai, Y. (2009). Scratch: Programming for Everyone. *Communications of the ACM*, *52*(11), 60–67. doi:10.1145/1592761.1592779

Román-González, M., Pérez-González, J., & Jiménez-Fernández, C. (2017). Which cognitive abilities underlie computational thinking? Criterion validity of the Computational Thinking Test. *Computers in Human Behavior*, *72*, 678–691. doi:10.1016/j.chb.2016.08.047

Rumelhart, D. E. (1994). Toward an interactive model of reading. In R. B. Ruddell, M. R. Ruddell, & H. Singer (Eds.), *Theoretical models and processes of reading* (pp. 864–894). International Reading Association.

Scherer, R., Siddiq, F., & Sánchez Viveros, B. (2018). The cognitive benefits of learning computer programming: A meta-analysis of transfer effects. *Journal of Educational Psychology*, *111*(5), 764–792. doi:10.1037/edu0000314

Serafini, F., & Gee, E. (2017). *Remixing multiliteracies: Theory and practice from New London to new times*. Teachers College Press.

Smith, R., Snow, P., Serry, T., & Hammond, L. (2020). The Role of Background Knowledge in Reading Comprehension: A Critical Review. *Reading Psychology*, *42*(3).

Sullivan, A. (2019). *Breaking the STEM Stereotype: Reaching Girls in Early Childhood*. Rowman & Littlefield.

Vee, A. (2017). *Coding Literacy: How Computer Programming Is Changing Writing*. The MIT Press. doi:10.7551/mitpress/10655.001.0001

Vogel, S., Hoadley, C., Castillo, A. R., & Ascenzi-Moreno, L. (2020). Languages, literacies, and literate programming: Can we use the latest theories on how bilingual people learn to help us teach computational literacies? *Computer Science Education*, *30*(4), 420–443. doi:10.1080/08993408.2020.1751525

Vygotsky, L. (2012). *Thought and language*. MIT Press.

Westlund, J., & Breazeal, C. (2015). The Interplay of Robot Language Level with Children's Language Learning During Storytelling. In *Proceedings of the Tenth Annual ACM/IEEE International Conference on Human-Robot Interaction Extended Abstracts*. ACM. 10.1145/2701973.2701989

Willis, A. I., & Harris, V. (2000). Political acts: Literacy learning and teaching. *Reading Research Quarterly*, *35*(1), 72–88. doi:10.1598/RRQ.35.1.6

Wing, J. M. (2006). Computational Thinking. *CACM Viewpoint*, 33-35. http://www.cs.cmu.edu/afs/cs/usr/wing/www/publications/Wing06.pdf

ADDITIONAL READING

Aguilar, R. (2014, July 30). *Your Call: Is coding the new literacy?* [Radio broadcast]. KALW. https://www.kalw.org/show/your-call/2014-07-30/your-call-is-coding-the-new-literacy

Bers, M. U. (2018). Coding as a Literacy for the 21st Century. https://www.edweek.org/education/opinion-coding-as-a-literacy-for-the-21st-century/2018/01

Bers, M. U. (2019). Coding as another language. In C. Donohue (Ed.), *Exploring key issues in early childhood and technology: Evolving perspectives and innovative approaches* (pp. 63–70). Routledge. doi:10.4324/9780429457425-11

DevTech Research Group. (2021). Coding as Another Language: Teaching programming as a literacy of the 21st century. https://sites.tufts.edu/codingasanotherlanguage/

Fedorenko, E., Ivanova, A., Dhamala, R., & Bers, M. U. (2019). The Language of Programming: A Cognitive Perspective. *Trends in Cognitive Sciences*, *23*(7), 525–528. doi:10.1016/j.tics.2019.04.010 PMID:31153775

Hassenfeld, Z. R., & Bers, M. U. (2019). When We Teach Programming Languages as Literacy. https://www.literacyworldwide.org/blog/literacy-now/2019/05/16/when-we-teach-programming-languages-as-literacy

Vee, A. (2013). *Ideologies of a New Mass Literacy.* https://vimeo.com/61820239

Vee, A. (2013). *Is coding the new literacy everyone should learn? Moving beyond yes or no.* http://www.annettevee.com/blog/2013/12/11/is-coding-the-new-literacy-everyone-should-learn-moving-beyond-yes-or-no/

KEY TERMS AND DEFINITIONS

Access: The ability, permission, or right to use, communicate, or approach something or someone.

Composition: A musical, artistic, written, or digital artifact created by a person or a group of people.

Grammar: The structure and system of a language.

Language: A socially and culturally constructed symbolic system of human communication conveyed using speech, gesture/manual signs, and writing.

Literacy: The ability to read, write, speak, and listen in a way that enables a person to communicate effectively and make sense of the world around them.

Programming Language: A set of commands, instructions, and symbols that humans can manipulate in order to communicate with computers.

Syntax: The set of rules, principles and processes of a language that govern the arrangement of words and phrases.

ENDNOTE

1 Note that constructionism is based on but slightly different from constructivism as it is used in language acquisition research; the latter refers to the assumption that children construct meaning collaboratively in an interplay between the individual and the environment, and specifically other speakers who use language around them.

Chapter 6
Computational Thinking and Life Science:
Thinking About the Code of Life

Amanda L. Strawhacker
Tufts University, USA

ABSTRACT

Life science and computer science share the educational goals of fostering students to engage in inquiry-based learning and solve problems through similar practices of discovery, design, and experimentation. This chapter outlines the pedagogical links among traditional life science and emerging computer science domains in early childhood education, and describes an educational intervention using the CRISPEE technological prototype. CRISPEE, designed by a research team of developmentalists, biologists, educators, and computer scientists, invites young children to use computational logic to model design processes with biological materials. Findings are discussed as they relate to new understandings about how young children leverage computational thinking when engaged in design-based life science, or biodesign.

INTRODUCTION

As part of my research at the DevTech Research Group, I (like all the researchers in our lab) have spent years collecting data about young children's engineering, technology, and programming learning by implementing and evaluating informal curricular interventions. In the 20 years since the creation of DevTech by Marina Bers, our collective research experience on running these camps and play sessions

DOI: 10.4018/978-1-7998-7308-2.ch006

has resulted in a cumulative wealth of knowledge about effective practices for introducing robotics, coding, and other STEAM-themed topics for the first time to 4- to 8-year-old learners. One of the very first activities that we like to play in our robot-themed camps is a game called, "*Robot or Not?*" The premise of the game is simple: the researcher shows a group of children a picture of an object, and asks "Is this a robot, or not?" If a child thinks yes, they jump up and down; if their answer is no, they stand still; and if they aren't sure or they need more information, they wiggle from side-to-side. In addition to being a fun game to get some energy out, *Robot or Not?* provides an opportunity for children have conversations about what makes something a robot. We found early on that children exploring robotics for the first time understandably hold a variety of assumptions and ideas about robots that range from precocious to erroneous. *Robot or Not?* offers a low-stakes playful settting to explore children's ideas, allowing researchers to address misconceptions and identify gaps in knowledge.

Conversations get especially rich when players disagree about whether something is a robot. For example, most children jump up and down when they see a picture of a famous robot character from a movie, and stand still for a picture of a dog, but a picture of a stuffed toy stitched to *look* like a robot is more ambiguous. When we reach the inevitable point in the game when children are uncertain, the researcher pauses to invite children to list characteristics that they think robots have, in order to agree on a shared definition. A common list includes the following criteria: Robots are made of metal or plastic; They have special parts like gears and motors that non-robots do not have; These special parts can move and make sounds automatically; Some robots are built to look like humans; All robots need an engineer or programmer to tell them what to do. This list may grow or change depending on the children in the group, but one criterion is common across every conversation that I've ever led or observed with children playing this game: robots are machines, and so they are definitely *not* alive. And yet, as advances in biotechnology and genetics change the very nature of what we mean by "alive", I find myself questioning this foundational assumption about machines that even young children understand, and wondering what it could look like to have that conversation in our early childhood STEAM camps.

Thus, in my doctoral thesis, I set out to explore the relationship between children's understanding of computational algorithms, and algorithms in the natural world, such as DNA—the genetic "code of life". I wanted to know if we could create tools, frameworks, and lesson activities to invite children to meaningfully engage with concepts from genetics and biology in a playful and developmentally appropriate way, just as we've seen successfully in early computer science education (Bers, 2020). The NSF-funded *Making the Invisible Tangible* project led at Tufts University and Wellesley College (CHS-1564019), attempted to explore the pedagogical connections linking computational thinking to engineering design and life science content

(Strawhacker, Verish, Shaer, & Bers, 2020a, 2020b, 2020c; Verish, Strawhacker, Bers, & Shaer, 2018). We sought to develop a suite of lesson activities and an educational technology prototype, called the CRISPEE kit, that could bring the real-world relevance and design creativity of coding into children's exploration of microbiology, a historically challenging field for young learners to break into.

In this chapter, I share our experiences testing the CRISPEE learning intervention with children. I first outline relevant pedagogical connections and priorities among traditional life science and more recent computer science domains, with a focus on the emerging STEM-integrated domain of *biodesign* (using engineering practices and methodologies to solve biological problems, using biologically-based building materials). Finally, I describe findings from an educational research intervention with the novel CRISPEE prototype, to explore whether and how young children can apply digital learning tools and computational thinking skills to explore biological content.

Science and the STEM Revolution

In the last few decades, the US education community has seen a boom in shifting tools, attitudes, and priorities surrounding STEM (or "I-STEM"), the catch-all term for integrated science, technology, engineering, and mathematics content. Whether the justification is military and federal concern for an economic pipeline of new STEM jobs needing to be filled (Vossoughi & Vakil, 2018), civic concern over the lack of diversity and representation in current STEM fields (Sullivan, 2019; Vossoughi & Vakil, 2018), or novel learning opportunities made by possible by innovations educational technologies (Bers, 2020; Kafai & Walker, 2020), STEM has become as ubiquitous in curriculum standards and school district outcomes as the classic learning goals of "reading, writing, and arithmetic" (Garrett, 2008; McComas & Burgin, 2020; Sanders, 2008; Sullivan, 2019).

It is beyond the scope of this chapter to delve into the debate surrounding how or why STEM has launched into national (and international) prominence (see Vossoughi & Vakil, 2018, for a critical depth examination), but one pronounced trend is that technology, computer science, and engineering have made sweeping advances as educational domains. This is partly due to advances in computational technology, which have contributed to a growing library of innovative, hands-on learning tools that make previously inaccessible concepts more tangible and accessible than ever before, especially for the youngest learners in PreK-2nd grade who most benefit from an integrated, hands-on approach to STEM learning (Chappell, et al., 2021; Kafai & Walker, 2020; National Research Council, 2000). In many ways, new tool development has been spurred by a focus on design-based learning approaches, rooted in the theory of constructionism from the field of computer science education, and learning pedagogies (e.g. "playground-style" technology use, Bers 2012) that engage learners in creating digital and computational artifacts to visualize learning – in other words, to use novel platforms and modalities to construct, share, and reinterpret ideas through design (Bers, 2012; Papert, 1993). But where does this leave science? Why has the "S" in STEM education seemed to maintain a business-as-usual learning model in spite of new integrated science standards (Bybee, 2010, 2014; NGSS Lead States, 2013), introduced during a renaissance of integrated STEM education? For the purposes of this chapter, I will focus on life science and biology (used more or less interchangeably) as I unpack these questions and explore potential future directions for early education.

Bringing K-12 Life Science into the 21st Century

The question of how to reckon traditional life science education with a changing modern world has been debated in the education community long before the push for "21st century skills". In the 1980s and 90s, researchers championed Science, Technology, and Society (STS), a movement to engage students in leveraging scientific knowledge in decisions about policy, social, and public life (e.g., Yager, 1996). As life science approaches a new threshold of change and innovation, these initiatives to are even more important to prepare future citizens for biology-based dilemmas of the 21st century. Novel design-based methods that leverage living materials and genetic "re-coding" to engineer solutions to human problems foreshadow new directions for life science as a field, and thus, new goals for STEM-integrated life science education and participation (Kafai & Walker, 2020; Walker & Strawhacker, 2021).

Recent decades have seen the emergence of crosscutting and speculative domains like *biodesign* (applying engineering practices to the design of biological materials to solve human problems), leading to such advances as foods bioengineered to be more shelf-stable (e.g., Arctic Apples, 2020), tactical clothing made with genetically-engineered spider silk (Cumbers, 2019), and more common bioengineered products like insulin, medicines, and vaccines (Nawla, 2014). Technology and biology are converging in a way that outpaces our ability to fully understand it, much less explain it to children. Still, in the face of such transformative advances in "real" life science practices, limiting biology education to the same observational tools and methods of 19th century naturalists feels jarringly outdated.

From an education perspective, critics of integrating life science with other STEM fields point to clashing epistemologies and philosophies of each discipline (e.g., Clough & Olson, 2016; McComas & Burgin, 2020; Zeidler et al., 2016). The argument goes that traditionally, the goal of life science is to observe and explain phenomena in the natural world, which inherently diverges from engineering and computer science goals of building things to solve problems (see McComas & Burgin, 2020 for a depth discussion of life science in relation to integrated STEM education). Further, when we introduce design into any discipline, we inherently introduce power dynamics about who decides which designs are valuable, who stands to benefit from designs, and which design challenges are not pursued (e.g. Calabrese Barton & Tan, 2019). If issues of ethics and access are left unaddressed at the early education level, emerging 21st century life science pedagogies risk perpetuating ingrained inequities and injustices (e.g. underrepresentation of minority groups) currently facing STEM fields like computer science (Vakil, 2018).

My motivation for exploring innovative life science education in this chapter aligns with a perspective voiced by my former doctoral advisor, Prof. Marina Bers, about computer science education: "the rationale for supporting the introduction of computer science starting in kindergarten shouldn't be the creation of the future workforce, but the future citizenry" (Bers, 2018, p. 500). Further I agree with colleague Prof. Justice Walker that a main challenge of life science and biology education today is that schools use "19th century practices, to teach 20th century concepts, to create 21st century citizens" (J. Walker, personal communication, February 15, 2021). I do not argue that we should stop teaching children about the wonders of the natural world, or the traditional concepts and methods that make up life science education as we now know it. However, if we accept that an implicit goal of biology education is to democratize scientific knowledge and methods for the benefit of cultivating an informed public, capable of making scientifically grounded, ecologically sustainable, and socially just decisions for our world, then we must prioritize pedagogies that enable students to engage with relevant, authentic, and current scientific information and approaches. Further, I argue that a key part of fostering children's engagement with ethical 21st century design-based life science, involves developing their computational thinking skills.

Computational Thinking in Life Science: The Case for Biodesign

While the debate rages among researchers about how to position K-12 life science education in an integrated STEM curriculum, pre-professional programs and universities find themselves facing a different challenge: how to train their early-stage life scientists to better understand computational thinking?

In the 1950s, when the discovery of the DNA double-helix had just launched a new field called *bioengineering*, STEM training programs recognized the need to offer more biology courses to their engineering students (Naik, 2012; Nebeker, 2002). Today, bioengineering progressively relies not only on natural biological processes, but also artificial ones, thanks to new advances in computational biology and nanotechnology (Naik, 2012). As the discipline advances, university educators like Rubinstein and Chor (2014) insist that it is time to address students' lacking computational thinking skills, which they see as critical for modern biology training:

Life sciences are going through a dramatic biotechnological revolution. [...] Life sciences curricula, however, have hardly been altered to reflect this revolution...[and] not enough emphasis is put on developing abstract and algorithmic thinking skills.

This gap presumably starts at the classroom, but it lingers later on. Biology in many institutes and labs is still primarily a descriptive science with little computational approaches being used on a daily basis. Computational approaches in this context are not the mere use of tools, but the integration of computational thinking and algorithms to experiments design; to data generation, integration, and analyses; and to modeling (Rubinstein & Chor, 2014, p. 1).

Others have noted this issue as well, and in response, the ecosystem of tools and technologies for biodesign, bioengineering, and biomaking is growing (as young field, these nascent terms are still evolving and converging but they all refer to biology experiences that leverage computational thinking and problem-solving through design). Most biodesign education programs are targeted at older students in high school or college (e.g. Kafai, Telhan, Hogan, Lui, Anderson, Walker, & Hanna, 2017; Kuldell, 2007), but many countries already mandate computer science and engineering education starting in Kindergarten (Cejka, Rogers, & Portsmore, 2006; Metz, 2007; Pretz, 2014). Life science has been taught to this age range for decades already, so why should we not introduce biodesign earlier?

Biodesign is currently viewed as too advanced for early education, and using current models for teaching about microbiology, that is certainly true. However, research suggests that young children may already hold preconceptions about genetics and biology, gleaned from popular culture and media aimed at children and young adults (Elmkesky, 2013; Venville, Gribble & Donovan, 2005). In my own interviews with over 100 children aged 4-9 years in the greater Boston area, I found that around 15% of my sample had already heard of concepts like "genes" and "DNA", and some could explain these and other advanced biology concepts (like viruses) with surprising accuracy (Strawhacker, Verish, Shaer, & Bers, 2020b).

There are also overlapping themes that young children explore in early childhood that are foundational to biodesign. Life science and engineering share similar methodologies for asking and answering questions (i.e., the scientific method) and building and testing solutions to human problems (i.e., the engineering design process). Both involve steps of ideating, designing (experiments or prototypes), iterating, and refining that young children practice starting in Kindergarten. Similarly, biology and computer science both rely on computational concepts of abstraction (e.g., of proteins within cells, or subroutines within coded instructions), modularity (segmentation of organs to make a body, or hardware parts to make a robot), and algorithmic logic (in genetic codes or computer codes) to understand and model how systems operate within hierarchical structures to function as a whole. These computational concepts may sound highly sophisticated, but educational coding tools like the ScratchJr programming language (www.scratchjr.org), KIBO robotics kit (www.kinderlabrobotics.com), BeeBot robot (www.terrapinlogo.com), Code-a-Pillar (www.fisher-price.com), and more all demonstrate how those concepts can successfully be introduced to children as early as preschool.

From a developmental perspective, integrated STEM is the preferred learning model for early childhood (Aldemir & Kermani, 2017; Wortham, 2006). Engineering brings creative agency and hands-on exploration to biology lessons, which can be abstract and overly-structured for young learners (Ostroff, 2016). Introducing novel scientific topics of biodesign brings real-world relevance and context to STEM explorations, connecting children's learning to topics in their broader community and society. New technologies offer children a chance to playfully explore topics that were previously too microscopic, invisible, or time-consuming to engage meaningfully.

Given these findings, that constraints for bringing biodesign into early childhood would appear to rest more on the side of pedagogical approaches and educational tools than on children's developmental capacity. In the following sections, I describe a research project that set out specifically to address this gap, by designing a learning intervention and tangible technology supports to engage children in applying computational thinking concepts to biodesign content.

Design of CRISPEE: A Tangible Tool and Learning Intervention to Model "Coding with Genes"

The NSF-funded *Making the Invisible Tangible* project attempted to explore the pedagogical connections linking computational thinking to engineering design and life science content (NSF grant no. CHS-1564019). The goal of the project, headed by Dr. Orit Shear of Wellesley University and Dr. Marina Bers at Tufts University, was to explore the viability of translating these integrated STEM themes to an early childhood context, designing tangible technologies and curricular supports as needed to meet young children's developmental needs. I participated as a doctoral student researcher to develop a biodesign learning tool and curriculum, together with a team of university student researchers trained in a variety of backgrounds, and expert consultants from fields of biology, ethics, and education. We implemented this learning tool with over 125 children, families, and teachers in school, museum, and makerspace settings, and used findings to iteratively redesign the technology and intervention (Strawhacker, Verish, Shaer, & Bers, 2020a, 2020b, 2020c; Verish,

Strawhacker, Bers, & Shaer, 2018). Throughout the rest of this chapter, I will share our experiences testing this technology and curriculum with children, and describe how these findings highlighted the role of computational thinking in design-based biology for early childhood.

The Tangible CRISPEE Technology Kit

Inspired by advances in learning and design frameworks for child-computer interaction (e.g., Antle & Wise, 2013; Horn, Crouser, & Bers, 2012; Lester, Rowe, & Mott, 2013), we set out to test whether young children could explore the computational and biological ideas we identified as learning goals if we provided them with tangible tools and story-based contexts to represent learning. After researching relevant early childhood learning standards and frameworks, we identified a list of cross-cutting STEM learning goals for our intervention (Strawhacker, Verish, Shaer, & Bers, 2020c). These goals can be summarized in three steps: (1) introduce or recall (depending on the child's experience) the model of coding languages in machines, and present genes as a kind of coding language for living things; (2) engage children in designing and testing their own gene codes (using technology to model this process); and (3) prompt children to apply their newfound gene-design knowledge to solve speculative and story-based problems. The result of the 4-year project exploring these learning outcomes is the CRISPEE kit prototype (see Figure 1), and accompanying curriculum for a 15-hour learning intervention.

CRISPEE is modeled loosely on DNA extractor/incubators that use the CRISPR/Cas-9 gene editing software – the most prominent technological advance to bring genetic engineering from "tinkering" to "cut and paste editing" (Doudna, 2015). CRISPEE lets children explore how genes can function like a coding language to determine the color of a bioluminescent (glowing) animal's light. The model, based on real processes of genes and bioluminescent proteins, relies on "gene blocks" (made with wood, felt, conductive Velcro, and resistors) that turn red, green, and blue glowing lights "on" or "off". The "on" light colors then mix according to light-color physics (in which a regular color wheel begins with different primary colors than in solid-color mixing) to determine the resulting light color. In addition to letting children explore fascinating naturally-glowing animals, the mixture of gene colors also lets children explore the visual arts concept of additive color-mixing with light. a less-explored companion to subtractive color-mixing with solids, like crayons and paints. Because this is a universal concept beyond biology, the additive primary colors (red, green, and blue) always mix to create the same secondary colors, whether working with CRISPEE, a children's light table, or cello-paper and flashlights.

Figure 1. The prototyped CRISPEE Kit for modeling biodesign of bioluminescent (glowing) animals' light color includes (1) a tangible interface for building and testing light codes, (2) various gene blocks with different color controls, (3) a picture book to introduce how CRISPEE works in a story context, (4) LED-interactive stuffed animals to display children's coded lights, and (5) speculative designs for logic-gate "biosensor" controls to code light changes under certain environmental conditions

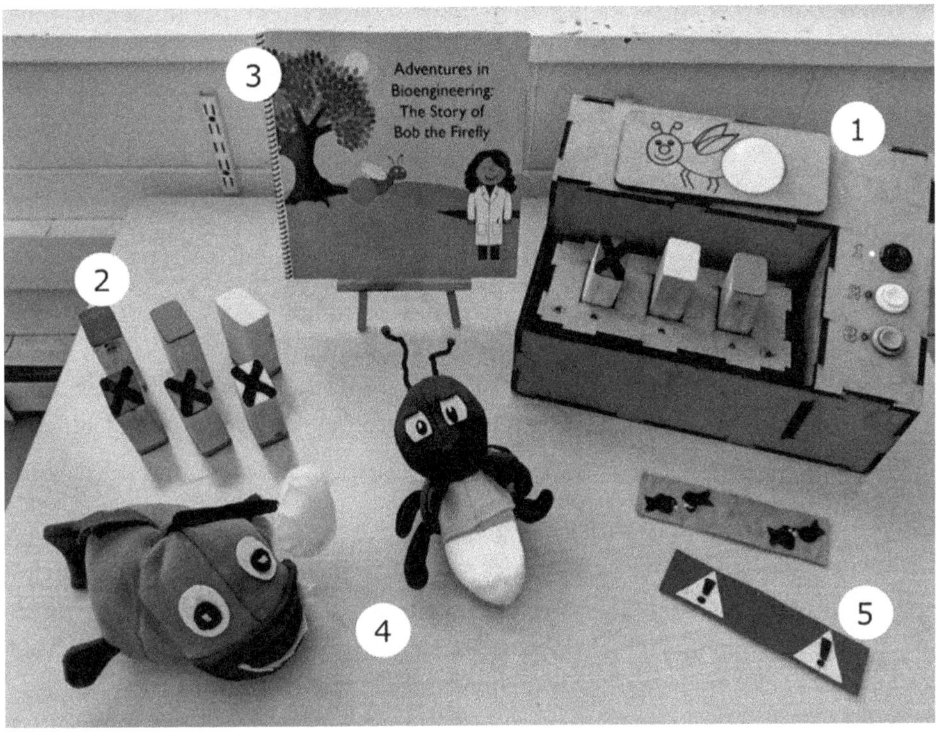

Figure 2. Three-step CRISPEE interaction
Source: Verish et al. (2018)

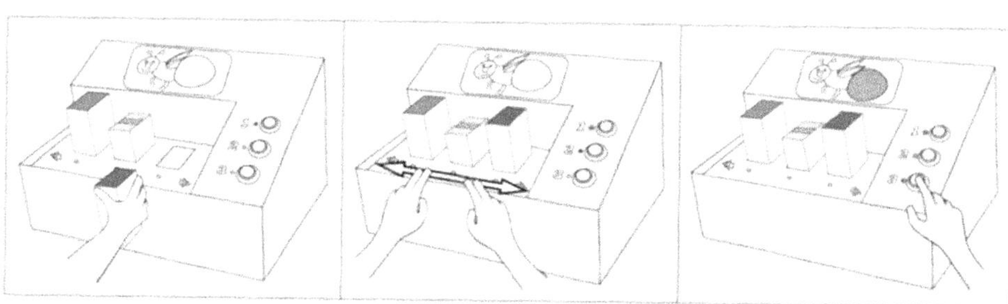

Figure 2 shows the interaction process to play with CRISPEE. First, a child selected a detachable faceplate showing the animal they want to use in their model. We provided four choices of naturally bioluminescent animals: firefly, zebrafish anglerfish, and jellyfish. Next, they selected three gene blocks to code for the light color in their animal model, which will determine which colors will be lit up, and which will remain silenced. I deliberately used computational words like "code" and "program" to describe their chosen block sequence, in order to highlight the design-based nature of this step. Finally, children shook the platform vigorously until several lights indicated CRISPEE was ready to test. Shaking was added for two reasons. First, it represents the real-life centrifuge process that DNA extractors use to tease apart DNA, allowing new genetic sequences (programmed by the engineer) to insert themselves into the genome. While we did not necessarily expect children to understand the details of this process, children in the curricular intervention viewed videos of this centrifuge process in laboratory CRISPR machines and made the connection to CRISPEE's action. Second, shaking reinforced the idea that the colors children selected would mix together, creating a single output light color. After shaking the platform, children finally pressed a button to test their new light, which glowed out of an oversized lightbulb framed by their animal faceplate.

The CRISPEE Curriculum: A Biodesign Curriculum for 5-8 Year Old Children

In addition to the technology kit, our research team developed, iteratively implemented, and refined a 15-hour NGSS-aligned curriculum intervention for 5-8 year old children to explore biodesign in a developmentally appropriate learning progression. Included in this curriculum were original learning supports, as well as suggestions for commercially- and freely-available resources.

Storytelling with picture books is an effective learning device to introduce young children to science topics that are typically too abstract (experimental methods), distant (e.g. outer space), or microscopic (e.g. cell biology) for children to meaningfully explore (Mantzicopoulos, & Patrick, 2011; Monhardt & Monhardt, 2006). I wrote the *Adventures in Bioengineering* picture book to introduce concepts like biodesign in the context of a developmentally-appropriate story, as well as to illustrate the interaction steps for using the CRISPEE tool (see Figure 3). In the story, a firefly with a genetic inability to glow is separated from his friends, who cannot locate him without his light. He seeks the help of a bioengineer, who works with Bob to reprogram his genes so he can glow and relocate his firefly community. The story presents concepts and vocabulary words (e.g., bioengineering, genes, bioluminescence) in-text and in a glossary, and narrates through a problem-design-solution process. This story was important for helping children understand a potential context for what kind of problems biodesign might be useful to address.

Figure 3. Cover art (left) and pages (center, right) from the Adventures in Bioengineering storybook, an original picture book developed to introduce a story-context for biodesign, as well as illustrate how to use the CRISPEE tool

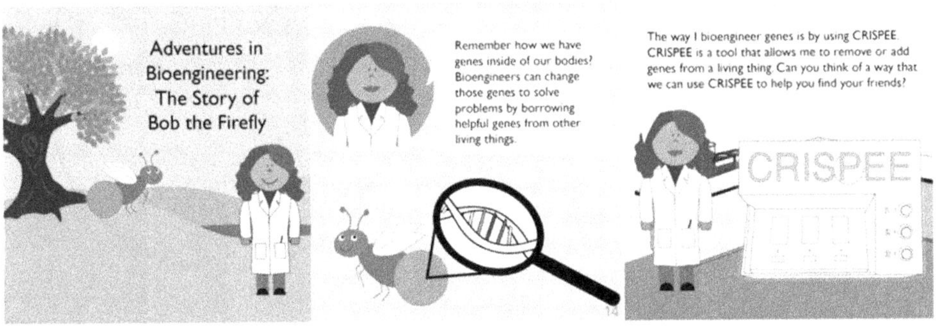

Part of the challenge of introducing biodesign for any age group is the ethically sensitive nature of work that involves changing the genes of living animals and organisms. In order to explore these questions without presenting any single perspective as "ethically correct," we created an anchor chart called the Ethical Design Process. This chart was directly inspired by the Engineering Design Process commonly taught in DevTech Research Group's camps and interventions, but included questions that biodesigners ask at each step of the design process to ensure that their solutions are as ethical and responsible as possible (see Figure 4). The questions identified on the chart, such as "What are the possible consequences [of the design]?" and "How can we make our solution less harmful?" were adapted from or inspired by transcript conversations with young children in my research sample. Questions were further refined with assistance from expert consultants, including a biodesign professor at MIT, and a philosophy of ethics professor at Wellesley College. While this chart represents only a narrow portion of the kinds of ethical work that ecological conservationists and biodesign scientists must conduct as part of their professional design initiatives, these questions provided enough provocation that children could meaningfully consider ethical consequences of their design steps while engaging with CRISPEE (discussed more in later sections).

Figure 4. The Ethical Design Process (left) was modeled directly on the DevTech Research Group's Engineering Design Process (right)

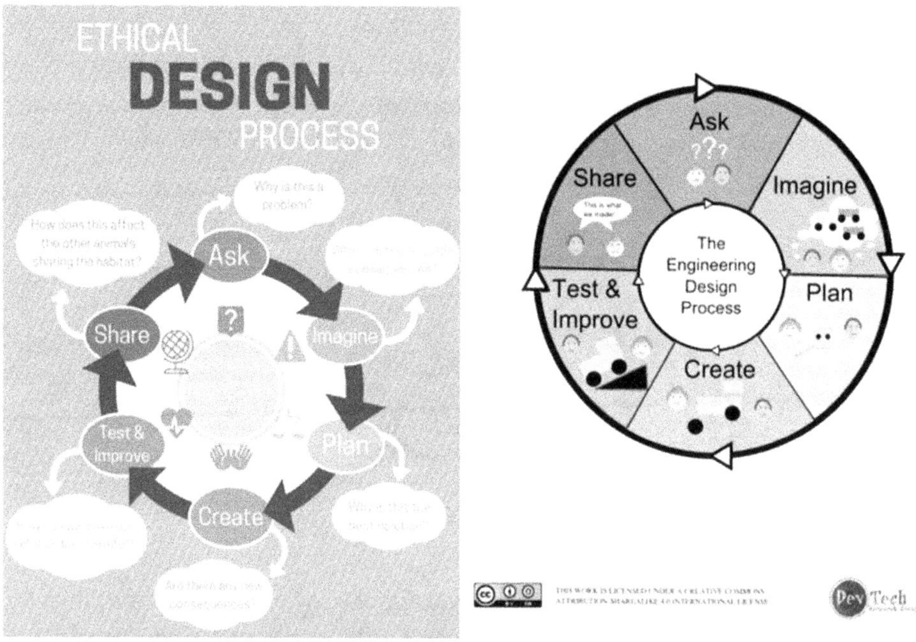

Children also explored a variety of familiar and traditional curricular supports, including biology- and engineering-themed picture books, microscopes and slides, child-sized laboratory equipment and gear (e.g. lab coats, beakers), researcher-created worksheets and games (e.g. word searches) packaged into individual Design Journals, and materials for light and color play (see Figure 5). In particular, children engaged richly with a light table purchased specifically because it used Red, Green, and Blue primary color knobs to control the color produced by the table. Children used this tool to explore light mixing, incorporating translucent marbles and toys into their play, and frequently commented that it mixed colors using the same rules as CRISPEE.

All of these materials were crafted or chosen specifically to support a concept relevant to our core learning goals for biodesign engagement. In the following sections, I describe examples from user testing and curriculum interventions involving CRISPEE and a small team of my research assistants. Each example shows how children naturally incorporated aspects of computational thinking into their biodesign play.

Figure 5. Children complete an original worksheet activity using biology-themed reference picture books

COMPUTATIONAL THINKING IN CHILDREN'S CRISPEE PLAY

Algorithmic Logic

CRISPEE was designed intentionally to leverage Kindergarten-appropriate physics (i.e., light color mixing) and biology topics, as well as to forefront computational thinking concepts. To make a successful glowing light with CRISPEE, children needed to understand a few algorithmic concepts. First, they needed to understand the binary nature of the blocks. For example, CRISPEE has two red blocks, one marked with an X for "turn red off", and the other marked with a solid color for "turn red on". They cannot be used at the same time, because (as the children often explained to each other), "CRISPEE doesn't understand if you tell it 'yes red' AND 'no red'." Second, children needed to add *one of each* color block in their sequence, or CRISPEE would terminate the test, because one color had too many inputs (the on-and-off problem again). Third, they had to understand that the colors they chose to turn "on" would mix together to create a new color, while the colors they turned "off" would remain silenced.

With just these simple coding rules, we found that children made connections to concepts of order and pattern. Younger children (aged 4 and 5 years) were more likely than older ones to use CRISPEE blocks to make a repeating color pattern (e.g., blue-red-blue) purely for aesthetic appeal, which researchers guided into an exploration of the rules for what blocks CRISPEE will "accept".

Other children mentioned technological toys and robot kits from their home or school, and used computational words like "program" and "code" before researchers introduced them. These children were more likely to test the same program multiple times in different sequences (e.g., red-green-blue, then blue-green-red) to see if block sequence would change the output light (see Figure 6). Even though we did not present CRISPEE as a robot or coding toy, these clues indicate that children understood CRISPEE as a computational tool for exploring sequencing, patterns, and algorithms.

Figure 6. A child tests a CRISPEE program with the state goal of seeing how the sequence of colors impacts the output light
Source: Strawhacker et al. (2020a)

Figure 6 shows a 7-year-old child testing a gene program with three On blocks (green-blue-red). His plan for the next program is laid out the table in front of him, with a reversed sequence of colors (red-blue-green) in Off blocks. Out of 28 total programs that he tested with CRISPEE, 15 were deliberately to explore the effect of changing the sequence of blocks in the same program.

Debugging and Hardware / Software

Although the debugging process and hardware/software relationships are separate computational thinking concepts, they were so related in children's explorations that I will discuss them together here. Children in my studies spent a good amount of time identifying and attempting to repair bugs (technical malfunctions) while using CRISPEE. This included debugging their CRISPEE codes to design a specific color they wanted (e.g., Strawhacker et al, 2020b), and assisting researchers in exploring actual bugs in the prototype (e.g., Strawhacker et al, 2020a). As a proof-of-concept research prototype, CRISPEE occasionally malfunctioned, which sparked curiosity in children about the hardware and how it worked. Children frequently requested to look inside CRISPEE, and found the internal mechanics as interesting as the actual interface.

Children also made scientific (evidence-based) observations about the CRISPEE kit during times when our research team had to repair the tool. For example, one child noticed that the wood interface "smells like a bonfire" (a byproduct of laser-cutting the parts), and he closely examined the moving platform to see if it "uses wheels, like my robot at home". He was especially curious about how CRISPEE could produce light when seemed be made of non-electronic materials. When examining the conductive Velcro on the underside of the programming blocks he asked, "is that stuff Velcro? How can it do stuff if it's just cardboard or wood?" In this instance, he was relating CRISPEE's interactions (e.g. lights, buttons) to its interface made of familiar, non-technical materials (e.g. wood, felt, Velcro), and and trying to reconcile this observation with his understanding of machines that require computational components to execute software code.

The children's curiosity inspired me to leave a laptop at the CRISPEE center with videos running to show how different parts of CRISPEE were made, which in turn led to further exploration about computational hardware. When I left a video running about laser-cutting to show why the CRISPEE wood smelled like it was burned, one girl asked, "What else can that laser thing cut? Can it cut glitter? Can it cut paper?" Watching researchers fix the CRISPEE prototype and finding new bugs became a favorite past-time during our CRISPEE curriculum interventions. One child even began collecting field notes in a hand-drawn bug log when she found a CRISPEE error, complete with drawings and labels of the specific malfunctioning CRISPEE code (see Figure 7).

Figure 7. A free-drawn "bug log" of a malfunctioning CRISPEE program, created by a 6-year-old girl to add to her Design Journal

One collaborative debugging episode occurred while children played with three different versions of the CRISPEE prototype. They were surprised to find that the program for a blue light, which everyone knew how to make, was returning a magenta color on one of the CRISPEEs. Four children and two researchers all worked together to solve the problem, and eventually discovered that CRISPEE blocks were not compatible across different prototype versions, a discovery that surprised even the lead engineer. Without the children's willingness to explore and test different solutions, it's unlikely that this issue would have been resolved, since most of the researchers gave up before the children did!

Design Process

Children were prompted to engage in steps of an ethical design process through large-group discussions, activities using their Design Journals, and games and songs involving the Ethical Design Process anchor chart. One activity, called Design a Helpful Animal, invited children to imagine a problem they could solve by biodesigning an animal, and then to consider positive and negative consequences of that design. Children identified diverse and interesting problems to solve, and offered creative solutions.

Several children wrote in their design journals about environmental problems they cared about. For example, during a circle conversation, a 5-year-old girl shared a memory of a trip she had taken to Florida, where she learned that the sea turtle population there was becoming threatened due to plastic pollution. The turtles were eating plastic bags, instead of their normal diet of jellyfish. For her design, this girl wanted to "give fox 'smell genes' to turtles", so that they could tell the difference between plastic bags and jellyfish. In addition to the positive consequence of saving turtle populations, she identified a negative consequence of turtles suddenly starting to hunt food that foxes eat, as a result of sharing their "smell gene".

Similarly, a 6-year-old boy spent over an hour imagining, sharing, and revising a design idea to help cheetahs (his favorite animal) by giving them "more genes" to be "smarter and faster". Originally it seems he just wanted to make cheetahs even faster (see Figure 8), although through conversation with two other intervention participants (a 7-year-old boy and a teacher, both of whom he knew from school) he was pressed to justify his design problem. When asked why cheetahs should be faster, the boy's answer focused on the dangers of poachers threatening the cheetah population. The friend and teacher both validated and extended his idea by offering vocabulary words (e.g., "endangered") to capture his concern for the cheetah's welfare (i.e., "why cheetahs are getting killed"). This example indicates the boy's conviction that a biodesign solution should involve serving or helping animals to escape harm, an ethical purpose that both his friend and his teacher readily understood and accepted.

Figure 8. A boy's design journal page describing his idea to enhance cheetahs with genes to help them escape poachers

Not all children directly engaged with ethical consequences of biodesign. For example, one 5-year-old girl, inspired by the circle-time story about turtles, drew pictures of plastic bags floating in an ocean of sea creatures, and wrote a line from the perspective of the animals: "Don't litter because I can die." Although her design doesn't suggest a bioengineered solution, it indicates that she was connecting biodesign to environmental maintenance and ecological stewardship. Other children wrote story-style narratives and focused more on individual animal characters than an ecosystem-level issue.

Surfacing Computational Thinking in Biodesign Education

Taking up the call from Rubinstein and Chor (2014) to examine the early beginnings of the gap in students' computational thinking preparedness for modern-day biology, this section reflects on the ways that the biodesign curriculum intervention and tangible CRISPEE tool organically fostered children's engagement with computational thinking. In the examples presented earlier, children explored algorithmic logic, debugging, hardware construction, and the engineering design process, although none of these concepts was explicitly introduced. Analysis of video transcripts suggests that these topics emerged spontaneously due to three main factors.

The first factor relates to the CRISPEE technology itself. Because CRISPEE was a computational object, children leveraged their understanding of machines to understand how CRISPEE works. Children drew on prior experiences with educational robotic kits, computer hardware, technology references in children's media (e.g., tv and books), and their family members' professions (e.g., doctor, software engineer) to make sense of CRISPEE's novel interactions and interface. This led to hypotheses when first engaging with CRISPEE about making color patterns with blocks, changing block sequences to explore the resulting light effect, and voicing theories like, "the blue blocks are maybe a different coding language than the green ones." Simply by using a technological tool, children were cued to engage their prior knowledge and experiences with technology, suggesting that computational concepts such as algorithmic logic may serve children in computer-mediated tasks, even when the task is not inherently related to computation.

The second factor contributing to computational thinking was the expertise of facilitators in the room during interventions, which emerged during tech malfunctions. At least one of the engineers who actually helped construct CRISPEE prototypes was present at each research session. This was a preventive measure, since I expected to the prototypes to fail occasionally. What I did not expect was that these technology breakdowns would inspire children's curiosity about hardware and software, and engage them in debugging practices of logging failures, observing engineers as they repaired prototypes, and persisting with tests to determine the nature of bugs. Children thrilled at the chance to take CRISPEE apart and put it back together again, and were self-motivated to help in any way they could with the debugging process. This finding points to the power of surprising tech interactions, including bugs and malfunctions, to engage children in authentic computational practices.

Third, the researcher-developed curricular tools, particularly the *Adventures in Bioengineering* storybook and the Ethical Design Process, supported children's engagement in iterative and reflective design cycles. In user tests of CRISPEE *without* the curricular components, children played for about 20-30 minutes before shifting to hardware explorations, discussions about science, or simply moving on to different activities. In contrast, children in the curricular interventions returned again and again to CRISPEE to try to realize their designs. For example, some children spent several days learning the logic of the gene blocks in order to recreate animals they learned about from the biology reference books we provided. When invited to create their own biodesigned solutions to a problem of their choosing, one child worked so hard on his design concept about endangered animals that he skipped free-play time and snack! This finding is most interesting to me, because it speaks to the ways that computational thinking processes like creative design, and learning domains like biology, can support children's engagement in ethical and altruistic play and learning. The diversity of children's engagement with ethical design offers valuable insight for future research in this area. Interestingly, none of the students in my sample made a biodesign to aid humans, but instead focused on ways to solve problems that animal might face. This surfaces another ethical challenge in biodesign education, since in reality, bioengineered organisms are primarily created to solve human problems. Perhaps children are so altruistic in their thinking at this age that it simply does not occur to them to view animals as a resource to serve human needs, or perhaps their attitudes are a product of the fact that anthropomorphic and empathic animals are commonly cast as protagonists in children's media (including the *Adventures in Bioengineering* picture book used in my studies). Future research should consider the impact of narrative framing to explain the purpose of biodesign work as it is actually practiced.

CONCLUSION

While researchers continue to debate the role of computational thinking in the biology classroom, findings from the CRISPEE research project suggest that biodesign may be a fruitful way to engage children as young as 5 years old in meaningful, relevant, and even ethical applications of computational concepts to the natural world. Even without a novel tool and curriculum, children may leverage computational logic to understand repeating patterns in nature, distinguish natural and human-made materials, and consider engineered solutions to environmental problems. As biodesign continues to emerge as a 21st century domain, I hope that future citizens and designers carry forward the propensity found in children, to leverage biodesign as a tool to engineer sustainable, environmentally responsible, and ethical solutions to problems that face humans and all organisms on our planet.

ACKNOWLEDGMENT

CRISPEE was developed as part of the *Making the Invisible Tangible* research project funded by the National Science Foundation (Grant No. 1564018) and pilot-tested at early childhood museums and learning settings around the Greater Boston area from 2017 to 2019. This project is a collaboration between the DevTech Research Group at Tufts University the Wellesley College Human-Computer Interaction Lab. Thank you tothe hundreds of children, families, and education professionals who contributed to the research, as well as the incredible student researchers and engineers at the DevTech and HCI Labs, without whom this research would not have been possible without.

REFERENCES

Aldemir, J., & Kermani, H. (2017). Integrated STEM curriculum: Improving educational outcomes for Head Start children. *Early Child Development and Care, 187*(11), 1694–1706. doi:10.1080/03004430.2016.1185102

Antle, A. N., & Wise, A. F. (2013). Getting down to details: Using theories of cognition and learning to inform tangible user interface design. *Interacting with Computers, 25*(1), 1–20. doi:10.1093/iwc/iws007

Arctic Apples. (2020). Retrieved from: https://www.arcticapples.com/

Bers, M. U. (2012). *Designing digital experiences for positive youth development: From playpen to playground.* Oxford University Press. doi:10.1093/acprof:oso/9780199757022.001.0001

Bers, M. U. (2019). Coding as another language: a pedagogical approach for teaching computer science in early childhood. *Journal of Computers in Education, 6*(4), 499-528.

Bers, M. U. (2020). *Coding as a playground: Programming and computational thinking in the early childhood classroom.* Routledge. doi:10.4324/9781003022602

Bybee, R. W. (2010). Advancing STEM education: A 2020 vision. *Technology and Engineering Teacher, 70*(1), 30.

Bybee, R. W. (2014). NGSS and the next generation of science teachers. *Journal of Science Teacher Education, 25*(2), 211–221. doi:10.100710972-014-9381-4

Calabrese Barton, A., & Tan, E. (2019). Designing for rightful presence in STEM: The role of making present practices. *Journal of the Learning Sciences*, *28*(4-5), 616–658. doi:10.1080/10508406.2019.1591411

Chappell, C., Dabholkar, S., Dilley, C., Heiland, M., Huang, A., Kuldell, N., Kurman, M., Legault, J., Scheifele, L., Scholze, A., Takara, C., & Tuck, E. (2021, April 8-12). *The BioMaker Ecosystem: Technologies, Spaces and Curricula for K-12 Making with Biology*. American Educational Research Association 98th Virtual Annual Meeting.

Clough, M. P., & Olson, J. K. (2016). Connecting science and engineering practices: a cautionary perspective. In L. A. Annetta & J. Minogue (Eds.), *Connecting Science and Engineering Education Practices in Meaningful Ways: Building Bridges* (pp. 373–385). Springer. doi:10.1007/978-3-319-16399-4_15

Cumbers, J. (2019). New This Ski Season: A Jacket Brewed Like Spider's Silk. *Forbes Magazine Online*. Retrieved from: https://www.forbes.com/sites/johncumbers/2019/08/28/new-this-ski-season-a-jacket-brewed-from-spider-silk/#2788fa63561e

Doudna, J. (2015, September). *How CRISPR lets us edit our DNA* [Video file]. Retrieved from: www.ted.com/talks/jennifer_doudna_we_can_now_edit_our_dna_but_let_s_do_it_wisely#t-686789

Garrett, J. L. (2008). STEM: The 21st century sputnik. *Kappa Delta Pi Record*, *44*(4), 152–153. doi:10.1080/00228958.2008.10516514

Horn, M. S., Crouser, R. J., & Bers, M. U. (2012). Tangible interaction and learning: The case for a hybrid approach. *Personal and Ubiquitous Computing*, *16*(4), 379–389. doi:10.100700779-011-0404-2

Kafai, Y., Telhan, O., Hogan, K., Lui, D., Anderson, E., Walker, J. T., & Hanna, S. (2017, June). Growing designs with biomakerlab in high school classrooms. *Proceedings of the 2017 Conference on Interaction Design and Children*, 503-508. 10.1145/3078072.3084316

Kafai, Y. B., & Walker, J. T. (2020). Twenty things to make with biology. Proceedings of Constructionism, 598-606.

Lester, J. C., Rowe, J. P., & Mott, B. W. (2013). *Narrative-centered learning environments: A story-centric approach to educational games. Emerging Technologies for the Classroom. 223-237.*

Mantzicopoulos, P., & Patrick, H. (2011). Reading picture books and learning science: Engaging young children with informational text. *Theory into Practice*, *50*(4), 269–276. doi:10.1080/00405841.2011.607372

Monhardt, L., & Monhardt, R. (2006). Creating a context for the learning of science process skills through picture books. *Early Childhood Education Journal*, *34*(1), 67–71. doi:10.100710643-006-0108-9

Naik, G. R. (Ed.). (2012). *Applied Biological Engineering: Principles and Practice*. BoD–Books on Demand. doi:10.5772/2101

National Research Council. (2000). *From neurons to neighborhoods: The science of early childhood development*. U.S. National Research Council.

Nebeker, F. (2002). Golden accomplishments in biomedical engineering. *IEEE Engineering in Medicine and Biology Magazine*, *21*(3), 17–47. doi:10.1109/MEMB.2002.1016851 PMID:12119874

NGSS Lead States. (2013). *Next Generation Science Standards: For States, By States*. The National Academies Press.

Ostroff, W. L. (2016). *Cultivating curiosity in K-12 classrooms: How to promote and sustain deep learning*. ASCD.

Papert, S. (1993). *The children's machine: Rethinking school in the age of the computer*. Basic Books.

Rubinstein, A., & Chor, B. (2014). Computational thinking in life science education. *PLoS Computational Biology*, *10*(11), e1003897. doi:10.1371/journal.pcbi.1003897 PMID:25411839

Sanders, M. E. (2008). Stem, stem education, stemmania. *Technology Teacher*.

Strawhacker, A., Verish, C., Shaer, O., & Bers, M. (2020c). Young children's learning of bioengineering with CRISPEE: A developmentally appropriate tangible user interface. *Journal of Science Education and Technology*, *29*(3), 319–339. doi:10.100710956-020-09817-9

Strawhacker, A., Verish, C., Shaer, O., & Bers, M. U. (2020a, April). Debugging as Inquiry in Early Childhood: A case study using the CRISPEE prototype. *Computational Thinking for Science Learning. Symposium. Annual Meeting of the American Educational Research Association (AERA).*

Strawhacker, A., Verish, C., Shaer, O., & Bers, M. U. (2020b). Designing with Genes in Early Childhood: An exploratory user study of the tangible CRISPEE technology. *International Journal of Child-Computer Interaction, 26*, 26. doi:10.1016/j.ijcci.2020.100212

Sullivan, A. A. (2019). *Breaking the STEM stereotype: Reaching girls in early childhood.* Rowman & Littlefield Publishers.

Vakil, S. (2018). Ethics, identity, and political vision: Toward a justice-centered approach to equity in computer science education. *Harvard Educational Review, 88*(1), 26–52. doi:10.17763/1943-5045-88.1.26

Venville, G., Gribble, S. J., & Donovan, J. (2005). An exploration of young children's understandings of genetics concepts from ontological and epistemological perspectives. *Science Education, 89*(4), 614–633. doi:10.1002ce.20061

Verish, C., Strawhacker, A., Bers, M. U., & Shaer, O. (2018). CRISPEE: A Tangible Gene Editing Platform for Early Childhood. *Proceedings of the Twelfth International Conference on Tangible, Embedded and Embodied Interaction (TEI).* 10.1145/3173225.3173277

Vossoughi, S., & Vakil, S. (2018). Toward what ends? A critical analysis of militarism, equity, and STEM education. In *Education at war* (pp. 117–140). Fordham University Press. doi:10.2307/j.ctt2204pqp.9

Walker, J., & Strawhacker, A. (Co-chairs). (2021, April 8-12). The Biomaker Ecosystem: Technologies, Spaces and Curriculum for K-12 Making with Biology [Symposium]. *American Educational Research Association* (Virtual Conference).

Wortham, S. C. (2006). *Early childhood curriculum: Developmental bases for learning and teaching.* Kevin M.

Yager, R. E. (1996). Meaning of STS for science teachers. *Science/technology/Society: as reform in science education*, 16-24.

Zeidler, D. L., Herman, B. C., Clough, M. P., Olson, J. K., Kahn, S., & Newton, M. (2016). Humanitas emptor: Reconsidering recent trends and policy in science teacher education. *Journal of Science Teacher Education, 27*(5), 465–476. doi:10.100710972-016-9481-4

ADDITIONAL READING

Brown, B. A. (2021). *Science in the city: Culturally relevant STEM education*. Harvard Education Press.

Duschl, R. A., Schweingruber, H. A., & Shouse, A. W. (Eds.). (2007). *Taking science to school: Learning and teaching science in grades K-8* (Vol. 500). National Academies Press.

Elmesky, R. (2013). Building capacity in understanding foundational biology concepts: A K-12 learning progression in genetics informed by research on children's thinking and learning. *Research in Science Education*, *43*(3), 1155–1175. doi:10.100711165-012-9286-1

Inagaki, K., & Hatano, G. (2006). Young children's conception of the biological world. *Current Directions in Psychological Science*, *15*(4), 177–181. doi:10.1111/j.1467-8721.2006.00431.x

Metz, K. E. (2006). The knowledge building enterprises in science and elementary school science classrooms. In *Scientific inquiry and nature of science* (pp. 105–130). Springer.

Nalwa, H. S. (2014). A special issue on reviews in nanomedicine, drug delivery and vaccine development. *Journal of Biomedical Nanotechnology*, *10*(9), 1635–1640. doi:10.1166/jbn.2014.2033 PMID:25992435

Walker, J. T., & Kafai, Y. B. (2021). The biodesign studio: Constructions and reflections of high school youth on making with living media. *British Journal of Educational Technology*, bjet.13081. doi:10.1111/bjet.13081

Walker, J. T., Strawhacker, A., Angleton, C., Allan, J., Konwar, A., Obayomi, O., & Kong, D. S. (Eds.). *2021. Proceedings of the Global Community Bio Summit (GCBS) 4.0*, Cambridge, Massachusetts. Retrieved from: www.biosummit.org

KEY TERMS AND DEFINITIONS

Biodesign: An emerging science movement that applies methods and approaches of creative and engineering to the design of living materials and systems.

Computational Thinking: Broadly, a set of cognitive skills, processes and concepts that involve expressing problems and their solutions in ways that a computer could also execute.

CRISPEE: A tangible technological prototype and suite of educational materials designed to engage children in playfully exploring biological algorithms (e.g., genetics) through the lens of computer programming.

Early Childhood Education: Education of children from birth through age 8 years.

Life Science: Any fields of science related to biology or the study of life and living systems.

Programming: Also called coding, computer programming is the process of designing and building an executable computer program to accomplish a specific computing result or to perform a specific task.

Sequencing: Arranging elements of a system in a particular order, e.g., commands in a computer code.

STEM Education: An educational approach that integrates domains of science, technology, engineering, and mathematics.

Tangible User Interface: A computer interface in which the user manipulates digital information using physical gestures and interactions.

Chapter 7
Computational Expression:
How Performance Arts Support Computational Thinking in Young Children

Amanda L. Strawhacker
Tufts University, USA

Amanda A. Sullivan
Tufts University, USA

ABSTRACT

In the past two decades, STEM education has been slowly replaced by "STEAM," which refers to learning that integrates science, technology, engineering, arts, and mathematics. The added "Arts" portion of this pedagogical approach, although an important step towards integrated 21st century learning, has long confused policymakers, with definitions ranging from visual arts to humanities to art education and more. The authors take the position that Arts can be broadly interpreted to mean any approach that brings interpretive and expressive perspectives to STEM activities. In this chapter, they present illustrative cases inspired by work in real learning settings that showcase how STEAM concepts and computational thinking skills can support children's engagement in cultural, performing, and fine arts, including painting, sculpture, architecture, poetry, music, dance, and drama.

DOI: 10.4018/978-1-7998-7308-2.ch007

INTRODUCTION

What kind of thinking does it take to write a story, paint a landscape, or put on a play? How does a poet know how to structure a stanza, or an architect know how to start designing a memorial statue? Creativity and inspiration are an important part of the artistic process, but equally important are cognitive and psychosocial traits that support many creative endeavors such as perseverance, logical reasoning, abstraction – in other words, computational thinking. Further, just as learners can and should explore diverse artistic mediums from an early age, young children's cognitive and psychosocial development can also benefit from early exposure to computational thinking. In chapter one of this book, Professor Marina Bers framed computational thinking as an expressive process that involves problem solving by thinking like a computer. In this chapter we delve into this" expressive process" to explore the communicative and creative potential that children can tap into when they explore foundational computational thinking skills.

In the past two decades, STEM education has been slowly replaced by "STEAM", which refers to learning that integrates Science, Technology, Engineering, *Arts*, and Mathematics. The added "Arts" portion of this pedagogical approach, although an important step towards integrated 21st century learning, has long confused policy makers, with definitions ranging from visual arts to humanities to art education, and more (Henderson, 2020). We take the position that Arts can be broadly interpreted to mean any approach that brings interpretive and expressive perspectives to STEM activities. In this chapter, we will discuss how STEAM concepts and computational thinking skills can support children's engagement in a range of liberal and performing arts, including painting, sculpture, architecture, poetry, music, dance, and drama. This chapter specifically hones in on how the performing arts (e.g., dance, drama, music, etc.) are connected to computational thinking. We present descriptive cases that are inspired by research conducted in early-learning settings over a period of several years to illustrate the many overlaps between supporting young children's exploration of drama and computational thinking.

BACKGROUND

STEAM in Early Childhood

Young children (ages 5-8 years) are at a critical stage in their cognitive, social, emotional, and physical developmental trajectories. Integrated educational STEAM experiences allow children to explore diverse approaches, perspectives, and mediums related to STEM, offering a divergence from the positivist "hard sciences" mindset of

20th century science, math, and construction education. Similar to how representation of diverse religious, cultural, ethnic, and gender backgrounds can support a child's developing identity, diverse modes of STEAM education support a child's developing identity as one who does – or is able to do – STEM work.

The arts help teachers engage more thoroughly with STEM, as well. Research shows that despite the global educational trend of bringing novel STEM domains (e.g., computer science, technology, robotics) into early childhood classrooms (Sheffield et al., 2018), pre-service and in-service early childhood educators around the world still indicate insecurity and mixed levels of confidence in their ability to meaningfully teach these subjects (e.g., Dong & Xu, 2020; Masoumi, 2020). In contrast, arts integration has a long history in education, and teachers generally report feeling confident and comfortable with bringing these domains into the early childhood classroom (Bresler, 2007; Hartman & Dani, 2020; Leung, 2020). Arts integration makes STEM more accessible to teachers by allowing them to leverage their experience in arts education to create more engaging and meaningful pathways into STEAM exploration for students.

With the growing focus on technology and STEM education, some critics of computers in education have expressed fear that technology may inhibit children's natural play and creativity (e.g., Cordes & Miller, 2000; Oppenheimer, 2003). The STEAM integration approach counters this by demonstrating that the flexibility inherent in art practices and how naturally this can be applied to STEM content, therefore making STEM more appealing to young learners as well as accessible for educators (Robelen, 2011).

Performing Arts in Early STEM Education

In the performing arts, including fields like theater, music, and dance, children literally embody the emotion or message they are aiming to present. Integrating STEM content in the performing arts offers children the chance to embody a STEM identity (e.g., by acting out the role of a scientist wearing a white lab coat), and even appropriate STEM practices into their "toolbox" for creative and expressive play (e.g., building and testing props for a dramatic performance).

Over a decade of research has demonstrated that the performing arts can positively benefit young children's learning across multiple STEM domains. For example, Ingram & Reidel (2003) found a significant positive link between in-school arts-integrated programming (as part of the Arts for Academic Achievement program) and standardized test scores. Similar research found that children in Chicago arts-integrated elementary schools performed better on tests than children in control schools (Catterall & Waldorf,1999). At the early childhood level, Erdoğan and Baran

(2009) reported that drama-infused math instruction for Turkish six-year-olds was linked to mathematics achievement test scores.

The performing arts offer playful and collaborative ways to integrate self-expression and creativity with traditional STEM content in ways that support foundational knowledge as well as self-confidence. Research funded by the US Department of Education indicates that teachers using arts-integrated strategies have significant positive effect on children's STEM learning (Ludwig & Song, 2016). In a randomized controlled study of an early childhood program titled "Early Childhood STEM Learning Through the Arts" created by the Wolf Trap Foundation for the Performing Arts, researchers found that students in the program outperformed peers in control schools on the Early Math Diagnostic Assessment (EMDA). Moreover, teachers in this study reported that music, movement, and drama was particularly beneficial for students who were shy, who had never been to school, or who were speaking another language (Ludwig & Song, 2016). In this way, the performing arts help teachers connect with children who may otherwise struggle socially. Taken together, this body of research speaks to the value of integrating the arts with other curricular areas in order to boost children's confidence, foster connections, and even master STEM learning content – in other words, it illustrates the benefits of adapting a STEAM approach.

Computational Thinking & STEAM

As we have seen, there is a growing body of work on the impact of the performing arts on STEAM education. The focus of prior work has mainly concentrated on linking the arts with mathematics achievement (e.g., Erdoğan & Baran, 2009). With the recent national and international focus on the importance of early childhood exposure to coding, computational thinking, and engineering, it is useful to look at how the performing arts can serve to foster new 21st century computing skills and mindsets.

Computational Thinking (hereafter, CT) has been described as the thought processes involved in constructing and/or decomposing the sequential steps of a task so that it can be carried out by a computer (Cuny, Snyder, & Wing, 2010; Aho, 2011; Lee, 2016). In chapter two of this book, Bers (2021) argues that CT is more than a style or category of thinking. She asserts that the ability to apply CT skills and practices for the purpose of self-expression through code and technology is equally important for young children to explore, which nicely aligns with creative, dramatic, and performing arts education perspectives.

In this chapter, we take the second approach, and further propose that when children engage in STEAM-integrated dramatic arts and performance, they leverage CT skills and practices to create a shared experience outside of themselves. Through the arts, students move from mastery or even self-expression through code to connection with their community and audience. In the process of creating shared cultural artifacts and experiences with technology, children not only hone their computational thinking skills and abilities, but also cultivate their creative capacity and enrich their identities as agents of transformative social experience. In this view, technology is not just a tool to practice with, a puzzle to solve, or even a language to learn, but instead becomes an extension of the child's expressive and artistic palette, alongside paints, costumes, and musical instruments. In the case vignettes presented later in this chapter, we offer real example cases of children exploring STEAM-integrated arts and drama experiences using performative platforms such as robotics kits and computer programming languages.

Computational Thinking & the Performing Arts: How Do They Relate?

Offhand, it may seem like the arts and CT have little in common. Surprisingly, though, the mindsets, skills, and practices from the two domains actually overlap quite a bit. Modularity and decomposition, for example, are CT skills that involves breaking a task into smaller, more manageable parts. Decomposition can also be applied to composing a piece of music with multiple movements or writing a play with multiple acts. Taking the music example, the composer could choose to work on each individual "movement" (i.e., a self-contained part of the musical composition)at a time and then break that down into even smaller sections such as the main melody and distinct harmonies.

At the early childhood level, it may be difficult for educators to decipher which CT and performing arts concepts are most important for young children to focus on. It is important to choose concepts from both domains that will reinforce foundational early childhood skills that are relevant across all classes and content matter. Sequencing (i.e., understanding that order matters), for example, is relevant to CT and drama but is also a foundational early math and literacy skill.

Bers (2020) describes 7 "powerful ideas" from CT that young children should focus on. These include: algorithms, modularity, control structures, representation, hardware/software, the design process, and debugging. Table 1 defines each of these powerful ideas and illustrates how each one relates to the performing arts as well as foundational early childhood skills.

Table 1. Mapping early childhood CT concepts with performing arts activities & early childhood skill

Bers (2020) Computational Thinking Concepts	Performing Arts Activities & Practices	Early Childhood Skills
Algorithms – A series of ordered steps taken in sequence.	**Story Arc** – Understanding that a play follows a sequence including a beginning, middle, and end.	**Sequencing** **Logical organization**
Modularity – Breaking down tasks and procedures into simpler, manageable units.	**Breaking Down Music** - Learning to play a song one verse at a time.	**Breaking down a large task**
Control Structures – Controlling the sequence in which a program is executed. Making decisions based on conditions.	**Character Reactions** – Exploring if/then reactions for characters through acting improvisations.	**Pattern Recognition**
Representation – Concepts can be represented by symbols.	**Music Notation** - Learning to read music notation and understanding symbols represent notes, rhythms, etc.	**Symbolic representation of letters & numbers**
Hardware/Software - Computing systems need both hardware & software to operate.	**Lighting & Sound Boards** – Understanding that lighting & sound boards rely on hardware & software.	**Recognizing objects that are human engineered**
Design Process – An iterative process used to develop programs & artifacts with multiple steps.	**Editing** – Iteratively editing a script or piece of music.	**Writing Process** **Scientific Method** **Editing/Revision**
Debugging – Fixing problems in our programs.	**Rehearsals** – Using the rehearsal process to troubleshoot a song that doesn't sound quite right, a scene that doesn't work, or a lighting cue that is off.	**Perseverance** **Problem-Solving**

THREE CASES OF STEAM ACTIVTIES TO SUPPORT COMPUTATIONAL THINKING

The previous section explored the powerful ideas of CT young children can begin to develop, as well as how and how they map on to the arts and traditional early childhood content. But what does this look like in practice? In this section, we will explore how these concepts can be brought to life with hands-on robotics kits, interactive coding applications, and in-person practices. We will present three case studies that include fictionalized vignettes (all school and child names are pseudonyms) that are inspired by actual early learning curricula and projects that successfully wove together the performing arts, technology, and CT. These vignettes are based on real observations and data collected in the settings described, but include fictionalized depictions of students.

The first case study will describe a project integrating robotics with music and dance performance in Singapore preschools. The second explores a curriculum that explores music and computer science. Finally, the third vignette describes a STEAM summer camp that integrated performance arts practices through rehearsals and a final performance for families.

Cultural Dances with KIBO

At Sunny Day preschool in Singapore, group partners Nadia, Siti, and Rizwan are all working hard on their Dancing KIBO robot project. "Can you pass me the motors?" Nadia asks in English, and Rizwan replies, "here you go" in Malay. Because of Singapore's multicultural community, most schools offer bilingual education in English, and one of the other three national languages: Mandarin, Malay, and Tamil. Although everyone learns English at Sunny Day, these three children are in the Malay "Mother Tongue" program. Today, they are excited to test the dancing KIBO project they have been working on for several weeks. They can't wait to show off their special Malay dance to their classmates in their school showcase!

Nadia is nearly done attaching two motors to their KIBO. Meanwhile, Siti is building their dance program while Rizwan checks their Design Plan Worksheet and gives Siti instructions. "Begin, Spin, Wait for Clap, Shake, End," he reads off the code they wrote yesterday, while Siti touches each block in the program in front of her. "It's all here! We're ready Nadia!" Nadia hands the robot to Rizwan, who carefully holds KIBO's flashing scanner light over the barcode on the Begin block. When the robot beeps loudly, Nadia and Siti shout "Next!" and Rizwan moves to the next block, until he has scanned all the way down the line. Nadia tests their robot, remembering to add a sound sensor to hear their clap. Finally, after a few tests, the team decides the robot is ready to decorate. Siti carefully attaches the decorated platform they had all worked on last week. She is careful not to crush or tear the tissue paper skirts of their robot dancer, and makes sure it is sturdy before she takes her hands away.

"Everyone, it's time to put on your costumes!" their teacher announces. All three children take out the silky skirts and hats they brought from home. They giggle and twirl in their special outfits, and take turns dancing next to KIBO. Finally, their teacher starts the music and nods to their group. Rizwan presses "Start" on the KIBO, and all three partners line up to dance along with their robot. "Spin, Clap, Shake" their teacher reads off their Design Plan and acts along in front of them. As the three children dance side-by-side with their robot, their classmates smile and nod to the music. Finally, they finish the song and take a bow to a room of applause. "Well done, Malay team!" their teacher says, and helps them carry their robot to a table while the Tamil group prepares to practice their dance.

Figure 1. KIBO Robot and sample wooden programming blocks

This case example was taken from a study conducted in 2015, when the government of Singapore announced a novel initiative called the "PlayMaker Progamme" to bring a maker-centered, screen-free approach to early childhood technology instruction (Bers, 2020; Chambers 2015; Digital News Asia 2015; Sullivan & Bers, 2017). Preschool classrooms across the country used novel technologies like BeeBot (a programmable floor robot) and Circuit Stickers (paper-thin conductive electrical components that can be connected to create programmable circuits). Additionally, several classrooms explored KIBO robotics described in the vignette above through a research collaboration with the DevTech Research Group at Tufts University (Sullivan & Bers, 2017).

KIBO is a robotics construction kit designed for children ages 4–7 years to practice early engineering, programming, and design skills (Sullivan & Bers 2015). KIBO was created by the DevTech Research Group through funding from the National Science Foundation and made commercially available through KinderLab Robotics. KIBO is a robotics construction kit that involves hardware (the robot itself) and software (tangible programming blocks) used to make the robot move (Sullivan and Bers, 2015). The kit contains easy to connect construction materials including: wheels, motors, light output, and a variety of sensors. KIBO is programmed to move using interlocking wooden programming blocks. These wooden blocks contain no embedded electronics or digital components. Instead, KIBO has an embedded scanner in the robot used to scan the barcode on each block one at a time (see Figure 1).

Children in the PlayMaker Programme study spent 7 weeks building, coding, and testing a KIBO robot, as well as learning and rehearsing a meaningful dance from their unique culture. Just as Rizwan checked their group's Design Plan to remember their KIBO program, children spent time matching their dance steps to KIBO's coding instructions, creating an algorithm that represented the familiar dance steps they wanted to perform. Coding with a specific order in mind connects to the powerful computer science idea of algorithms, or the concept that the sequence of coded instructions is important for the final product. Children felt intrinsically motivated to develop and stick to specific algorithms because they knew it would make their dancing robots perform in a way that matched their initial choreography and creative vision.

Children also explored the distinction between KIBO's software, or coded instructions, and robotic hardware, the physical pieces that are required to make the robot move. Nadia knew that her KIBO needed special hardware parts, motors and a sensor, to act out all the instructions in their software (Spin, Wait for Clap, and Shake). The concept that a machine or computer will only act out the instructions in its coding software is a complex idea for young children. In comparison, the idea that your body can only act out a dance if your mind learns the steps is easy to understand. By exploring coding through the metaphor of a musical performance, children can embody the hardware/software relationship themselves. They easily understand the relationship between software instructions and hardware as they act out the dance steps alongside their robots, perhaps even listing their coding instructions out loud as they move. Papert, a pioneer of computational thinking in children, called this *body syntonic* learning, and argued that it is a natural way for young children to learn about their world (Papert, 1980).

In this example, children explored coding software and robotic hardware through the lens of song and dance performance. Children's cultural connection to their dance, and their commitment to accurately showing the steps of the dance, added meaning and creativity to an otherwise straightforward (and common) STEM activity to code a robot to move in certain direction. In the next vignette, we meet a child who is exploring concepts of sequencing and patterns through code, in the context of musical composition.

Coded Musical Compositions

Livvy is 7 years old and is learning how to use the ScratchJr programming app on her family's tablet. She has been practicing how to record sounds into coding blocks, make characters, and write short lines of code. Today, she wants to make a music video for her little brother of his favorite song, *5 Little Speckled Frogs (see lyrics below).* She tries to record the whole song, but her tune is interrupted! ScratchJr sound record blocks can only hold about 1 minute of song. She erases the recording and thinks about the lyrics of her song:

"*Five Little Speckled Frogs*" Song Lyrics
Five *little speckled frogs,*
Sat on a speckled log,
Eating some most delicious bugs.
Yum!
One jumped into the pool,
Where it was nice and cool,
Then there were ***four*** *speckled frogs.*
Glub glub!
Four *little speckled frogs,*
Sat on a speckled log,
Eating some most delicious bugs.
Yum!
One jumped into the pool,
Where it was nice and cool,
Then there were ***three*** *speckled frogs.*
Glub!
Three *little speckled frogs*
(song continues)...

After singing through the song in her head, she says out loud, "most of this song is the same part over and over. Just the number of frogs changes." She thinks a little longer, and then begins to record three new sounds, shown in Table 2. Then she organizes her blocks into a code, with her verse block repeating in between the counting down blocks (see Figure 2). Soon, she finishes her first page and proudly replays it again and again. Just as she is ready to move onto the next part of the song, she hears her mom calling her to come eat lunch. She runs to the table beaming, and spends lunch telling her family all about the parts she wants to add to her song project after she finishes eating.

Livvy was engaging in the computational thinking skill of Algorithmic logic and sequencing when she worked on her ScratchJr song project. By creating her song through code instead of simply singing it, she was forced to consider the repetitive patterns in the song. Music also supports Livvy's intrinsic motivation to accurately code the sequence. She and her audience (her little brother) know what the song sounds like, and if there are changes in the sequence it becomes a musical reinterpretation – but not the old familiar classic.

Table 2. Livvy's sound block recordings

Sound Block	Livvy's Recording
Block 1 (Verse):	*Sat on a speckled log,* *Eating some most delicious bugs.* *Yum!* *One jumped into the pool,* *Where it was nice and cool,*
Block 2:	***Five** little speckled frogs*
Block 3:	*Then there were **four** speckled frogs.* *Glub!*
Block 4:	***Four** little speckled frogs*
Block 5:	*Then there were **three** speckled frogs.* *Glub!*

Figure 2. Livvy's ScratchJr Program, which reads "Start on Green Flag, then Play Recordings in this order: 2, 1, 3, 4, 1, 5"

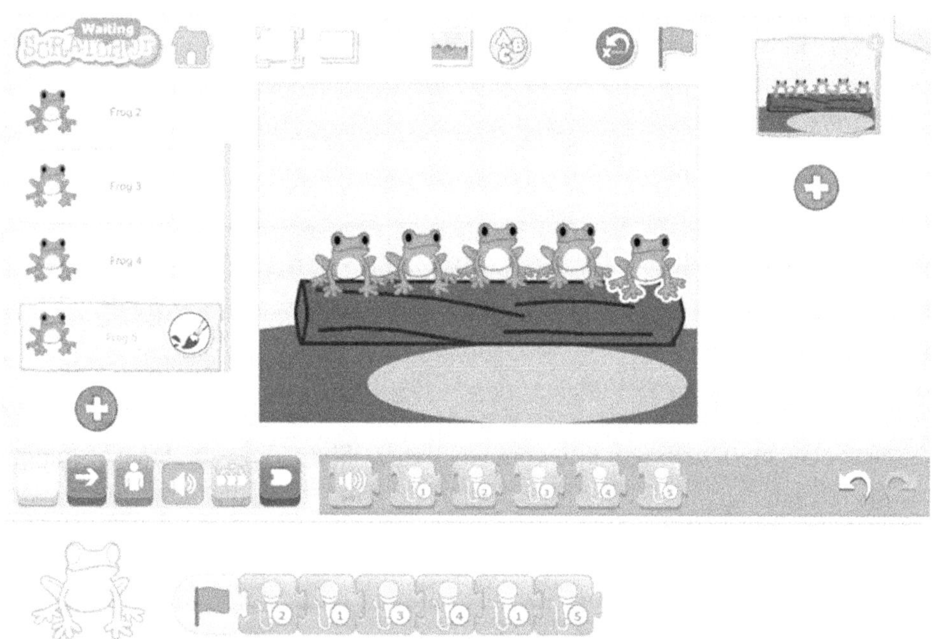

Musicians and computer scientists both share a very important understanding about the value of algorithmic sequencing. Even people who have never coded before can understand why it would be important for a computer or machine to carry out instructions in a specific order to function correctly. We need sensor-automated sinks to wait to be activated until they turn on the water faucet, and a robot coded to move straight 3 times and then right will end up in a very different place than if it turned right 3 times and then moved straight. Similarly, the sequence of notes, phrases, and stanzas in a musical score is important to creating the mood and emotion that the composer is trying to evoke, in addition to recreating the rhythm and cadence of familiar songs accurately. The shared emphasis on sequencing and patterning is part of the reason that computer science researchers have recently begun to explore AI (artificial intelligence) algorithms to aid in music composition (e.g., Moruzzi, 2017).

When Livvy breaks up the parts of her song into different sections and pieces them together again through code, she is also practicing modularity, the practice of decomposing problems into smaller, constituent parts. This is distinct from breaking music phrases into stances or couplets, which is more similar to a literacy or poetry task, but instead requires Livvy to recognize the specific lyrics of the song that change through each iteration of the verse. As Livvy works through multiple rounds of testing and refining her project, she may also choose to test specific parts of the program individually, rather than launching the whole song from the beginning, which is also a form of modularity. In early childhood, modularity is a challenging but important skill that sets the foundation for larger and more complex tasks like writing a lengthy story, exploring complex math concepts, and more.

Livvy's vignette highlights the importance of using algorithmic logic, modularity, and debugging to explore musical composition with a technological tool. Moreover, this vignette demonstrates a simple way that educators can infuse music with computational thinking through a freely available coding application. the third and final example in this chapter, we consider how children can apply the same mindset of debugging and refining a "work in progress", to the task of rehearsing and executing a culminating performance about their completed STEAM projects.

Reflecting on Computational Thinking and Performing Arts

"...And now, I'd like to introduce the Robot Campers to share with you their robotic Zoo Animals! Go ahead and say hello, campers!" Calvin's camp counselor announces and the room full of parents, grandparents, siblings, and babysitters erupts in applause. He is a little nervous, but when his counselor, Lisa, gives him a big smile and two thumbs up, he stands up with his friends and waves at his family in the crowd. Lisa moves to the side of the stage area, so Calvin can still see her but she is not in the way of the audience, and she holds up the poster of the Engineering Design Process for the campers to see. Then she holds up her fingers to silently count down "3 – 2 – 1 – GO!" and the campers start singing "the Design Process

Song" at the same time. Lisa points to the poster and mouths the words along with them. Everything is going exactly like they practiced yesterday in rehearsal. Calvin's nervousness starts to fade and soon he is smiling and singing loudly. When they get to the end of the song, he doesn't even need anyone to point at him to remind him when to sing his big ending – this is even better than rehearsal!

Next, he gets ready to show-and-tell about his robot. Just like in a school play or performance, each camper has lines to say at particular times. When it is Calvin's turn, Lisa hands him his decorated robot and he holds it out for everyone to see. "This is my Zebra. Zebras are my favorite animal because they're kind of like horses but with better colors." All the adults laugh and Calvin looks up, confused, but Lisa nods for him to keep going. She also puts her hands around her mouth as if she might shout, to remind him to talk louder. He keeps going, raising his voice to say "My three zebra facts are: Zebras are herbivores because they eat grass, Zebras live in Africa, and many Zebras together is called a Dazzle. I programmed my Zebra with many forwards because zebras can travel very far, and I used a light sensor to make it only move when it's light outside, because Zebras sleep at night and are awake during the day. It took me 8 tries to make the sensor work right. Also, I used black and white streamers for the Zebra stripes." Calvin places his robot on the ground and presses the "Play" button to show his moving Zebra robot. When it stops moving Lisa starts clapping, and then the whole audience claps so loudly that Calvin jumps. He sees his Mom and brother standing up, clapping loudest of all. "Thank you, Calvin! Go ahead and take a bow!" Lisa says. Calvin bows like they practiced that morning, and carried his robot back to his seat.

After the showcase, Calvin walked back to the bus stop with his family, holding his new Robot Engineer certificate with a picture he remembered posing for earlier that week, of him holding his Zebra robot. He was so proud that he asked his mom if they could hang the certificate on the refrigerator at home. "Are you still sad you didn't get to take the robot home?" his brother asked. "No," Calvin said, "this is even better," and he held up his certificate, beaming.

Calvin's experience at his robot showcase is a common scene from the STEAM-themed camps run at the DevTech Research Group. In addition to being a powerful way to commemorate the time and effort that children put into their STEAM projects and invite adults to understand the work they did at camp, performances like these are wonderful opportunities to practice performance and dramatic arts. Calvin's several rehearsals prepared him for large-group public speaking, communicating through song, gestures, and speech in front of an audience, and coordinating with a team of other performers to make sure their performance came across as a clear, cohesive presentation. The teacher moves made by counselor Lisa are some classic supports that drama teachers use to guide children to focus and remember their parts, and remind children that they are not alone. This kind of practice is important for children's perseverance and executive functioning, especially during the potentially overwhelming experience of performing for a crowd.

Another goal of helping children rehearse and refine a staged performance is that it is another way to practice the Engineering Design Process. This is a multi-step practice that professional and novice engineers, designers, and builders engage in when working through a design project. Educators and researchers use various versions of this cycle (e.g., see Lachapelle & Cunningham, 2007, and Milto, Portsmore, McCormick, Watkins, and Hynes, 2020), but almost all of them move through a process of asking a question, brainstorming and testing different solutions, and iterating on designs until the designer is satisfied. The design cycle commonly used at DevTech STEAM camps is depicted in figure 3, notably an infinity loop to signify to children that the process is never fully "done," and you can constantly jump around to any stage in the process as you work. This is akin to rehearsing a staged performance: there are always changes and improvements that can be made.

Figure 3. The engineering design process

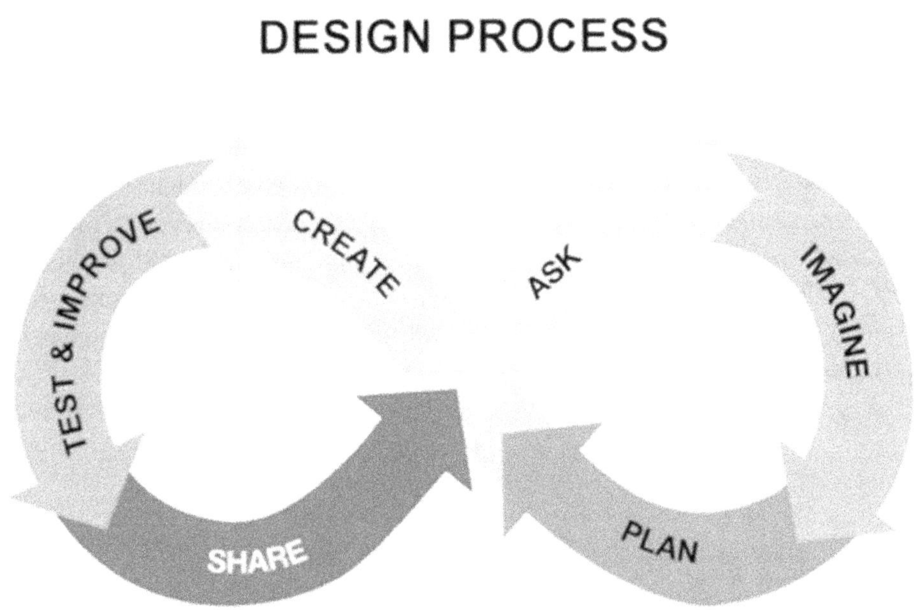

Calvin's references to improving on his rehearsal point to another important practice that he surely explored when building his zebra robot as well: Debugging. Debugging, the practice of identifying and resolving problems in your work, is a popular skill to build when creating a STEM project. It's especially useful when the program or machine is meant to solve a problem, or is expected to perform in a very specific way. However, this process is also useful for dramatic performances, when timing, coordination, and precise execution of lines and actions is important to expressing a clear, unified message to an audience. Just as Calvin spent "8 tries" reworking and iterating on his robotic sensor, he also spent several rehearsals learning his lines for the showcase performance. While debugging with technology can improve his competence as a technician and builder, debugging his performance can support his developing confidence in sharing and expressing his design ideas.

Summer Camp STEAM Showcase

The vignettes above demonstrate ways that children can engage with computational thinking through a STEAM integration approach, with dramatic, performance, and visual arts playing as important a role in children's creative design work as engineering, logic, and technology. In addition to highlighting children's creative and expressive potential through computational media, these stories also depicted adults – parents, teachers, and counselors who put in considerable time and effort "behind the scenes" to support children's positive early STEAM experiences. In this section, we unpack some practical steps that adults can take to increase children's chance of success with STEAM-themed performing arts explorations.

We have argued throughout this chapter that computational thinking is just one among many skillsets that children can leverage during creative STEAM play, but we have not yet described what other types of learning adults might look for. In 2012, Bers outlined the Positive Technological Development framework, a model of psychosocial behaviors that children can practice through engagement with technology (Bers, 2012). This framework drew on extensive research in latent character trait development to yield six behavioral indicators of children's psychosocial engagement, also called the *6 C's* (see figure 4).

Figure 4. Positive Technological Development Framework, including (left) general character assets that children can develop, (center) six corresponding behaviors (i.e. "6 C's") that children exhibit when using technology to indicate they are developing that asset, and (right) example classroom practices that can support positive behaviors through technology use

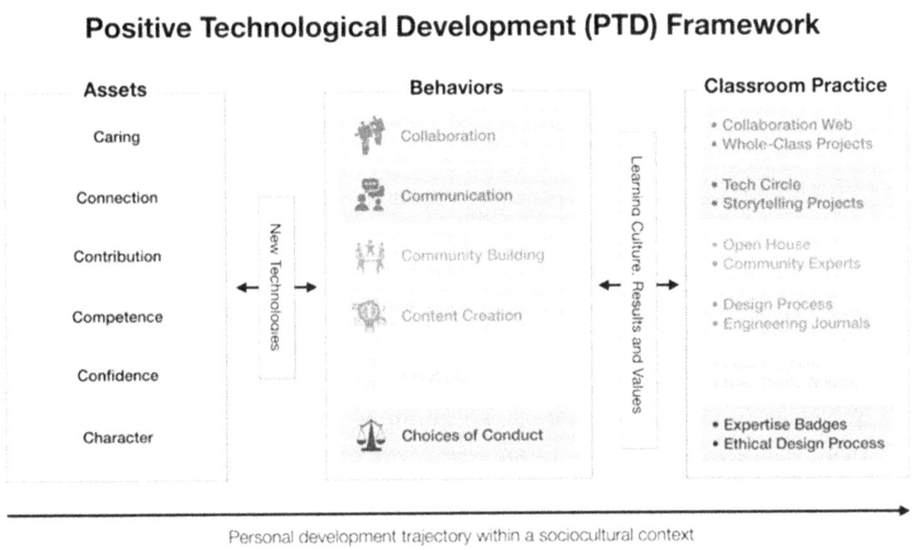

What does this framework mean for CT teaching and learning? The 6 C's can be a guiding framework to remind adults of the positive psychological and prosocial outcomes that children stand to gain through developmentally appropriate, meaningful, child-directed technology exploration (Bers, 2012). Below, we outline teaching and mentoring practices that specifically relate to integrating STEAM and the performing arts. In each example, we use the 6 C's to highlight aspects of psychosocial development that are supportive of CT learning:

- ***Support Children in Giving and Receiving Constructive Feedback.*** An important aspect of both art and engineering is respecting the iterative, flexible, and sometimes unpredictable nature of the design process. In our curricular work with children, we often encourage children to pause and celebrate the process through mid-point project shares and requests for help and feedback, typically in medium or large groups. In addition to helping children course-correct in the event of technology bugs or challenges, these conversations give children opportunities to practice *communicating* their work and ideas in a way that others can understand. Children who are not presenting ideas also have learning opportunities, and must practice how to provide feedback or opinions in a *collaborative* and constructive way, without hurting the presenting child's feelings. Adults can help children develop empathic and constructive *choices of conduct* by offering a simple structure for feedback, such as naming one aspect of another child's project that they admire or enjoy for every suggested change. This type of exercise is common in performing arts classes after children perform a scene or monologue, but less common in coding or computer science classes. However, it is especially helpful for young children because it allows them to observe their peers' reactions to their words and choices, which important for their developing social and perspective-taking skills.

- ***Implement Rehearsals.*** While we appreciate the worry of over-practicing and taking the joy out of an expressive experience, rehearsing for a performance is an important part of the creative process. In STEM, the arts, and life, there is always room to "improve," and learn from past failures. Giving children opportunities to experience failure in low-stakes settings bolsters their confidence ahead of larger, more intimidating performances. When adults skip these practice opportunities, children may feel unsure about their ability to express their creative selves for "real" events and audiences, leading to stifled creativity and, in the worst-case scenario, stage fright and fear of failure. Instead, adults can liken rehearsals to the "Test & Improve" step of the engineering design cycle (see Figure 3). This positions rehearsals as a necessary step toward *content creation*, rather than letting small failures discourage a child from completing their performance. Importantly, rehearsals also offer low-stakes environments for children to practice respectful listening as audience members. Adults can invite these children to offer feedback and encouragement (e.g., through applause) to help them feel involved in rehearsing for more passive parts of a performance. Positive presentation of failures can help children outline the next steps for debugging their work, allowing space for them to explore *creative* solutions, and ultimately make their work even better than they originally thought it could be!
- ***Communicate with Education Stakeholders.*** Children may have wonderfully rich performing arts STEAM experiences in the context of a camp or classroom, but how can we support the transfer of these skills to other settings and learning contexts? Adults can set children up for STEAM success (in the performing arts or in any STEAM program) by communicating ideas and strategies to keep the learning going with other adults and caregivers in the child's learning environment, including home and school. Regular notes sent home to family members, classroom teachers, or other counselors can outline key concepts, activities, and vocabulary words from the day's STEAM lesson, and can even prompt adults with questions or ideas to connect to

relevant experiences at home or in the local community. These can easily be drafted ahead of time, and shared in a timely way at pick-up and drop-off, or classroom transitions. Connecting STEAM learning to children's daily experiences (e.g., at the grocery store, while watching TV, or even while eating dinner) can cement CT concepts that they will carry into later academic settings. Perhaps more importantly, this kind of communication among a child's entire *community* of mentors and educators serves to recruit them as STEAM-allies within the child's learning network.

- ***Be a STEAM Role Model.*** Adults leading STEAM activities have an opportunity to model all of the 6 Cs by diving in and taking part in the creative process alongside children! In addition to being a fun and dynamic way to lead activities, this method allows children to see first-hand how an experienced STEAM practitioner works through challenges using perseverance, positivity, curiosity, and passion. Many learning approaches, especially in early school settings, attempt to "manage" children's attention, motivation, and contribution to activities (Rogoff, Paradise, Arauz, Correa-Chávez, & Angelillo, 2003). By comparison, research has shown that when children are positioned as responsible contributors to a shared task in which adults also take part, they exhibit more intrinsic motivation to observe and participate (e.g., Rogoff, Paradise, Arauz, Correa-Chávez, & Angelillo, 2003). Studies also show that regardless of the teaching approach, children are constantly observing and learning from adults around them, which leads to schemas of social understanding (e.g., Sullivan, 2019). For example, if their female teacher never engages in an engineering project of her own, or their male caregiver insists that he is not creative, this contributes to children's developing understanding (and stereotyping) of gender roles (Sullivan, 2019). Adults can foster a more equitable STEAM narrative by engaging as collaborators, creators, and experimenters alongside the children they are mentoring. This can lead to more agency and freedom of creative expression in children, an important aspect in their arts-integrated learning. Finally, adults can telegraph expectations and values to children through their praise and affirmations. Research shows the importance of praising effort, rather than output, for fostering a resilient growth mindset in children (Dweck, 2008; Sullivan, 2019). For example, telling children that you admire their pretty decorations cues them to attend to visual output of their work, which shifts focus from the creative process. Over time, this kind of praise can lead to perfectionism or performance anxiety in children. Instead, admiring the effort and time they must have put into their detailed decorations sends the message that the process of designing is more important than the product, and reinforces children's perseverance and initiative.

CONCLUSION

In the professional world, the line between the arts and computing is becoming increasingly blurry. Fields like film and video game design, for example, rely heavily on both artistic and computing abilities. Colleges like Yale are now offering majors in Computing and the Arts, melding computer science with drama, music, and more. Why, then, should we teach these fields in isolation at the early childhood level? Both CT and performing arts activities offer young children new and engaging ways to build problem-solving skills, express themselves creatively, and build self-confidence. By integrating the performing arts with CT, it also becomes possible for parents and educators to reach children who are not typically drawn to STEM or computing. Through these integrative experiences, children learn that computational skills can be applied to the arts, and in fact, *support* them in furthering their artistic endeavors.

REFERENCES

Bers, M. (2020). *Coding as a Playground: Programming and Computational Thinking in the Early Childhood Classroom* (2nd ed.). Routledge Press. doi:10.4324/9781003022602

Bers, M. U. (2012). *Designing digital experiences for positive youth development: From playpen to playground*. Oxford University Press. doi:10.1093/acprof:oso/9780199757022.001.0001

Bresler, L. (Ed.). (2007). *International handbook of research in arts education* (Vol. 16). Springer Science & Business Media. doi:10.1007/978-1-4020-3052-9

Catterall, J. S., & Waldorf, L. (1999). Chicago Arts Partnerships in Education: Summary evaluation. In E. B. Fiske (Ed.), *Champions of change: The impact of the arts on learning* (pp. 47–62). Arts Education Partnership. Retrieved from https://artsedge.kennedy-center.org/champions/pdfs/ChampsReport.pdf

Chambers, J. (2015). *Inside Singapore's plans for robots in pre-schools*. GovInsider. Retrieved from: https:// govinsider.asia/smart-gov/exclusive-singapore-puts-robots-in-pre-schools/

Cordes, C., & Miller, E. (2000). *Fool's gold: A critical look at computers in childhood*. Academic Press.

Digital News Asia. (2015). *IDA launches $1.5m pilot to roll out tech toys for preschoolers*. Retrieved from: https://www.digitalnewsasia.com/digital-economy/ida-launches-pilot-to-roll-out-tech-toys-forpreschoolers

Dong, C., & Xu, Q. (2020). Pre-service early childhood teachers' attitudes and intentions: Young children's use of ICT. *Journal of Early Childhood Teacher Education*, 1–16. doi:10.1080/10901027.2020.1726843

Dweck, C. S. (2008). *Mindset: The new psychology of success.* Random House Digital, Inc.

Erdoğan, S., & Baran, G. (2009). A study on the effect of mathematics teaching provided through drama on the mathematics ability of six-year-old children. *Eurasia Journal of Mathematics, Science & Technology Education, 5*(1), 79–85. Retrieved from https://www.ejmste.com/v5n1/EURASIA_v5v1_SErdogan.pdf

Hartman, S. L., & Dani, D. (2020). Full STEAM Ahead: Creating Interdisciplinary Informal Learning Opportunities for Early Childhood Teacher Candidates. *Journal of STEM Teacher Education, 54*(1), 3. doi:10.30707/JSTE54.1/MNCB7975

Henderson, A. (2020, July 21). *So Why Is There An "A" In STEAM?* [Blog post]. Retrieved from https://amt-lab.org/blog/2020/5/so-why-is-there-an-a-in-steam

Ingram, D., & Riedel, E. (2003). *What does arts integration do for students?* University of Minnesota, Center for Applied Research and Educational Improvement.

Lachapelle, C. P., & Cunningham, C. M. (2007, March). Engineering is elementary: Children's changing understandings of science and engineering. *ASEE Annual Conference & Exposition*, 33.

Leung, S. K. (2020). Teachers' belief-and-practice gap in implementing early visual arts curriculum in Hong Kong. *Journal of Curriculum Studies, 52*(6), 857–869. do i:10.1080/00220272.2020.1795271

Ludwig, M., & Song, M. (2016). *Evaluation of professional development in the use of arts-integrated activities with mathematics content: Findings from the evaluation of the Wolf Trap Arts in education model development and dissemination grant.* American Institutes for Research. Retrieved from https://education.wolftrap.org/sites/default/files/Full%20WT%20AEMDD%20Report_Final_Jan-2015updated%20with%20date%2Bappendix.pdf

Masoumi, D. (2020). Situating ICT in early childhood teacher education. *Education and Information Technologies*, 1–18.

Milto, E., Portsmore, M., McCormick, M., Watkins, J., & Hynes, M. (2020). *Novel Engineering, K–8: An Integrated Approach to Engineering and Literacy.* NSTA Press.

Moruzzi, C. (2017, November). Creative AI: Music composition programs as an extension of the composer's mind. In *3rd Conference on" Philosophy and Theory of Artificial Intelligence*. Springer.

Oppenheimer, T. (2003). *The flickering mind: The false promise of technology in the classroom, and how learning can be saved*. Random House Incorporated.

Papert, S. (1980). *Mindstorms: Children, Computers and Powerful Ideas*. Basic Books.

Robelen, E. W. (2011). STEAM: Experts make case for adding arts to STEM. *Education Week, 31*(13), 8.

Rogoff, B., Paradise, R., Arauz, R. M., Correa-Chávez, M., & Angelillo, C. (2003). Firsthand learning through intent participation. *Annual Review of Psychology*, 54. PMID:12499516

Sheffield, R. S., Koul, R., Blackley, S., Fitriani, E., Rahmawati, Y., & Resek, D. (2018). Transnational examination of STEM education. *International Journal of Innovation in Science and Mathematics Education (formerly CAL-laborate International), 26*(8).

Sullivan, A., & Bers, M. U. (2015). Robotics in the early childhood classroom: Learning outcomes from an 8-week robotics curriculum in pre-kindergarten through second grade. *International Journal of Technology and Design Education*. Advance online publication. doi:10.100710798-015-9304-5

Sullivan, A., & Bers, M. U. (2017). Dancing robots: Integrating art, music, and robotics in Singapore's early childhood centers. *International Journal of Technology and Design Education*. Advance online publication. doi:10.100710798-017-9397-0

Sullivan, A. A. (2019). *Breaking the STEM stereotype: Reaching girls in early childhood*. Rowman & Littlefield.

ADDITIONAL READING

Bers, M. U. (2012). *Designing digital experiences for positive youth development: From playpen to playground*. Oxford University Press. doi:10.1093/acprof:oso/9780199757022.001.0001

Bers, M. U. (2017). *Coding as a playground: Programming and computational thinking in the early childhood classroom*. Routledge. doi:10.4324/9781315398945

Cohrssen, C., & Garvis, S. (Eds.). (2021). *Embedding STEAM in Early Childhood Education and Care*. Palgrave Macmillan.

Henderson, A. (2020, July 21). So Why Is There An "A" In STEAM? [Blog post]. Retrieved from https://amt-lab.org/blog/2020/5/so-why-is-there-an-a-in-steam

Ingram, D., & Riedel, E. (2003). *What does arts integration do for students?* University of Minnesota, Center for Applied Research and Educational Improvement.

Robelen, E. W. (2011). STEAM: Experts make case for adding arts to STEM. *Education Week*, *31*(13), 8.

Sullivan, A., & Bers, M. U. (2017). Dancing robots: Integrating art, music, and robotics in Singapore's early childhood centers. *International Journal of Technology and Design Education*. Advance online publication. doi:10.100710798-017-9397-0

Sullivan, A., Strawhacker, A., & Bers, M. U. (2017). Dancing, drawing, and dramatic robots: Integrating robotics and the arts to teach foundational STEAM concepts to young children. In Khine, M.S. (Ed.) Robotics in STEM Education: Redesigning the Learning Experience. (pp. 231-260). Springer Publishing.

KEY TERMS AND DEFINITIONS

Arts Education: A field of educational research and practice informed by investigations into learning through arts experiences.

Computer Programming: The process of designing and building a stepwise list of instructions (program) for a computer or machine to carry out.

Creative Expression: A broad spectrum of using artistic to engage in storytelling and idea-sharing, sometimes related to expressing aspects of selfhood (e.g., identity, personal experiences). Methods of creative expression include dance, writing, theater, drama, acting, singing, music, broadcasting, digital design, and scriptwriting.

Dramatic Arts: The art of the writing and production of plays; drama.

Early Childhood Education: Education of children from birth to age 8 years.

KIBO: A screen-free programmable robotics kit for young children with blocks, sensors, modules, and art platforms.

Performing Arts: All forms of creative activity that are performed in front of an audience, such as drama, music, and dance, etc. Also called Performance Arts.

ScratchJr: A programming application for young children to create games, animations, stories, and more.

STEAM Education: An approach to learning that integrates science, technology, engineering, the arts, and mathematics for encouraging student inquiry, dialogue, and problem-solving.

Section 3
Contexts

Chapter 8
Fostering Computational Thinking in Homes and Other Informal Learning Spaces

Madhu Govind
Tufts University, USA

ABSTRACT

This chapter provides theoretical and practical insights for fostering children's computational thinking (CT) in homes and other family-friendly spaces such as libraries, museums, and after-school programs. The family context—the kinds of roles, interactions, and opportunities afforded by parents, caregivers, and siblings—is essential for understanding how young children learn and engage in CT. This work is informed by research on how everyday activities and educational technologies (and the contexts in which they are used) can be designed to promote opportunities for CT and family engagement. This chapter discusses ways to support children's CT by co-engaging family members in collaborative coding activities in homes and other informal learning spaces.

INTRODUCTION

Parents and caregivers have long played an important role in children's early learning and development (NAEYC & Fred Rogers Center, 2012; Rideout, 2014). From birth onwards, children rely on their caregivers to provide them with the care they need to be happy and healthy, and to grow and develop in positive ways. Caregivers of today's increasingly digital and global landscape are tasked with exposing their young children to an emerging set of skills that enable them to successfully navigate

DOI: 10.4018/978-1-7998-7308-2.ch008

the constantly evolving and technology-rich society in which they live. One set of these so-called "twenty-first century competencies" is computational thinking, the focus of this book.

In Chapter 1, Bers describes computational thinking (CT) as a way of thinking in new ways that invites creativity, collaboration, and critical thinking, among other important skills. Prior to the 1980s, CT was not a term that was used outside of the field of computer science (CS), let alone in the homes of young children and families. In many ways, CT is still largely situated within the CS discipline. However, as technology continues to grow and young children are increasingly exposed to a wide range of technological tools, CT is being treated more like the "universally applicable attitude and skill set" Wing (2006) and others purported it to be. In line with this perspective, CT can be applied to a variety of problem-solving situations that do not necessarily require the act of coding or manipulating digital technologies but can be supported when partnered with them. As such, CT represents a set of skills that can be learned and fostered through young children's everyday play and learning activities, many of which occur in informal spaces in the presence of family members.

It is important to acknowledge that informal spaces, as described in this chapter, have less to do with physical location (though that can be a distinguishing factor) but rather more to do with the kinds of interactions and learning opportunities afforded to children and other individuals occupying that space. For instance, an after-school chess club with local community members might take place in a formal school setting, but the nature of the activity and the participants involved in the activity make it more of an informal learning context. Callanan, Cervantes and Loomis (2011) summarize five key dimensions of informal learning: 1) non-didactive, 2) highly socially collaborative, 3) embedded in meaningful activity, 4) initiated by learner's interest or choice, and 5) removed from external assessment. Accordingly, the informal learning spaces discussed in this chapter refer to environments that invite multiple pathways for attaining and transmitting knowledge, promote social and collaborative interactions, and engage children in meaningful and self-driven activities for the sake of enrichment, not evaluation. Such spaces include children's homes, museums, libraries, community centers, after-school enrichment programs, and other spaces that are accessible to young children and their caregivers.

The goal of this chapter is to provide theoretical and practical insights for fostering children's CT in homes and other family-friendly informal learning spaces. The family context, which is comprised of the various roles, interactions, and opportunities afforded by parents, caregivers, siblings, and extended family members, is essential for understanding how young children learn and engage in CT. This work is informed by research on how everyday activities and educational technologies (and the contexts in which they are used) can be designed to promote opportunities for CT and family engagement. After presenting an overview of this literature, I discuss one workshop model for engaging families in collaborative computing and CT activities in informal learning spaces. These workshops, called Family Coding Days, utilize two coding platforms for young children shared in other chapters of this book: the ScratchJr app and the KIBO robotics kit. The chapter concludes with some practical insights for families, educators, and practitioners when facilitating such events like Family Coding Days or other family-oriented opportunities that aim to strengthen young children's development of coding and CT skills.

THE ROLE OF CAREGIVERS IN SUPPORTING CHILDREN'S COMPUTATIONAL THINKING

Long before formal schooling begins, parents and caregivers (hereafter referred to as "caregivers" to synthesize the various terms used to identify the primary adults who care and provide for a child) guide young children's participation in culturally valued activities and practices (Barron, Martin, Takeuchi, & Fithian, 2009; Lave & Wenger, 1991; Rogoff, 1999; Vygotsky, 1978). These activities and practices serve to promote children's development in a variety of cognitive domains, including but not limited to early literacy and language, numeracy, scientific thinking, artistic and musical abilities, and competency with technology, all of which are connected in some capacity to CT. This section summarizes the literature on caregivers' roles as gatekeepers, facilitators, co-learners, and co-designers of children' coding and CT experiences.

Although children have some agency in how they explore the world around them, the role that caregivers play, either implicitly or explicitly, in guiding children's participation in cultural and social activities is critical. For instance, caregivers may provide access to particular toys or media in the home, or they may arrange for the child to engage in informal learning opportunities outside of the home at local libraries or other community spaces (Rideout, 2014). Caregivers may also engage more directly in these activities by attending these informal learning experiences with their children, co-viewing media, or playing with toys and games together (Takeuchi & Stevens, 2011). Regardless of the level of involvement caregivers may have, one thing is clear: caregivers act as "gatekeepers" (The Toy Association, 2019, p. 5) of children's early play and learning experiences, and thus, of children's earliest experiences with multiple dimensions of CT.

Various frameworks and definitions of CT have been proposed over the years (see Bers' Chapter 1). This chapter focuses on seven powerful ideas of CT that are developmentally appropriate for young children: algorithms, modularity, control structures, representation, hardware/software, design process, and debugging

(Bers, 2018). These seven powerful ideas encompass CT concepts and approaches that children can exhibit through everyday activities around the home and other informal spaces. Table 1 illustrates common activities in the home that support CT. One example of a CT concept that can be fostered through daily activities is algorithms. Caregivers may support children's algorithmic thinking by structuring daily schedules in a logical and step-by-step fashion, such as through morning and evening routines. Algorithmic thinking is also involved in library and museum settings when families use maps to navigate from one space to another or plan out how they will visit multiple exhibits over a span of several hours.

Another example of a CT concept that can be fostered in informal settings without technological devices is control structures. In programming, control structures are complex sets of code that determine whether and how other pieces of code are executed (e.g., repeat loops, events, and conditionals). In the broader sense of CT, however, control structures can be described as a way of understanding whether and how everyday activities should be performed. Let's look at these two statements as an example: 1) If it is cold outside, then I will wear a jacket, but if it is warm outside, then I will leave my jacket at home. 2) Because I have two feet, I perform the action of putting on my sock and then my shoe two times: first for my left foot and the second time for my right foot. The first statement identifies a situation for which the outcome will be determined by an external event, the weather. The second statement represents an activity that involves a step-by-step process that is repeated a total of two times. These examples illustrate how families may already be encouraging or supporting children's understanding of conditionals and repeat loops but may not be aware of it, possibly because they may lack an understanding of CT or because the activities do not necessarily involve technology. However, as Relkin and Strawhacker detail in Chapter 3, there are many unplugged activities that can help promote children's CT and can be facilitated by parents or other family members.

Caregivers may also support young children's CT development through more explicit technology-mediated activities that engage children in viewing, manipulating or producing computational artifacts. Computational artifacts are anything created by a human using a CT process and a computing device, such as a program, image, video, audio, presentation, or web page file (K-12 CS Framework, 2016). Although children may be exposed to viewing computational artifacts from a very young age using televisions or mobile devices (Rideout, 2014), the act of manipulating or producing them might be a novel experience. Thus, in addition to their roles as gatekeepers and resource providers, caregivers can also serve as facilitators of children's technology-related experiences. As facilitators, caregivers not only provide access to opportunities, but they also may be present to provide verbal, emotional, physical, or cognitive scaffolding to help children understand difficult concepts or guide their learning (Neumann, 2017; Yelland & Masters, 2007).

Table 1. Examples of activities in the home that support computational thinking (CT)

CT Powerful Idea (Bers, 2020)	Activity Example	Activity Description
Algorithms	Following daily routines and schedules	From the moment children wake up and to the time they get ready to fall asleep, children often follow a set of routines that help them understand everyday events and procedures. By co-constructing schedules with children, guiding them to choose the order of activities, or asking children to remember the steps for doing something, caregivers reinforce the idea of algorithms or sequencing.
Modularity	Setting up the dining table for a meal	Each place setting (plate, cup, fork, spoon, knife, etc.) represents a module that can be duplicated. Caregivers may demonstrate how to arrange one place setting, and children may follow along and arrange the remaining dinnerware, displaying their understanding of decomposition (breaking down a large task into smaller chunks) and modularity (recreating the module for the other place settings that day or on a different day).
Control Structures	Choosing what to wear based on the activity or weather	When explaining to children why it would be appropriate to wear a coat before going outside in the cold weather, caregivers are introducing children to the idea of decision-making and cause and effect. These foundational concepts promote understanding of control structures such as loops and conditionals, which are used to direct which actions take place and the order in which they take place.
Representation	Sorting items by size, color, or other attributes	Many objects in homes, including young children's toys and games, have distinct attributes, such as color, size and shape. Some of these attributes signify important meanings. For example, the red button on a TV remote and the red button on a video camera can both be used to record something. When children recognize and learn these associations, they are engaging with the idea of symbolic representation.
Hardware/ Software	Using a remote to turn on/off devices	A computing device works by having physical components (hardware) that are given instructions using programs (software). When children use remotes to turn on/off devices or to adjust the volume, brightness, or other settings on the device, children are engaging with the concepts of hardware and software.
Design Process	Building structures with blocks or other materials	Children's play activities often involve blocks, building bricks, or other sturdy objects that can be stacked on top of one another to create different structures. Children may be provided design challenges that invite them to think, plan, and test their ideas and then determine whether their creations successfully met the challenge. Through the process of building, testing, and improving on their ideas, children participate in an iterative design process.
Debugging	Finding or fixing something that is missing, broken or incorrect	It is not uncommon for a young child to misplace something around the home or to act displeased when their favorite toy is broken. The act of finding solutions to problems can support children's debugging skills when they engage with computing devices. For instance, children may be guided to retrace their steps, tinker and explore how broken parts can be glued back together, and persevere through challenging tasks.

Source: (IGI, 2021)

The role of caregivers as facilitators is particularly salient in library and museum settings (Campana, Haines, Kociubuk, & Langsam, 2020; Ehsan, Ohland, Dandridge, & Cardella, 2018; Swartz & Crowley, 2004). Caregivers may explore these spaces with their children, pointing out different features of exhibits and encouraging children to interact with the objects or individuals in the space. For example, a study by Ehsan et al. (2018) examined five families' interactions with a computer-based coding game at an engineering exhibit in a science center. The authors found that although children could engage in CT independently while playing the game (citing evidence of children's algorithmic thinking, debugging, and decomposition, among other CT skills), parents were able to provide complementary CT competencies that enabled children to advance to higher game levels. The authors conclude that children's engagement in CT can differ depending on the presence and involvement of parents, and that the role of parents may evolve as children gain more experience and competence with technology.

Particularly in the context of novel technologies such as tablet-based apps and creative computing tools, caregivers may also assume the roles of co-learner and co-designer. The terms *participatory learning* and *joint media engagement* (JME) are often used to characterize how caregivers can co-engage in activities and learn alongside their children (Clark, 2011). Takeuchi and Stevens (2011) define JME as "spontaneous and designed experiences of people using media together… [such as through] viewing, playing, searching, reading, contributing, and creating, with either digital or traditional media" (p. 9). Whereas many parent-child interactions might involve the parent teaching or guiding the child in accomplishing a complex task ("zone of proximal development", Vygotsky, 1978), novel technologies can disrupt traditional roles of teacher and learner and enable families to participate more collaboratively and actively (Barron et al., 2009; Connell, Lauricella, & Wartella, 2015; Takeuchi & Stevens, 2011).

As gatekeepers, facilitators, learners, and designers, caregivers play a variety of roles in promoting young children's CT development. Accordingly, a number of programs and interventions have been aimed at bringing children and their families together in informal spaces to engage in collaborative STEM (science, technology, engineering, and mathematics), computing, and making activities. Some examples of these programs include Family Creative Learning, CS is Elementary, and Be a Scientist Family Science Program. The Family Creative Learning program consists of a series of workshops for school-age children and their families to learn and co-create projects using the Makey Makey invention kit and Scratch programming language (Roque, 2016; Roque, Lin & Liuzzi, 2014). The CS is Elementary program, formerly known as Family Code Night, provides free event kits for schools to host large-scale family learning events for K-5 students and families using Code.org and unplugged activities (Pearce & Borba, 2017). The Be A Scientist Family Science Program brings children and parents together for five-week workshops to engage in hands-on science and engineering-related activities (Pierson, Momoh, & Hupert, 2015).

To provide a closer look at one family workshop model and the ways in which it can support children's coding and CT skills, this next section describes a research project at the DevTech Research Group at Tufts University called Family Coding Days. This project, like the aforementioned programs, explored how families with young children engage in unplugged and technology-supported coding and CT activities. The technologies used in this project were the ScratchJr programming application and the KIBO robotics kit, both of which are introductory block-based programming languages developed by the DevTech Research Group that expose young children to foundational coding and CT concepts. Due to the interface differences between ScratchJr (screen-based application) and KIBO (tangible robotics set), the project also explored the ways in which the type of interface may influence how families interact with the technology and with one another during the workshops.

FAMILY CODING DAYS

The Family Coding Days project at the DevTech Research Group first originated in the early 2000s as Project Inter-Actions, an exploration of intergenerational learning with robotics. Children between the ages of 4-7 and their parents attended a series of five-week workshops, during which they were introduced to programming using LEGO bricks. The project revealed a number of interesting findings about the ways in which children and parents learn about technology and engage with powerful ideas such as sequencing, looping, and debugging (Beals & Bers, 2006; Bers, 2007; Bers, New & Boudreau, 2004). In particular, the project revealed how these workshops could generate a multigenerational "community of practice" (Lave & Wenger, 1991) that encourages families to engage with each other and with new knowledge and skills by producing creative computational artifacts.

About a decade later, the creation of newer coding technologies for young children such as ScratchJr and KIBO made it possible to explore family dynamics around creative computing with a fresh perspective. Anecdotal evidence from informal community outreach events and pilot workshops at local early childhood centers and museums indicated that families having the opportunity to code together with ScratchJr and KIBO could learn together, too. Thus, these 1-2-hour workshops called "Family Coding Days" were developed and implemented with over 100 families with young children. A detailed protocol was devised with recruitment strategies, facilitation tips, sample workshop agendas, activity prompts, and instructions for research data collection, enabling interested event facilitators around the country to host their own Family Coding Day workshops in informal learning spaces. A total of 14 workshops (nine ScratchJr and five KIBO) were facilitated in after-school programs, libraries, and museums in the local New England area and in several cities across the United States. The research sample of Family Coding Day attendees included mostly mothers holding an associate degree or higher with children in the 5-7-year range. Over half of families did not have any prior experience with the technology (ScratchJr or KIBO) before attending the event, and about one-third had children who had some exposure to the technology through school.

Each Family Coding Day workshop utilized one type of coding technology (ScratchJr or KIBO) and consisted of three general types of activities: learning about the technology, co-creating a coding project, and sharing projects with peers. Families were together for most of the workshop, except for the introductory learning activity. Children's introduction to the coding technology included off-screen games followed by a play-based tutorial, whereas parents' introduction consisted of a step-by-step tutorial followed by an open discussion. When families were rejoined for the co-creating activity, facilitators provided families with sample prompts for their projects (e.g., "Program a ScratchJr character or KIBO robot to perform a dance, be an animal, or act out a scene from a favorite book or movie") but encouraged families to come up with their own creative ideas. As families worked on their projects, facilitators walked around to assist and observed how families interacted with one another and with the technology.

This next section presents a summary of findings from Family Coding Days, focusing particularly on findings related to children's CT engagement. Then, two illustrative case studies are presented from a small follow-up study, in which individual parent-child dyads were videotaped while working together on a ScratchJr or KIBO project together for 20 minutes. Although the context of these individual case studies was different from the multigenerational community gatherings of Family Coding Days, the case studies provided unique insight into the kinds of parent-child CT interactions that can take place around creative computing activities.

Findings from Family Coding Day parent surveys indicated that parents reported a significant increase in their own coding interest, as well as in their children's. Parents also observed children actively exploring CT concepts of algorithms, design process, and debugging. In addition, parents reported engaging collaboratively with their children (as opposed to the activity being child- or adult-directed) and particularly enjoyed watching their children take initiative as planners of their coding projects. Parents provided a variety of cognitive, affective, and technical scaffolding support for their children. For example, observed parental behaviors included asking their children questions about their projects, offering suggestions and words of encouragement, and showing children where to click or press buttons (Govind, Relkin, & Bers, 2020; Relkin, Govind, Tsiang & Bers, 2020). No significant differences in workshop outcomes, as reported by parents, were found between families who attended ScratchJr workshops and families who attended KIBO workshops, suggesting that families experienced similar forms of collaboration and joint engagement regardless of interface (Govind & Bers, 2020).

The follow-up study of parent-child dyadic interactions with ScratchJr and KIBO shed insight into how families were able to engage collaboratively to create their computational artifacts. For example, families positioned themselves strategically so that the tool was equally accessible. In addition, if one person was using the tool without sharing, the other person would ask to take a turn or make verbal suggestions to contribute actively to the project. Consistent with our previous study findings, families' interactions and behaviors during this 20-minute activity further clarified parents' role as coaches and children's role as planners. The following case studies illustrate this coach-planner dynamic and how parents, even without having explicit knowledge of computer programming or CT, were engaging in conversations and behaviors that seemed to enhance children's display of CT.

KIBO Case Study: Eye of the Tiger

Jordan (child) and Caroline (parent) are playing with the KIBO robotics kit during a KIBO family coding play session. Caroline asks whether Jordan would like to make KIBO into an animal or something else, to which Jordan responds "animal… a tiger!" Knowing that her son has had extensive experience with KIBO from school, Caroline lets Jordan take the lead, remarking, "you might have to lead the way because I don't know how to do this." Jordan goes to the crafts table and begins looking through the various colors of construction paper. They work together to find all the orange paper in the stack. Seeing the variety of colors to choose from, Jordan thinks about making a rainbow instead, but Caroline encourages him to stick with the tiger idea.

Caroline begins drawing a face on the tiger and prompts Jordan with questions about a tiger looks like so that she can draw it properly. She asks, "What color should the stripes be? Does a tiger have whiskers? Does a tiger have eyebrows? Does a tiger have eyes?" Jordan gets excited about their project after seeing the tiger drawing. He asks his mother for help to tape the drawing onto KIBO. They bounce ideas off of one another to figure out how to ensure the drawing is sturdy and upright so that it doesn't fall off the robot. Caroline makes the connection that "it's like putting a character on a parade float". Jordan and Caroline finish taping the tiger body onto KIBO. As a finishing touch to their decorations, Caroline offers a suggestion to make a tiger tail out of a rolled-up piece of orange construction paper. Jordan considers the suggestion but instead decides to cut a zigzag design on a rectangular strip of paper and tapes it to the back of the KIBO robot. They run out of tape, which prompts them to move onto the next part of their project: programming KIBO.

Caroline asks Jordan for ideas for their program. Jordan at first seems distracted but then walks over to the blocks and confidently says, "Forward, backward, lights—these are the eyes blinking—and play sounds." Jordan points out that there are three different sounds that can be recorded using the sound recorder module, so he records the first two tiger growling sounds. Then Caroline asks for the third turn and records her own growling sound. Jordan independently assembles and scans the full program, which contains a repeat forever loop, and presses the green triangle on KIBO to play the program. He remarks to his mother, "Look, I'm scanning these blocks again on purpose because I want this [forward] to go more than once." Caroline expresses her excitement to see their project come alive and gives her son a high-five. When they showcase their KIBO tiger project to the researcher, Jordan recalls his favorite part as creating the tiger's tail. Caroline replies that she enjoyed that too but that her favorite part was recording the growling sounds. Figure 1 displays Jordan and Caroline's collaborative creation, a KIBO tiger.

Figure 1. Jordan (child) and Caroline (parent) co-create a KIBO project: a robotic tiger that moves around, blinks its eyes, and makes growling sounds
Source: Reprinted from Govind (2019)

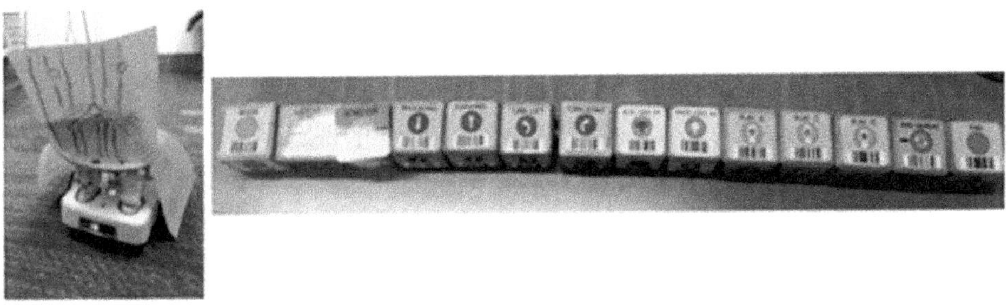

CT Connections

There are many examples of CT skills displayed by Jordan in this case study. To illustrate a few of them, I will focus on the following three powerful ideas of CT: algorithms, representation, and design process. Jordan displays his understanding of algorithms by successfully assembling and scanning the KIBO program. He demonstrates to his mother how the blocks of the program represent the tiger's actions and even explains how scanning the block multiple times was not an accident but rather done purposefully to repeat the robot's actions. When discussing ideas about how to program their KIBO tiger, Caroline and Jordan choose blocks that would represent different aspects of a tiger. The light blocks, as Jordan remarks, represent the tiger's flashing eyes. The recorded sounds represent the tiger growling. These associations made between the characteristics of a tiger and the actions that KIBO can do (i.e., move, turn on its lightbulb, and play sounds) display Jordan's understanding of the computational idea of representation. Finally, Jordan and Caroline engage in an iterative design process to create their KIBO decorations. They imagine what a tiger's face looks like, create their drawing, use tape to secure the decoration to the robot, and improve their design by adding a tiger tail. These design choices are not independently made by Jordan but rather facilitated by Caroline through asking questions and offering suggestions.

ScratchJr Case Study: Magic and Mystery

Shaan (child) and Bina (parent) decide they are going to create a play, maybe something about "dragon avengers" or perhaps a wizard story. They are participating in a ScratchJr family coding play session. Shaan opens a new ScratchJr project and begins scrolling through the backgrounds. Because they are making a play, Shaan selects a stage background and announces that their theme will be "magic and mystery." Bina agrees with enthusiasm and points out the wizard as they scroll through the page of ScratchJr characters. Shaan selects the wizard and navigates to the paint editor tool to customize the wizard as Gandalf from *Lord of the Rings*. Bina sounds out the name Gandalf while Shaan uses his finger to type "Gandolf the great wizard" as the character's name. There is extensive trial and error involved in customizing their Gandalf character, requiring Bina to show Shaan how to use the "undo" feature on the paint editor tool.

Once their first character is set, Shaan switches back and forth between the character panel and paint editor to design the other characters for their project. Laughing at their silly creations, Shaan and Bina create the following *Lord of the Rings* characters: Frodo, Treebeard, and Legolas. There seems to be no suitable character that would be easy to modify into Treebeard or Legolas, so Shaan makes the Treebeard character from scratch. Bina points out they don't have a lot of time left, so instead of creating a completely new character for Legolas, Bina encourages him to use the fairy and pretend that the wings and heels are "where his bow and arrow are."

With only a few minutes left to finish their projects, Bina suggests to Shaan that perhaps they should move on to programming their characters. Shaan agrees and decides immediately that he wants to record a story. He selects the green "Record Sound" block and starts recording, "Once upon a time, there were these... uh hold on a second… there were these four fighters..." Shaan plays back his recorded sound and blushes with embarrassment about his pause in the middle. Bina offers the suggestion of practicing what to say before recording. They notice they only have about a minute left, so Shaan instead navigates to the blue motion blocks. He swiftly drags the blocks into the programming area and snaps them together to create a dance. He asks Bina how he can get the character to dance and play his recorded story at the same time. Before she could respond, Shaan exclaims that he remembers how to do that and adds the "Start on Green Flag" block at the beginning of both sets of code. He then decides to use a "Repeat Forever" block for the movements so that the characters would keep dancing forever. Bina exclaims and asks, "Forever?! Why don't we just make it repeat a couple of times?" Shaan shrugs and proceeds to test his program, excitedly watching the character jumping around the screen. As Bina calls over the researcher to look at their project, Shaan copies the dance code to the other characters so that they are all dancing together on the stage. Figure 2 displays Shaan and Bina's collaborative creation, a *Lord of the Rings*-themed ScratchJr play.

CT Connections

In this case study, Shaan leads most of the activity, but Bina is fully present and engaged. Their back-and-forth exchanges and giggles throughout the course of the activity exemplifies the nature of their playful interactions and collaborative experience. To illustrate the ways in which CT is fostered in this example, I focus on the following powerful ideas: modularity, debugging, and control structures. Having chosen a magical and mysterious theme with many different characters, Shaan and Bina display the concepts of decomposition and modularity in their project. They break down their project idea (the large, complex novel and book series *Lord of the Rings*) into individual characters that they could then customize and program. At the end of their project, Shaan's ability to copy the code for one character to the others demonstrates his understanding that this set of code represents a "dance" module that could be used in multiple places. Debugging is also present when Shaan wonders how he can get the recorded sound and movement blocks to play at the same time. Although he attempts to ask his mother for help, he remembers that he forgot to include the "Start on Green Flag" block to the beginning of his code. Finally, Shaan's decision to use the "Repeat Forever" block rather than the orange repeat loop block demonstrates his understanding of the difference between infinite and finite loops, thus indicating his ability to navigate control structures.

These two case studies highlight some of the ways in which caregivers may promote young children's engagement in CT. As a practical tip, caregivers seeking strategies to support children's CT learning might consider asking the following kinds of questions while co-engaging in collaborative activities with children: "Can you show or tell me what you're doing? What are the steps we need to follow here? That looks interesting – what do you think will happen if we try this?" Such questions open the door for children to display their coding knowledge, ask and answer questions in return, and engage in deeper CT practices. The next section explores additional practical strategies for families and facilitators.

Figure 2. Shaan (child) and Bina (parent) co-create a ScratchJr project with multiple Lord of the Rings characters dancing on a stage and a "Legolas" character narrating the story
Source: Reprinted from Govind (2019)

SETTING FAMILIES AND FACILITATORS UP FOR SUCCESS

In addition to CT concepts and practices that are encompassed in Bers' (2018) framework of the seven powerful ideas, Brennan and Resnick (2012) identified that children participating in creative computing activities can also have CT perspectives. CT perspectives are defined as "children's evolving understandings of themselves, their relationships to others, and the technological world around them" (p. 10). One of these three perspectives is *connecting*, which describes how children recognize—and perhaps grow to appreciate—the power of creating with and for others. The Family Coding Days project and the related work shared in this chapter aptly point to this very aspect of CT. When children co-create computational artifacts or co-engage in unplugged CT activities with family members—often the closest people to young children both physically and social-emotionally—the opportunities to grasp, apply, and extend their understanding of CT are amplified.

Models that engage young children and families in collaborative activities are well-aligned with the principles of connected learning. Ito and colleagues (2013) define connected learning as "broadened access to learning that is socially embedded, interest-driven, and oriented toward educational, economic, or political opportunity" (p. 4). Collaborative computing activities, such as the ones described in the Family Coding Days project, are *interest-powered* and *production-centered*, inviting parents to co-design robotic creations or digital stories that are personally meaningful and interesting to their children and to themselves. These activities are also *peer-supported* and have a *shared purpose*, welcoming various opportunities for collaboration, feedback, and community building. Finally, these activities are *academically oriented* and *openly networked*, offering children the opportunity to learn new skills and connect their learning across different settings.

As recommendations for future research and practice, I offer the following practical tips and considerations for families and facilitators seeking to promote young children's coding and CT engagement:

- **Tools for consuming versus creating:** The JME literature highlights the many ways that children of today's increasingly global and technology-rich society are interacting with the technological tools around them. However, not all technology is the same. Some tools are made for consuming (e.g., televisions); others are made for creating (e.g., conductive play dough). Thus, the ways in which families can foster children's CT through those technologies should not be the same either. In any space that is considered to be "family-friendly", caregivers might consider the following questions: What might my child do or say while they are navigating this space? What might I be doing or saying while navigating this space with my child? How does the technology and the technology-mediated activity enable us to co-engage in CT? Facilitators and designers of family-friendly environments might examine the variety of technological tools in their spaces and think carefully about the kinds of interactions those tools might provide.
- **Leveraging community resources:** In addition to the kinds of tools, it is crucial to think about how the community will utilize the space and the resources needed to make the opportunity accessible and engaging for all attendees. Facilitators might consider the following questions: Are there enough robotics kits or materials for all families? How are we supporting bilingual families or families with children with special rights with this tool and activity? What resources may be needed to enhance accessibility and

inclusion? Facilitators should be encouraged to leverage the existing resources within communities when planning workshops, activities, and other events intended to support families' coding and CT engagement.

- **Facilitating bidirectional home-school connections:** One of the primary goals of this chapter is provide insight on how families can drive their own learning process about understanding what CT is and how it can be fostered through unplugged and technology-mediated activities. As computer science education becomes an increasingly important national and international priority in schools and other formal learning settings, continuing children's coding and CT learning in informal settings through family engagement initiatives will be increasingly salient. Stakeholders who play a role in facilitating children's informal and formal learning experiences might consider the following questions: What activities might parents already be doing in homes and informal learning spaces that foster children's CT, and how can we empower parents to recognize and extend those activities? What technological tools might be introduced in school settings and how is that learning being shared with families? Facilitating bidirectional home-school connections is essential to developing young computational thinkers in homes and other family-friendly spaces.

CONCLUSION

Computational thinking is a set of thought processes that can be learned and fostered through young children's everyday play and learning activities, many of which occur in informal spaces in the presence of family members. These activities do not necessarily require the act of coding or manipulating digital technologies but can be supported when partnered with them. This chapter bridges together the theory, research, and practice around families supporting children's coding and computational thinking through unplugged and technology-supported activities in the home and other family-friendly informal learning spaces such as community centers, museums, and libraries.

ACKNOWLEDGMENT

This research was supported by the National Science Foundation [grant number DRL-1118664] and the Scratch Foundation.

REFERENCES

Barron, B., Martin, C. K., Takeuchi, L., & Fithian, R. (2009). Parents as Learning Partners in the Development of Technological Fluency. *International Journal of Learning and Media*, *1*(2), 55–77. doi:10.1162/ijlm.2009.0021

Beals, L., & Bers, M. (2006). Robotic Technologies: When Parents Put Their Learning Ahead of their Child's. *Journal of Interactive Learning Research*, *17*(4), 341–366.

Bers, M. (2007). Project Inter-Actions: A multigenerational robotic learning environment. *Journal of Science and Technology Education*, *16*(6), 537–552. doi:10.100710956-007-9074-2

Bers, M. (2018). *Coding as a playground: Programming and computational thinking in the early childhood classroom*. Routledge Press.

Bers, M., New, B., & Boudreau, L. (2004). Teaching and learning when no one is expert: Children and parents explore technology. *Journal of Early Childhood Research and Practice*, *6*(2).

Brennan, K., & Resnick, M. (2012). Using artifact-based interviews to study the development of computational thinking in interactive media design [Paper presentation]. The meeting of the American Educational Research Association, Vancouver, BC, Canada.

Callanan, M., Cervantes, C., & Loomis, M. (2011). Informal learning. *Wiley Interdisciplinary Reviews: Cognitive Science*, *2*(6), 646–655. doi:10.1002/wcs.143 PMID:26302414

Campana, K., Haines, C., Kociubuk, J., & Langsam, P. (2020). Making the Connection: Computational thinking and early learning for young children and their families. *Public Libraries*, *59*(4).

Clark, L. S. (2011). Parental Mediation Theory for the Digital Age. *Communication Theory*, *21*(4), 323–343. doi:10.1111/j.1468-2885.2011.01391.x

Connell, S. L., Lauricella, A. R., & Wartella, E. (2015). Parental Co-Use of Media Technology with their Young Children in the USA. *Journal of Children and Media*, *9*(1), 5–21. doi:10.1080/17482798.2015.997440

Ehsan, H., Ohland, C., Dandridge, T., & Cardella, M. (2018). Computing for the Critters: Exploring Computational Thinking of Children in an Informal Learning Setting. *Proceedings of IEEE Frontiers in Education Conference*. 10.1109/FIE.2018.8659268

Govind, M. (2019). *Families That Code Together Learn Together: Exploring family-oriented programming in early childhood with ScratchJr and KIBO Robotics* [Unpublished master's thesis]. Tufts University, Medford, MA, United States.

Govind, M., & Bers, M. U. (2020). Family Coding Days: Engaging Children and Parents in Creative Coding and Robotics. Proceedings of Connected Learning Summit.

Govind, M., Relkin, E., & Bers, M. U. (2020). Engaging Children and Parents to Code Together Using the ScratchJr App. *Visitor Studies*, *23*(1), 46–65. doi:10.1080/10645578.2020.1732184

Ito, M., Gutiérrez, K., Livingstone, S., Penuel, B., Rhodes, J., Salen, K., Schor, J., Sefton-Green, J., & Watkins, S. C. (2013). *Connected Learning: An Agenda for Research and Design*. Digital Media and Learning Research Hub.

K–12 Computer Science Framework. (2016). *K-12 CS Framework*. http://www.k12cs.org

Lave, J., & Wenger, E. (1991). *Situated learning: legitimate peripheral participation*. Cambridge University Press. doi:10.1017/CBO9780511815355

National Association for the Education of Young Children (NAEYC) & Fred Rogers Center for Early Learning and Children's Media. (2012). *Technology and Interactive Media as Tools in Early Childhood Programs Serving Children from Birth through Age 8*. https://www.naeyc.org/files/naeyc/file/positions/PS_technology_WEB2.pdf

Neumann, M. (2017). Parent scaffolding of young children's use of touch screen tablets. *Early Child Development and Care*, *188*(12), 1654–1664. doi:10.1080/03004430.2016.1278215

Pearce, J., & Borba, S. (2017). *What Is Family Code Night?* https://www.naesp.org/blog/what-familycode-night

Pierson, E., Momoh, L., & Hupert, N. (2015). *Summative Evaluation Report for the Be A Scientist!* Project's Family Science Program. https://iridescentlearning.org/wp-content/uploads/2014/01/BAS-2015-Eval-FINAL-3.pdf

Relkin, E., Govind, M., Tsiang, J., & Bers, M. (2020). How Parents Support Children's Informal Learning Experiences with Robots. *Journal of Research in STEM Education*, *6*(1), 39–51. doi:10.51355/jstem.2020.87

Rideout, V. J. (2014). *Learning at home: Families' educational media use in America*. A report of the Families and Media Project. The Joan Ganz Cooney Center at Sesame Workshop.

Rogoff, B. (1999). Cognition as a collaborative process. In Handbook of child psychology. New York: Wiley.

Roque, R. (2016). Family Creative Learning: Designing Structures to Engage Kids and Parents as Computational Creators. In K. Peppler, Y. Kafai, & E. Halverson (Eds.), *Makeology in K-12, Higher, and Informal Education*. Routledge.

Roque, R., Lin, K., & Liuzzi, R. (2014). Engaging Parents as Creative Learning Partners in Computing. *Exploring the Material Conditions of Learning*, *2*, 687–688.

Swartz, M. I., & Crowley, K. (2004). Parent Beliefs about Teaching and Learning in a Children's Museum. *Visitor Studies*, *7*(2), 5–16.

Takeuchi, L., & Stevens, R. (2011). *The New Coviewing: Designing for Learning through Joint Media Engagement.* The Joan Ganz Cooney Center at Sesame Workshop.

The Toy Association. (2019). *STEM/STEAM Formula for Success*. https://www.toyassociation.org/ta/research/reports/stem-steam/toys/research-and-data/reports/stem-steam.aspx?hkey=6e80262f-1fea-4b37-a5e2-9679ec26f048

Vygotsky, L. S. (1978). *Mind in Society*. Harvard University Press.

Wing, J. M. (2006). Computational thinking. *Communications of the ACM*, *49*(3), 33–35. doi:10.1145/1118178.1118215

Yelland, N., & Masters, J. (2007). Rethinking scaffolding in the information age. *Computers & Education*, *48*(3), 362–382. doi:10.1016/j.compedu.2005.01.010

KEY TERMS AND DEFINITIONS

Caregiver: A person, often a parent or adult family member, who is responsible for the overall supervision, care, and well-being of a child.

Informal Learning: Any form of education that takes place outside of a structured setting or is embedded within daily life experiences.

Interface: A device that enables a person to communicate with a computer.

KIBO: A screen-free programmable robotics kit for young children with blocks, sensors, modules, and art platforms.

Scaffolding: A range of instructional techniques used to support a person in the learning process.

ScratchJr: A free block-based programming application for young children.

Unplugged: Describes activities such as games and puzzles that aid the teaching and learning of computer science but without the use of technology.

Chapter 9
Makerspaces as Learning Environments to Support Computational Thinking

Amanda L. Strawhacker
Tufts University, USA

Miki Z. Vizner
Independent Researcher, USA

ABSTRACT

Makerspaces are technology-rich learning environments that can uniquely support children's development. In education communities, makerspaces have become sites to take up explorations of personally-motived problem solving, and have been tied to 21st century learning outcomes of perseverance, creativity, persistence, and computational thinking. Elsewhere in this book, Bers described computational thinking as the set of skills and cognitive processes required to give instructions for a specific task in such a way that a computer could carry it out. But Bers also argued that the purpose of computational thinking is to cultivate a fluency with technological tools as a medium of expression, not an end in itself. Computational making is part of this expression. This chapter explores the ways in which tools, facilitation, and the physical environment can support children's engagement with powerful ideas of computational thinking through making.

DOI: 10.4018/978-1-7998-7308-2.ch009

INTRODUCTION

Makerspaces have become sites to take up exploration of personally-motived problem solving, and have been tied to 21st century learning outcomes of perseverance, creativity, persistence, and--particularly because of the emphasis on creation with digital tools--computational thinking (e.g. Campbell, Heller, & Goodman, 2018; González-González & Arias, 2019; Iwata, Pitkänen, Laru, & Mäkitalo, 2020). Elsewhere in this book, Bers described computational thinking as the set of skills and cognitive processes required to give instructions for a specific task in such a way that a computer could carry it out. But Bers also argued that the purpose of computational thinking is to cultivate a fluency with technological tools as a medium of expression, not an end in itself. Makerspaces provide informal learning spaces in which this can happen through computational making.

While the concept of computational making is nascent, we use the term in this chapter to refer to any creative making or design endeavor in which makers (anyone who creates or tinkers) leverage computational thinking skills (e.g. as outlined by Bers, 2020), to achieve their creative goals. We do not propose that computational making is a learning domain or a standard in itself. In much the same way that picking up and putting down weights isn't the reason one goes to a gym as much as maintaining overall health and wellness, and mastering grammar and syntax isn't the reason to learn a new language as much as self-expression and communication, we argue that making is not necessarily an educational goal itself. Rather, we pose in this chapter that an educational goal of making is to support the maker in developing a suite of psychosocial behaviors and character traits, including (but not limited to): confidence to tackle unstructured problems, competence with a range of physical and digital tools and creative practices, critical thinking skills to evaluate problems and imagine logical solutions, and creativity and agency to determine which problems to address that are personally or communally meaningful. By extension, an educational goal of computational making is to cultivate those same skills and behaviors, but in the context of projects that incorporate digital and technological tools, computational thinking skills, and disciplinary practices from fields such as computer science and engineering. Maker educators are familiar with the phrase, "children should be creators, and not just consumers of their own digital experiences" (Smith, 1982). By providing tools, community, and an environment to enable computational making, a makerspace supports children in creating, rather than consuming.

In the following chapter, we explore research on the educational affordances of makerspaces, with a focus on opportunities for computational thinking and making in the early years. Following this, we describe practice-based examples of computational making inspired by real children and events from our experiences designing and evaluating early childhood makerspaces. We use these examples to illustrate how computational making supports learning in areas such as computational thinking and creative agency. Finally, we share evidence-based principles for designing and facilitating a makerspace to support computational making, and conclude with a reflection about the impact of computational making on children's ability to create their own artifacts and ideas.

BACKGROUND

What is a Makerspace?

Soldering irons, 3D Printers, and robots are some examples of tools that people commonly feel are too complicated for most education settings, especially early childhood ones. In the early 2000s, the maker movement, branded by Dale Dougherty's *Make Magazine,* began to rise in popularity with the promise to democratize tools, expertise, and learning models of production (Dougherty, 2012). His conceptualization of makerspaces was widely interpreted to be in the same spirit of public libraries and their promise to democratize knowledge and literacy (Lakind, Willett, & Halverson, 2019). As Doughterty defines them, makerspaces are any places where people get together to *make*. The content of what is made can vary. It is the mindset of the community, the affordances of the available tools, and the intended purpose of the space that defines the unique identity of a makerspace. While the environment, community expertise, facilitation, and tools may be very different across spaces, all makerspaces--from very high-tech industry prototyping sites, to school-based maker labs, to at-home garages--are places where people can go to make things, and develop their making skills.

Makerspaces and the Environment-as-Teacher

Early childhood educators trained in the Reggio-Emilia approach, a pedagogical philosophy rooted in supporting children to direct their own learning, often say that "the environment acts as a third teacher" (Strong-Wilson, 2007). In other words, the environments that we inhabit inform the ways we think about and engage with the world. Anita Olds, a leading expert in designing North American childcare centers, wrote that "our attitudes and beliefs are the legacy we leave our children. Our thoughts, as reflected in our designs, in turn shape children's beliefs about themselves and life" (Olds, 2001).

As designers of computationally-enriched spaces and tools for young children, we have researched makerspaces in order to learn how the affordances of the environment can support children's learning, specifically in areas of computational thinking and making. Affordances are properties of objects or environments that guide users to the actions they can take, simply through their design and interface (Jones, 2003). For example, a door handle could be designed to invite a user to push, twist, or pull, depending on the affordances (e.g., shape, size, location) of the handle. In 2018, we conducted research on the differential impact of the "three teachers" identified in the Reggio-Emilia pedagogy (Strawhacker & Bers, 2018b). Specifically, we explored children's interactions with peers, facilitators, and the makerspace environment and materials, and looked for differences in key areas of Positive Technological Development (Bers, 2012), six behaviors that children can exhibit when using technology that are shown to correlate with psychosocial development in children. We found that each of these three "teachers" supported children's positive technology engagement in related but unique ways. Children and facilitators supported more social behaviors such as "making pro-social choices of conduct" and "building community connections," and the makerspace itself—through material affordances of the furnishings, materials, and tools, as well as on-display projects and provocations—supported children's creativity and content creation (Strawhacker & Bers, 2018b).

If children's spaces, tools, and experiences have such a strong impact on learning, we have an opportunity to set intentions about what we want these environments to communicate to children. Since beginning our work in 2015, we have maintained a goal of developing makerspaces to engage children in playful and self-directed computational thinking and making, in order to empower them as future citizens and leaders.

What Can Makerspaces Teach Us?

In his theory of loose parts for children's play, Nicholson wrote that "our education and culture condition us to believe that creativity is for the gifted few... and this is a culturally induced and perpetuated lie." It is this lie that the maker movement responds to, by promoting a vision of community that is participatory, egalitarian, and innovative (Hlubinka et al., 2013). Making can be successful for young children, as well as older youth in middle school, high school, and college (Blikstein, 2013; Halverson & Sheridan, 2014; Marsh et al., 2019). We chose to focus on early childhood because we know that children form stereotypes about themselves at a very young age (Martin & Ruble, 2004, Sullivan, 2016), and so we must ask what educational tools can we use to combat these harmful biases in the early years? Papert (1980) asserted that knowledge is constructed by the individual, through interactions with the real world and with digital worlds. In other words, learning happens by making. He also posed that the making should be personally meaningful, as evidenced by children's willingness to persevere on a project for its own sake in spite of challenges, a concept he termed "hard fun" (Papert, 2002). We align with his conception of learning, and maintain that learning happens through playful making.

Play in the early childhood years is dominated by make-believe fantasy play. For young children, this is serious work; through play, children develop new tools for thinking, being, and symbolizing (Scarlett, 2005). The maker movement boasts "If you can imagine it, you can make it" (Hlubinka et al., 2013). We know that, as with play, children's motivations to make, experiment, and explore come from need to understand the world around them, and one of their most trusted tools in the pursuit of this exploration is their imagination. A space explicitly designed to foster making and learning through playful, imaginative making seems developmentally appropriate, if not necessary.

In our own research, we intentionally designed spaces to interrogate the societal lie of "gifted creativity" by designing rooms with easy access to tools and materials, open-ended play provocations, and facilitators trained to let children take the creative lead on their own playful projects. The major distinction that separates makerspaces from other learning environments is that they offer learners a way to engage with a wider range traditional and novel tools, technologies, practices, and forms of expression than are typically acceptable in formal learning environments, for example, because of mess, limited space, or impractical arrangements (e.g. too few electrical outlets). Today, there are many technologies that children may encounter in their daily lives, including tools from their parent's generation (like phones and pagers), their grandparent's generation (like LEGO bricks and polaroid cameras) and from many generations before that (like pencils and paper). A makerspace offers children a place where they can form their own community, and engage with the tools and skills that will become part of their own generation. Today, that includes programming, robotics, and engineering (Bers, 2008).

Computational Making in Early Childhood

Given the wide range of applications and audiences for makerspaces, we set out to investigate how to design makerspaces that specifically engage young children in computational thinking and making. In order to do this, we revisited classic theories of how children learn by doing and making. Developmentalists have long known that between ages 5-7 years, young children undergo dramatic transitions in almost every domain, including cognitive, social, emotional, and physical changes (Sameroff & Haith, 1996). As their capacity for new understanding grows, patterns emerge in how children learn, and hands-on play, physical exploration, personal motivation, and social connection all play a heightened role in children's knowledge-gathering about their world. Famed developmental theorist, Piaget, is credited as the first to propose the idea of *constructivism*, the theory that children construct their own knowledge as they attempt to understand their own interactions with the world, such as building with block manipulatives to form early conceptions of gravity and spatial reasoning (Wadsworth 1996). Because of the importance the environment on children's learning, constructivist pedagogies argue for cultivating a learning environment full of provocations and opportunities to test, build, and explore with a variety of available materials. To support children's free exploration and idea construction, teachers can support learning by acting as facilitators and co-explorers (rather than gatekeepers) of knowledge (e.g. Bada & Olusegun, 2015; Hein, 1991).

Pertinent to makerspaces, new approaches have emerged to describe learning in technologically-enhanced and digital environments. Papert (1980) proposed *constructionism*, an adaptation of constructivism concerned with the unique meta-cognitive learning opportunities afforded by computer programming. He is perhaps best known for his idea that "you can't think about thinking without thinking about something", emphasizing the importance of models, representations, and tools to aid in knowledge construction (Papert, 2005). He argued that when children explore computer coding, they create digital "microworlds," or virtual spaces where they have programmed all the rules and behaviors in that world. When creating (and testing, and breaking) these programmatic rules, children can learn about how the "real" world works, and also about their own computational thinking processes (Papert, 1980). Further, he argued that by constructing artifacts to represent their thinking in the digital world, children can share ideas with peers, and learn from communal experiences and explorations.

How can a Makerspace Facilitate Computational Making?

What specific elements of the makerspace afforded opportunities to explore computational thinking and making? In the following section, we describe elements of space, materials, and facilitation that could be used in any space to engage young children in developmentally-appropriate computational activities.

Research on makerspaces for early childhood has revealed that the developmental needs of child makers have important consequences for the purpose and expectations of the making that occurs there. For example, kindergarten educators who use makerspaces with their students report that they prioritize learning goals of sensory and motor exploration, confidence-building, and agency in choosing and pursing projects (Strawhacker & Bers, 2018b). These goals differ from the purpose of spaces where professional makers develop prototypes for commercial use, or even educational spaces for older student makers to cultivate STEM skills and practices (Gravel, Bers, Rogers, & Danahy, 2018). In a study of the unique affordances of facilitators and the environment in Early Childhood Makerspace at Tufts University, the physical environment (including tools, materials, furniture, and layout of space) accounted mainly for children's engagement in content creation and creativity (Strawhacker & Bers, 2018b).

If the goal of an early childhood makerspace is to afford opportunities for children to explore making, it is important to understand how young children create and explore artifacts in the world. Researchers have described two distinct but equally valid ways of making (i.e., engaging in the design process), both of which are demonstrated in this vignette (Turkle & Papert, 1990; Papert & Harel, 1991; Worsley & Blikstein, 2013; Resnick, 2006; Vizner & Strawhacker, 2016). One popular approach is *bricolage*-style making (sometimes called "tinkering") in which the maker "mucks about" and eventually comes to a design conclusion by artfully or randomly bringing together objects and ideas (Beltagui, Sesis, & Stylos, 2021; Hatton, 1989). There is also a more top-down structured approach to design, where the creator follows a defined set of steps to find a solution to their problem (e.g., Papert & Harel, 1991; Vizner & Strawhacker, 2016). In order to support children's engagement in computational thinking, makerspaces can provide provocations for engaging in both of these ways of making.

Besides the materials and tools inside the room, children's space designers know that the room itself – the decor, furniture, arrangement of space – is important for helping children feel safe and secure enough to explore creative play. To help children focus on making, the space should feel as modular and flexible as the loose parts within it. Figure 1 highlights several design elements to guide the design of flexible, creative makerspaces. These guidelines were informed by interviews and observational research with kindergarten educators in diverse early childhood settings (see Figure 1; Strawhacker & Bers, 2018a; 2018b). For example, makerspaces should offer tools and materials that children can access and use on their own with minimal guidance. To support collaborative making, larger furniture or equipment should be too heavy for one child to lift, but safely movable by pairs or groups of children. In general, overly bright and colorful furnishings can be fatiguing in spaces where children prefer to focus (e.g. work tables), so opt for neutral, muted tones or natural materials, and provide clean, open work spaces (Olds, 2001). Instead, save that attention-grabbing decor for areas to display children's made works, which should also be labelled as often as possible with photos, captions, and other indicators to humanize and contextualize the children who represent the makerspace community. Finally, the factor mentioned by almost every educator in our studies was the importance of children's freedom and agency within the space, particularly the freedom to explore methods and materials that would be difficult to explore in a classroom, such as large-scale, messy, or technologically-mediated activities (2018a; 2018b).

Figure 1. Guidelines for designing early childhood makerspaces (Strawhacker & Bers, 2014)

Principle	Example
1. **New technologies** should let children explore making with contemporary forms, tools, and ideas	Offer developmentally appropriate robotics, film-making equipment, or circuitry kits
2. **Materials** should be visible, accessible, and inviting	Store materials in glass jars or wire baskets on low, uncrowded shelves.
3. **Furnishings** should be child-sized and functional for children's needs	Adjust wood work tables to approximately 51cm height, use wide floor areas for work
4. Elements of the room should promote **exploration and risk**	Learn about and practice safe use of tools like hot glue guns and sharp knives
5. The space should **reflect the children** who use it	Display children's work and pictures at child-height
6. Facilitation and space design should **"say yes" to children**	Before telling children what not to do, learn why they are doing it
7. **Building and sharing ideas** is as important as the finished product	Let children make mistakes and test ideas instead of correcting

Computational Making in Practice: Three Vignettes

Bers breaks down computational thinking (the skills, practices, and processes that comprise computational making) into the following powerful ideas. In the coming vignettes, we describe how children explored one or more of these concepts and processes through their making, as well as how the space afforded that exploration (see Table 1).

Two of the following vignette examples came from the Early Childhood Maker Space (ECMS) developed by the authors at Tufts University. The authors developed this space in 2017, with support from the broader Tufts community as a working laboratory to investigate—amongst other things—computational thinking and making. The ECMS became a space where engineers, educators, researchers, and children could come together to co-create the space and tools together. The third vignette came from the Kindergarten Makerspace (also developed by the authors) at the International School of Billund, a progressive international children's school in Denmark (Gravel, Bers, Rogers, & Danahy, 2018). Both spaces were supported by the LEGO Foundation, the Eliot-Pearson Curriculum Lab, Tufts University's DevTech Research Group, and Tufts' Center for Engineering Education and Outreach.

Table 1. Bers' Seven Powerful Ideas of Computational Thinking

Powerful Idea	Description
CONCEPTS	
Algorithms	Series of ordered steps in a sequence to solve a problem or achieve some goal. Sequencing is an important skill in early childhood. It is at the core of being able to tell a story, tie one's shoes, and make a peanut butter and jelly sandwich.
Modularity	The decomposing of a complex task or procedure into more manageable sub-parts, and an understanding that sub-parts can be put together to make a more complex entity. The ability to use sub-parts from a solution to one problem with other sub-parts to solve a new problem.
Control structures	The initiation and order of execution of a set of commands. This includes repeats, loops, conditionals, events and nested structures. Making decisions based on a set of conditions. For example, when a button is pressed, do some action or if it is dark out, turn on a light. Identifying patterns and using structures such as repeats and loops to execute them efficiently.
Representation	Symbolism develops in early childhood. The ability to represent concepts as symbols is important for computational thinking. The formal languages of computer science are representations of the programmer's thoughts organized in such a way that a machine can understand them
Hardware/software	Hardware and software are parts of a system. Software is used to control hardware. Hardware is built to interpret software and do some action. Depending on the hardware, this may be interpreting large data sets (computer) or navigating a maze (robot).
PROCESSES	
Design Process	A cycle with no explicit beginning or end where a child: asks questions, imagines, plans, creates, tests and improves, and shares their work. Engaging with and iterating through these actions is design process
Debugging	A systematic approach to isolating and addressing problems within an existing piece of work. For example, one might step through a program to find an error or check all connections on a piece of hardware.

Source: (Bers, 2012; Vizner, 2017)

All vignettes were inspired by or adapted from real events that took place in these makerspaces. All names presented are pseudonyms. All three vignettes illustrate how young children can engage in computational making, and emphasize how elements of space like tools, furnishing, and facilitation came together to support children's freedom and agency to pursue computational making.

Vignette 1: Exploring Representation with big-KIBO

In this vignette, we see a child exploring the computational concept of representation, both in his imaginative play and representation of a toy robot as a non-robotic vehicle (at times, a truck and a racecar), and his exploration of how to represent his plan for the robot's actions using its block coding language. The robot described in this vignette, called big-KIBO, is an experimental prototype of a scale version of the commercially-available KIBO robot. KIBO is an educational robot kit designed at DevTech and now produced and distributed through KinderLab Robotics, Inc. (for more information about the KIBO robot you can visit www.kinderlabrobotics.com). Intended to engage young children in developmentally-appropriate experiences with coding, computer science, and engineering, KIBO is programed with a tangible programming language consisting of wooden blocks with symbols for pre-readers (e.g. arrows), capitalized words for early-readers (e.g. FORWARD), and a barcode for the robot. In order to program KIBO, the child holds the robot over their chain of blocks—the program—and scans each block one at a time. Throughout this chapter, we will use written words (e.g. FORWARD) to refer to block-based instructions in the KIBO language.

One new and exciting tool that came out of the ECMS effort was a life-size rideable KIBO robot appropriately named big-KIBO (see Figure 2). Big-KIBO was built by the author to support his thesis work (Vizner, 2017) on investigating the role of scale in early childhood robotics. Big-KIBO is a replica of KIBO that is 109 times larger. It has a steel frame with a wooden top and acrylic see-through panels that allow children to see its "guts", the electrical components that make it work. It is powered by recycled wheelchair motors, and is strong enough to carry several full grown adults. Big-KIBO uses the same programming blocks as KIBO, due to its size children lift each block up to big-KIBO instead of holding it over their program.

As part of the author's thesis work (Vizner, 2017), Vizner invited two groups of three kindergartners with their classroom teacher to play with big-KIBO and/ or KIBO in the Early Childhood makerspace. As we see in the following vignette, children were invited to free-play with big-KIBO while Vizner facilitated these maker activities as a participant researcher.

Noah and two of his Kindergarten classmates are taking their second visit to the Early Childhood Maker Space today, and he is excited to continue building a tow-truck. The first time he visited the space, he tried to make his truck with a regular-sized KIBO, and was disappointed that it couldn't carry heavy objects, and that he couldn't ride the truck himself. Today, Noah begins to build with big-KIBO and states that he wants to design a feature to "help big-KIBO tow things." He uses duct tape, pipe cleaners and a foam block to construct a rudimentary towing hook, which he calls a "spoiler", and tests it by towing a stuffed animal (see Figure 3). He works on his "spoiler", and the design evolves to take on additional functions. "I'm going to make KIBO into a tow-race-car, a race car that tows things," he explains. He also decides to add antennas to "help big-KIBO climb." In the transcript excerpt below, Noah's regular classroom teacher asks him about his construction:

Figure 2. Big-KIBO and coding blocks in the ECMS at Tufts University
Source: Reprinted from Vizner (2017)

Teacher: Do you want to tell me what that thing is?

Noah: It's a spoiler that helps tow things [a spoiler is a stabilizing component typically found on the rear of a racecar]

Teacher: Oh, it tows things

Matt [another child]: Spoilers usually make [it so] cars don't spin

Teacher: Why does it need a spoiler

Noah: So that KIBO can go fast

Figure 3. Noah constructs his towing hook on big-KIBO, and tests it to see if it will tow his stuffed animal

Once he is satisfied with his creation Noah begins build a program for big-KIBO: BEGIN, SHAKE, SHAKE, FORWARD, FORWARD, END. In this transcript segment, Vizner (researcher) asks Noah about his program:

Researcher: What does it do?

Noah: This one is rock climbing [pointing at the first SHAKE]. This one is rock climbing [pointing at the second SHAKE]. This one is racing [pointing at the first FORWARD]. This one is racing [pointing at the second FORWARD].

Noah: But I need more racing [He gets up and looks for more FORWARD blocks]

In this example, Noah knows that each block represents a command to the robot and an action that it will do. Each block is a bit (i.e., a small unit of information) of information. Noah has also encoded an additional personally meaning bit of information to each block, i.e., climbing and racing. This makes each block a nested set of representations which big-KIBO is expected to carry out. By turning big-KIBO into a race car with a "spoiler", and using programming blocks to symbolize his robot racing, climbing, and towing, Noah is engaging the computational thinking concept of representation. The original KIBO robot kit supported his engagement with computational thinking, but Noah's engagement in computational making was much richer with big-KIBO. We argue that this can be attributed to the affordances of the makerspace, e.g., through experimental materials like the big-KIBO prototype, which was impractically sized for home or classroom use, as well as the freedom of the permitted activities in the makerspace that allowed him to pursue his pretend play goal of creating a ride-able racecar/tow-truck.

The Early Childhood Maker Space supported a "make anything" mindset through the tools housed there (crafts, novel robots, etc.), which allowed Noah to engage in a rich play and making experience with a computational thinking concept, motivated by his own interests. In the following example, we see a child whose making engage him in the computational thinking process of debugging.

Vignette 2: Engaging in Debugging with a CNC Mill

Another experience that took place in the ECMS was a movie-making camp for 6-8 year olds. The child participants produced stop-motion animation films from start to finish, including script-writing, prop and set creation, filming, and editing. The following vignette highlights children engaging in a lengthy and iterative debugging process in order to prepare the props for their films using a CNC mill. Debugging is a critical component of computational thinking, and reflects that the child maker (1) has a clear design goal in mind, and (2) engages in a process of iteratively refining their design to better achieve their goal. In this vignette, the child is working for the first time with an unfamiliar machine, a CNC (computer numerical control) milling machine (colloquially referred to as a CNC), and needs to explore and reuse the device many times before creating the exact construction he is working on. A CNC is a digital fabrication tool that uses blades on a rotating cylinder controlled by a computer to make precise cuts that form a 3D object (described in more detail below). The work flow for using this otherwise industrial tool was adapted for young children and supervised by an adult engineer and fellow researcher, allowing them to engage in the same process of trial and iteration that adult engineers using the CNC use to refine their designs.

Bobby, 7 years old, has been interested in medieval history ever since his parents took him to the renaissance festival last spring. When Bobby's mom told him he would be going to film making camp this summer, he immediately knew he wanted to make a movie about knights. He decided that the star of his film would be a knight who rides a horse and carries a sword. The first step for making his film was to create the props and characters. His camp counselor explained that they would be making all of the props themselves out of foam, using a big machine called a CNC. The counselor explained that a CNC is like a cutting machine that can make 3D objects, and uses pictures to make a program of where to cut. If Bobby drew a picture of each of the props he needed for his movie on a piece of paper, the computer could translate it into code that would tell the CNC how to cut out the same shape. This CNC uses a fast spinning blade which is precisely controlled by the computer program to move up/down, left/right, and front/back in order to cut the props out of foam. As it works the CNC creates waste material called swarf—Bobby chuckled at the sound of this word—which is cleaned by a vacuum as it works—Bobby also liked the idea that the machine cleans up as it works, so there is no clean-up time! Finally, his counselor introduced a visiting engineer/researcher, who would be helping the kids make CNC-created props for their movies. To make his props, Bobby drew a picture of a knight, a picture of a horse, and a picture of a sword (see Figure 4). Each picture filled up an entire piece of paper.

The engineer scanned Bobby's drawings and used a computer software to show how the CNC would cut his shapes out of the pink foam. Bobby put on safety goggles so he could watch as the machine cut the foam, and his sword slowly began to take shape. When the machine finished, Bobby excitedly removed his props from the CNC—but there was a big problem. The sword and horse and knight were all the same size! Bobby wanted the knight to be bigger than the sword, and smaller than the horse. At first, he was upset. It took him a long time to draw each of the pictures and he wasn't sure if he could do it again. Just as he was getting so frustrated that he wanted to quit, the engineer explained to Bobby that he didn't have to make new pictures, but instead they could program the computer to change the scale and size of the drawings he already had. In other words, Bobby could try to debug his design without losing all his earlier hard work. This is where Bobby got to experience the "rapid" part of rapid prototyping. Bobby worked with the engineer to change the size of his props, and even played around with scale factors, which adjusted the size of the props relative to each other, all on the computer. This debugging process took several tries. On his first re-scaling attempt, the horse came out way too small and knight was too big, which made Bobby laugh at "the giant and his tiny pony" (see Figure 5). On his second try, the sword was too small and the horse was huge. Bobby and the engineer worked on the computer and tried rescaling again and again until he was finally happy with the way all three props looked together.

Figure 4. Bobby's sword drawing, ready to be scanned into the computer

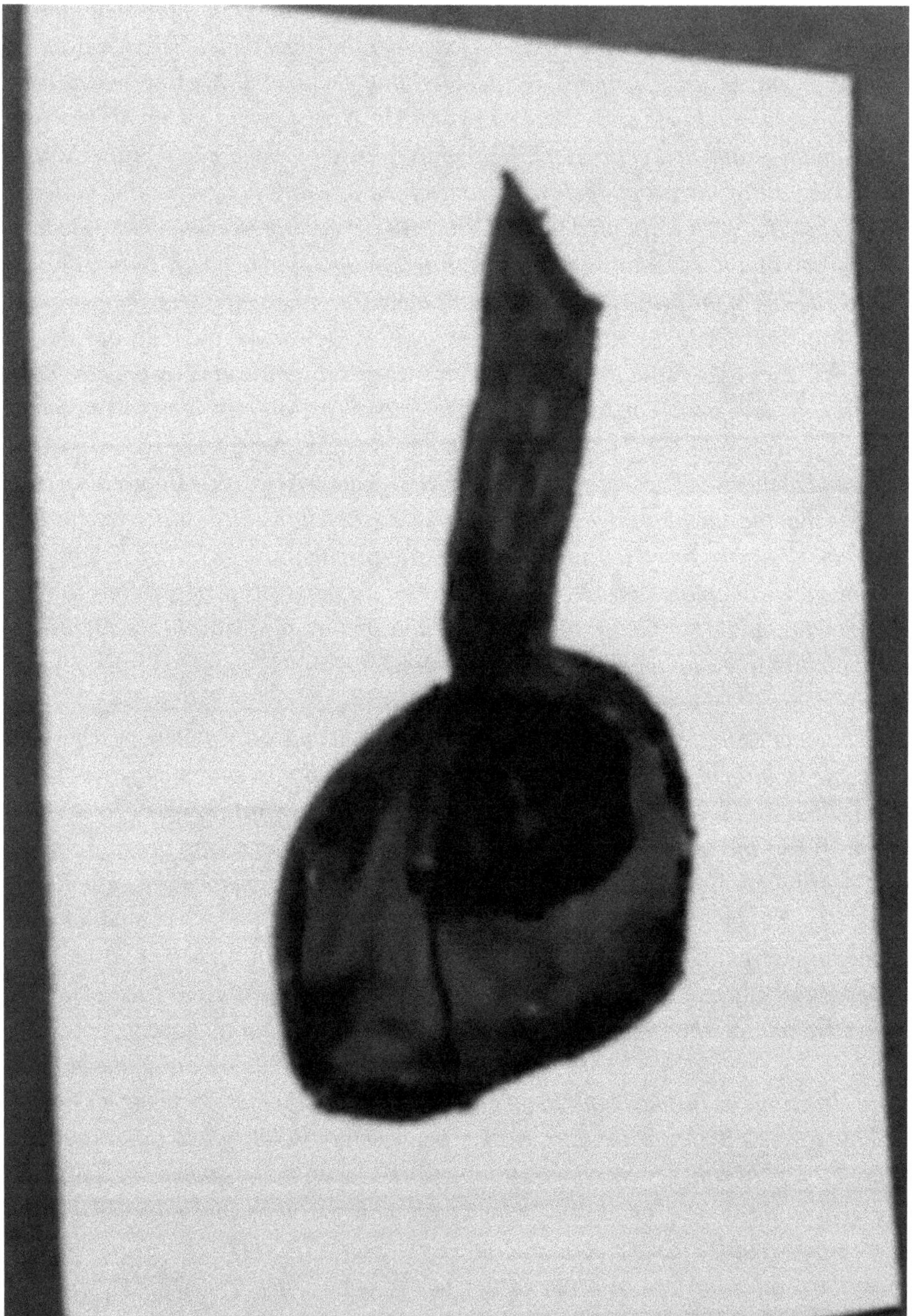

Figure 5. On his first resizing attempt, Bobby's knight was too big relative to his horse

It was finally time to film. Bobby sets up the knight and sword next to the horse and got ready to turn on the camera. He pressed "record", looked through the camera lens, and discovered that his parts were too big to fit in the frame! "No problem," Bobby said confidently, and walked back to the computer, ready to work on even more debugging. While the engineer watched, Bobby typed a few numbers into the computer to apply a 20% scaling factor to all of his props. He knew that by scaling their size uniformly instead of individually, the props would all keep the same relative size, and look the way he wanted when they were propped next to each other. He loaded a final piece of foam into the CNC and pressed "run." At the end of the cutting cycle, he lifted the lid to the CNC, and triumphantly pulled out his props. He now had the perfect cast for his film!

Bobby's iterative process highlights the computational thinking practice of the debugging process. Within the context of making, Bobby was not focused on the many trials it took him to perfect his design, but instead was motivated to keep working so that he could realize his ultimate vision of a movie about a medieval knight. While working towards the goal of having appropriately scaled parts he created, critiqued, and refined several prototypes of his props, engaging in multiple tests to improve his design. In Bobby's debugging process, we saw him engage in both bricolage and top-down design processes. When he arbitrarily sized each part, he took the bricolage approach. When he scaled them uniformly, he took a step-by-step approach. The makerspace, and the facilitators, machines, and equipment within it, afforded Bobby the opportunity to engage with novel tools and processes that he would be unlikely to encounter in any other setting designed for his age group. In the very beginning of the process, Bobby very nearly gave up on his first design, and would have missed out on a rich debugging and refining process that finally resulted in him proudly using perfected designs of his original prop idea for his self-made movie This example further highlights the flexibility of digital fabrication and rapid prototyping tools, and how they can support young children in computational making and thinking.

Vignette 3: Using the Design Process During Storybook-Inspired Making

This final vignette took place at the Kindergarten Makerspace at the International School of Billund. In this example, the affordances of the environment and the moves of the teacher supported children's engagement in computational play. Specifically, the children in this vignette engaged in a lengthy engineering design process to create a personally-meaningful KIBO robotics project, inspired by their favorite classroom storybook.

Abbie and Mathilde are partners in Ms. Maria's Kindergarten class. They are excited, because today they get to visit the new makerspace again. The makerspace is located next to the woodshop room, music room, and computer lab – but unlike all of those spaces, the makerspace has a whole room just for Kindergarteners!

When they first visited the space last week, they couldn't help running around the wide open room. The floor was cushy with mat tiles, and there were low benches, cushions, and "wobble chairs" located around the room. When they were sitting in a circle, a classmate shouted, "this room is like a gym!" Their teacher, Ms. Maria, smiled and said, "That's because we need lots of space to build your big ideas out of blocks and crafts – and of course, robots!" She brought out a box from behind her chair, and took out a KIBO robot. She passed it around the circle, showing children the batteries and wires inside through the clear panel on the back, the motors and wheels that move on its sides, and the green button that turns on its scanner. After she introduced KIBO, she divided children into pairs, gave them a robot, and let them explore how to build and code with it. Ever since that first visit, Abbie and Mathilde have been talking about what kind of a robot they would make the next time they visited the makerspace.

Figure 6. Children used furniture in the makerspace to build a "cave" environment as part of their robot obstacle course, inspired by their classroom storybook, Going on a Bear Hunt

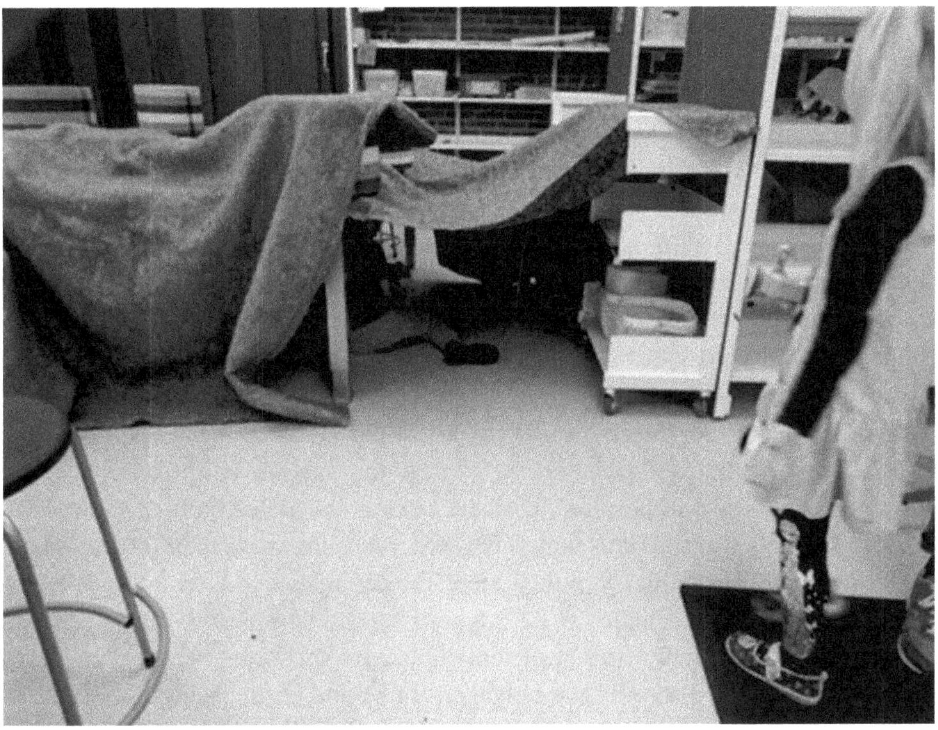

Today, Ms. Maria says there is a special activity waiting for them. When they arrive, they see all the chairs and cushions have been moved to the side of the room, and there are ramps, stools, and masking tape paths all around the floor. Today they will be making obstacle courses for their KIBO robots to navigate. Mathilde gets started ideating right away, saying "maybe we can make KIBO climb a hill!" She balances a ramp against a low stool. "I know, I know! We should make KIBO go on a Bear Hunt!" Abbie exclaims, referring to the story they have been reading at library time, and Mathilde eagerly agrees. The partners brainstorm about how to make a robot that acts out their favorite parts of the story, while Ms. Maria lends them her copy of the Bear Hunt book for inspiration. They are most excited to build the

"gloomy cave". The girls build a cave-exploring robot and test out several less-safe dark corners of the room before Ms. Maria suggests that they transform a tall table with blankets and stools to be a spooky cavern for KIBO to explore (see Figure 6).

While testing their KIBO, Abbie frowns. "It's too dark in here for the KIBO to help us find any bears," she says. Mathilde gets a flashlight from the materials wall to attach to their robot, but the flashlight is too heavy even when they try to use tape and rubber bands. They speak to Ms. Maria, who shows them a new KIBO part called a lightbulb, and some blocks to program different colorful lights for the lightbulb to shine. They add some light blocks to KIBO's program, and are confused when it still doesn't shine—they can't figure out what's wrong with their program. They continue to troubleshoot by testing different program sequences until Mathilde has the idea to try attaching the lightbulb to their KIBO. Suddenly, KIBO's light shines in many colors and the girls are delighted with their glow-in-the-dark cave-exploring robot! It wasn't the program after all, but the robotic parts and hardware that needed to be changed. Soon the whole class is having fun in their KIBO cave, and Abbie and Mathilde happily explain to anyone who wants to know how they got KIBO to shine its light in many colors.

In this vignette, the materials and furniture are used in a way that's rarely allowed in a classroom setting, and they become a part of the children's playful engineering design process. They are inspired to create a project modeled on their Bear Hunt storybook, and brainstorm how their KIBO robot will showcase their favorite cave-exploring scene. They quickly build a robot and begin testing different sites to use for their cave location. Finally, at the suggestion (and permission) of the teacher, they end up creating their own test site out of blankets and tables, which spurs them to engage in a longer process of iteratively refining and testing ways to make their robot shine a light. Finally, when they are satisfied with their final design, their enthusiasm spreads to others in the class who want to join in the fun, evidence that they are engaging in the positive technological behaviors of community building and communication through their robotic making project (Bers, 2012). The girls engage in the last step of the design process, sharing with the community, by inviting their classmates to play in their self-made cave, and sharing their new technical knowledge with others who are curious about the lightbulb hardware and software that completed their design. Rather than restricting their play or requiring "proper" use of this furniture, the teacher in the vignette considered the needs of the children and responded by simply granting them permission to explore the space in a new, creative way that also met her requirements for safety. Through the combination of space and facilitation aligned to support children's creativity, Abbie and Mathilde were able to pursue a rich engineering design process, completely motivated by their own creative ideas. In the process, they explored the computational thinking process of the design cycle, as well as concepts of hardware and software (when changing the code and parts to make their lightbulb shine) and debugging (when troubleshooting and testing to make their robot flash its lightbulb). Because of the affordances of the makerspace materials and facilitation that encouraged hands-on, child-directed play, Abbie and Mathilde were able to engage in rich computational making.

CONCLUSION

Throughout this chapter we've argued for the importance of children's environments on their learning. We also expanded on the nascent concept of computational making as making that leverages computational thinking skills and practices, and proposed that makerspaces are learning sites uniquely able to engage children in computational making. Finally, we maintained that our primary goal in pursuing this work was to empower children to become computationally fluent future leaders and citizens.

What Lies Ahead for Computational Making and Educational Makerspaces?

If agency and computational fluency are the values that we hope to impart to young children, then supporting the development of computational thinking and making are imperative. As our world becomes increasingly mediated by digital experiences, computational making can empower children to "be protagonists in their own learning" (Kuh, 2014). In other words, environments that promote computational making and thinking afford opportunities for children to create and produce artifacts of their own learning, rather than "consume" ideas through passive digital experiences.

While the educational opportunities inherent in makerspaces are vast, there is still much work to be done to realize the maker movement's goals of democratized knowledge, equipment, and expertise (Dougherty, 2012). Community makerspaces still struggle to empower makers to engage their own cultural funds of knowledge to impact designed solutions (Calabrese Barton & Tan, 2018). In a survey of 30 educational makerspaces in K-12 settings around the U.S., researchers from Drexel identified cultural inclusion as the biggest gap missing from school makerspaces—a serious omission, given that community-building and inclusive collaboration are explicit goals of the maker movement (Kim, Edouard, Alderfer, & Smith, 2018). As the maker movement enters its second decade, we hope to see an emphasis on shared culture-building take priority over less community-oriented maker goals such as access to expensive and highly technical equipment. We have shown above, in other chapters in this book (unplugged activity chapter), and in previous studies that computation making does not explicitly rely on high tech tools.

Educators can prioritize community-centered making in their own early childhood makerspaces by attending to the design guidelines introduced earlier in this chapter (see Figure 1). Specifically, displaying children's work, and even images of children working in the space at child's eye-level can improve a child's sense of connectedness to their maker community, even if they do not always play or engage with others in the space. Further, selecting materials, including reference books, tools, and craft materials that represent a diversity of maker cultures and histories, can help children connect to what Dale Dougherty calls the shared human act of making, passed on from centuries and even millennia of ancestors who were tool makers and users (Dougherty, 2012). Creating a space that allows young children to create comfortably and freely sets the stage for computational making.

Additionally, our prior work suggests that makerspaces can be productive sites for Positive Technological Development, a pedagogical approach that forefronts development of psychosocial and character traits through engagement with technology (e.g. Strawhacker & Bers, 2018b). Makerspace designers should incorporate practices and technologies that support these positive social and independent behaviors, in order to support children in meaningful and community-engaged computational making.

ACKNOWLEDGMENT

This research would not have been possible without the direction of Dr. Marina Bers, Dr. Brian Gravel, Dr. Chris Rogers, and Dr. Bruce Johnson of Tufts University. This work is made possible with generous funding from the LEGO Foundation, Tufts' Center for Engineering Education and Outreach, and the Evelyn G. Pitcher Curriculum Resource Lab. Finally, we thank the many children, families, educators, and makers who contributed to this research.

REFERENCES

Bada, S. O., & Olusegun, S. (2015). Constructivism learning theory: A paradigm for teaching and learning. *Journal of Research & Method in Education*, *5*(6), 66–70.

Beltagui, A., Sesis, A., & Stylos, N. (2021). A bricolage perspective on democratizing innovation: The case of 3D printing in makerspaces. *Technological Forecasting and Social Change*, *163*, 120453. doi:10.1016/j.techfore.2020.120453

Bers, M. U. (2012). *Designing digital experiences for positive youth development: From playpen to playground*. Oxford University Press. doi:10.1093/acprof:oso/9780199757022.001.0001

Bers, M. U. (2020). *Coding as a playground: Programming and computational thinking in the early childhood classroom*. Routledge. doi:10.4324/9781003022602

Calabrese Barton, A., & Tan, E. (2018). A longitudinal study of equity-oriented STEM-rich making among youth from historically marginalized communities. *American Educational Research Journal*, *55*(4), 761–800. doi:10.3102/0002831218758668

Campbell, L. O., Heller, S., & Goodman, B. (2018, March). Fostering computational thinking and student engagement in the literacy classroom through pop-up makerspaces. In *Society for Information Technology & Teacher Education International Conference* (pp. 3750-3754). Association for the Advancement of Computing in Education (AACE).

Dougherty, D. (2012). The maker movement. *Innovations: Technology, Governance, Globalization*, *7*(3), 11–14. doi:10.1162/INOV_a_00135

González-González, C. S., & Arias, L. G. A. (2019). Maker movement in education: maker mindset and makerspaces. In J. L. Jurado, C. A. Collazos, y L. F. Muñoz (Eds.), Ingeniería colaborativa, aplicaciones y usos desde la perspectiva de la Interacción Humano-Computador [Collaborative engineering, applications and uses from the perspective of Human-Computer Interaction]. Editorial: Universidad San Buenaventura de Cali. Colombia.

Gravel, B. E., Bers, M. U., Rogers, C., & Danahy, E. (2018). *Making engineering playful in schools*. The LEGO Foundation.

Hatton, E. (1989). Lévi-Strauss's bricolage and theorizing teachers' work. *Anthropology & Education Quarterly*, *20*(2), 74–96. doi:10.1525/aeq.1989.20.2.05x0841i

Hein, G. (1991). *Constructivist learning theory*. Institute for Inquiry. http://www.exploratorium.edu/ifi/resources/constructivistlearning.html

Iwata, M., Pitkänen, K., Laru, J., & Mäkitalo, K. (2020). Exploring potentials and challenges to develop twenty-first century skills and computational thinking in K-12 maker education. In *Frontiers in Education, 5(87), 1-16*. doi:10.3389/feduc.2020.00087

Jones, K. S. (2003). What is an affordance? *Ecological Psychology*, *15*(2), 107–114. doi:10.1207/S15326969ECO1502_1

Kim, Y. E., Edouard, K., Alderfer, K., & Smith, B. K. (2018). *Making culture: A national study of education makerspaces*. Drexel University.

Lakind, A., Willett, R., & Halverson, E. R. (2019). Democratizing the maker movement: A case study of one public library system's makerspace program. *Reference and User Services Quarterly*, *58*(4), 234–245. doi:10.5860/rusq.58.4.7150

Marsh, J., Wood, E., Chesworth, L., Nisha, B., Nutbrown, B., & Olney, B. (2019). Makerspaces in early childhood education: Principles of pedagogy and practice. *Mind, Culture, and Activity*, *26*(3), 221–233. doi:10.1080/10749039.2019.1655651

Olds, A. R. (2001). *Child care design guide*. McGraw-Hill.

Papert, S. (1980). *Mindstorms: Computers, children, and powerful ideas*. Basic Books.

Papert, S. (2002). Hard fun. *Bangor Daily News*, 2.

Papert, S. (2005). You can't think about thinking without thinking about thinking about something. *Contemporary Issues in Technology & Teacher Education*, *5*(3/4), 366–367.

Sameroff, A. J., & Haith, M. M. (1996). *The Five to Seven Year Shift: The Age of Reason and Responsibility*. The University of London.

Smith, M. (1982). *Creators not consumers: Rediscovering social education*. NAYC.

Strawhacker, A., & Bers, M. U. (2018a). Makerspaces for early childhood education (principles of space redesign) & Maker values of early childhood educators, organizing a grassroots space. In B. E. Gravel, M. U. Bers, C. Rogers, & E. Danahy (Eds.), *Making engineering playful in schools* (pp. 18–29). The LEGO Foundation.

Strawhacker, A. & Bers, M. U. (2018b). Promoting Positive Technological Development in a Kindergarten Makerspace: A Qualitative Case Study. *European Journal of STEM Education*, *3*(3), 9. doi:10.20897/ejsteme/3869

Strong-Wilson, T., & Ellis, J. (2007). Children and place: Reggio Emilia's environment as third teacher. *Theory into Practice*, *46*(1), 40–47. doi:10.1080/00405840709336547

Wadsworth, B. J. (1996). *Piaget's theory of cognitive and affective development: Foundations of constructivism*. Longman Publishing.

ADDITIONAL READING

Curtis, D., & Carter, M. (2014). *Designs for living and learning: Transforming early childhood environments*. Redleaf Press.

Honey, M., & Kanter, D. E. (Eds.). (2013). *Design, make, play: Growing the next generation of STEM innovators*. Routledge. doi:10.4324/9780203108352

Maker Education Initiative. (2015). *Youth Makerspace Playbook*. https://makered.org/wp-content/uploads/2015/09/Youth-Makerspace-Playbook_FINAL.pdf

Martin, L. (2015). The promise of the maker movement for education. *Journal of Pre-College Engineering Education Research (J-PEER), 5*(1), 4.

Martinez, S. L., & Stager, G. (2013). *Invent to learn: Making, tinkering, and engineering in the classroom*. Constructing Modern Knowledge.

Meehan, R. J., Gravel, B. E., & Shapiro, B. A. (2014). Card-sorting task to establish community values in designing makerspaces. *Poster presented at FabLearn.*

Nicholson, S. (1971). How not to cheat children: The theory of loose parts (Links to an external site.). *Landscape Architecture*, *62*, 30–34.

Resnick, M., & Robinson, K. (2017). *Lifelong kindergarten: Cultivating creativity through projects, passion, peers, and play*. MIT press. doi:10.7551/mitpress/11017.001.0001

Sheridan, K., Halverson, E. R., Litts, B., Brahms, L., Jacobs-Priebe, L., & Owens, T. (2014). Learning in the making: A comparative case study of three makerspaces. *Harvard Educational Review*, *84*(4), 505–531. doi:10.17763/haer.84.4.brr34733723j648u

Vossoughi, S., & Bevan, B. (2014). Making and tinkering: A review of the literature. *National Research Council Committee on Out of School Time STEM*, *67*, 1–55.

KEY TERMS AND DEFINITIONS

Constructionism: A learning theory arguing that learners can construct novel ideas through engaging with creative technological platforms, such as coding environments.

Debugging: In computer programming and software development, debugging is the process of finding and resolving "bugs" (errors) within computer programs, software, or systems.

Early Childhood Education: Education of children from birth through age eight.

Engineering Design Process: A common series of steps that engineers use in creating functional products and processes.

Environment-as-Teacher: A concept from the Reggio Emilia pedagogy that states, after the peers, educators, and self, the environment is a teacher (imparts information, values, and opportunities to explore) for young children.

KIBO: A screen-free programmable robotics kit for young children with blocks, sensors, modules, and art platforms.

Makerspaces: Collaborative spaces where people gather to get creative with DIY projects, invent new ones, and share ideas (also called hackerspaces, hackspaces, and fablabs).

Representation: The use of signs that stand in for and take the place of something else. It is through representation that people organize the world and reality through the act of naming and assigning meaning to its elements.

Chapter 10
Coding, Computational Thinking, and Cultural Contexts

Libby Hunt
Tufts University, USA

Marina Umaschi Bers
Tufts University, USA

ABSTRACT

This chapter examines the relationship between coding, computational thinking, and the contexts in which those concepts are learned. It recounts a pilot study where a 12-week robotics curriculum was taught in kindergarten classrooms at eight interfaith and secular schools in Boston, United States of America and Buenos Aires, Argentina. In this chapter, the authors explore how teachers and students drew from their socio-cultural environments to inform the language of computational thinking and support the internalization of computational concepts and, in turn, how computational thinking was used as a tool for deeper exploration of cultural traditions and beliefs, meaning-making, and creative expression.

INTRODUCTION

As computer technologies become ubiquitous in society, the call for incorporating STEM/STEAM education into early childhood classrooms grows louder on a global scale (Bers, 2019; Modan, 2019; K-12 Computer Science Framework Steering Committee, 2016), with computational thinking and coding lessons starting as early

DOI: 10.4018/978-1-7998-7308-2.ch010

as kindergarten (Sano, 2019; Seow, et al., 2019; Toikkanen and Leinonen, 2017). The kindergarten classroom is just one of many socio-cultural environments (e.g., home, school, faith-based settings) that play an integral role in a child's developmental process, contributing to the lens through which children view and understand the world. What does computer science education look like through this lens? How does cultural context shape the way a robotics-based program is taught and learned?

These ideas were explored as part of an interfaith project called *Beyond STEM: The Development of Virtues in Early Childhood Education Through Robotics*, led by Professor Marina Bers and funded by the Templeton World Charity Foundation. While the primary focus of the study related to the ways a robotics curriculum could support character development, this chapter focuses specifically on the relationship between coding, computational thinking, and the contexts in which those concepts are learned.

This chapter explores the role socio-cultural environments can play in informing the language of computational thinking and the internalization of computational concepts, and, in turn, how computational thinking can be used as a tool for deeper exploration of cultural traditions and beliefs, meaning-making, and creative expression. In it, we will demonstrate a number of ways that robotics and computational thinking can be used to help children strengthen their connections to their own cultural communities, faith-based or otherwise, and learn about others.

BEYOND STEM PROJECT OVERVIEW

Eight schools, four in Boston in the United States and four in Buenos Aires in Argentina, participated in the study. Six schools were representative of a major monotheistic religion in each country: Judaism, Catholicism, and Islam, and two schools were secular. These schools were selected for having a solid mission statement citing commitment to values education and character development — elements we hoped would be fostered throughout the project.

The project was implemented in 12 kindergarten classrooms — 5 in Boston and 7 in Buenos Aires — over a twelve-week period. The numbers are different because schools in Argentina are bigger and have more than one classroom per grade; this was the case for only one Boston school. A total of 224 children participated: 64 in Boston and 160 in Buenos Aires.

Although an increasing number of developmentally appropriate technological tools are available for children, this study utilized KIBO, a screen-free robotics kit designed for children ages 4-7 that supports cognitive development, creative problem solving, fine-tuning motor skills, and social engagement in a playful and developmentally appropriate manner (Bers, 2018; Lee, Sullivan, and Bers, 2013).

KIBO was selected because of its hands-on nature, as well as its open-ended platform that allows children to experiment, express themselves, and share experiences while actively practicing computational thinking skills.

The KIBO curriculum was developed by the DevTech Research Group into two different versions to align with frameworks and standards in computer science for both Massachusetts and the City of Buenos Aires. It consisted of twelve lessons, from foundational aspects of robotics and programming (e.g., sequencing) to more complex concepts like control structures, repeats, and conditionals. The curriculum followed the Coding as Another Language (CAL) approach (Bers, 2019) and built upon the theoretical framework of Positive Technological Development (PTD) (Bers, 2008; 2012; 2018) that promotes six positive behaviors ("the 6 C's") through the use of technology: content creation, creativity, communication, collaboration, community building, and choices of conduct. Each lesson featured a variety of activities — warm-up games, design challenges, creative projects, "technology circles" for peer-reflection, and plenty of free-play — that correlated to one of more of the "C's."

A team of researchers in Boston and Buenos Aires were present in the classrooms for each lesson. Using research instruments created by the DevTech Research Group, they collected data relating to the students' computational thinking, coding, and robotics knowledge, while also observing the ways individual teachers adapted the lessons to better meet the needs of their own classrooms, their school culture, and their faith.

Through this project, teachers and children in these two international cities developed technical skills through a robotics curriculum, while simultaneously fostering positive character traits, exploring their own cultures/beliefs, and learning about others. The experience was designed so teachers could present the robotics curriculum concurrently with discussions about what it means to be human, how to treat one another, and how to be citizens of the world.

Professional Development

In order for teachers to successfully integrate computational thinking and coding lessons into their classrooms, it is essential that they themselves understand the foundation of computational thinking. This does not require high-level expertise — only the willingness to learn (Govind & Bers, 2019). To ensure this, 31 educators and 5 administrators from the participating schools attended a one-day professional development workshop in their respective city. Of those participants, 24 had no prior coding or robotics experience. Here they learned about the concept of coding as a new literacy, the robotics curriculum, and the goals of the proposed project, and were given the chance to connect with one another.

Most importantly, however, the teachers were given the opportunity to engage with the KIBOs themselves. The teachers became students. They were encouraged to play, to make mistakes, to try and try again, and to ask for help when they needed it. This is the embodiment of the Design Process, a cycle of six steps: *ask, imagine, plan, create, test and improve,* and *share,* that they would later use to teach their students how an engineer approaches a problem.

Once the teachers had mastered the basics of the KIBO robotics kit, they gathered to discuss the challenges they had faced working with the KIBOs, the different values reflected in their individual faiths and school environments, and the ways in which the KIBO curriculum could be used to enhance the teaching of those values to their students.

Lastly, the teachers were asked to put their new skills to the test and create KIBO projects that were representative of their school, their faith, and their values. How the teachers chose to represent their schools varied — some told stories while others focused more on recognizable symbols — but at the core of each lay a commonality that was present throughout the professional development workshops. Amid religious differences, each participant of the workshop was an educator, and each program represented what they hoped their students would gain from their school's style of education. At the root, what each teacher hoped for was that their students feel inspired to learn, safe to wonder, and free to play.

CONCEPTUALIZING COMPUTATIONAL THINKING IN THE CLASSROOM

Engaging With the Design Process

The first lesson of the KIBO curriculum begins with a simple but essential question: "What is an engineer?" When posed with this question, students had plenty of guesses. A student at the Catholic school in Argentina said that engineers work with different types of energy, while another claimed engineers "fix ceilings so they don't fall." Others boasted having an engineer in their family, like one child in the Boston secular school who announced, "I'm an engineer. My dad is an engineer... An engineer is fixing problems and if they make mistakes, they try another time." Already, through the simple sharing of ideas based on their individual experiences, students were beginning to engage in two of the 6 Cs of Positive Technological Development framework- collaboration and communication (Bers, 2012).

Ultimately, the conversation steered to the understanding that engineers build robots (among other things). The students were all familiar with robots from different movies they'd seen, books they'd read, or even appliances in their homes. However, none had really stopped to consider: "What makes a human different from a robot? What makes something a robot at all? How does a robot do things?" As students began to conceptualize these somewhat philosophical questions, we saw the first instances of how teachers would come to use their different cultural contexts to make sense of computational thinking.

In each classroom, teachers had pictures of different items: a tree, a flower, a vacuum, a cell phone, and had their students determine which category the item belonged to, using cultural or faith-based worldviews to inform their category options. For the children at one secular school that prioritizes the appreciation of nature in its overall mission, the distinction was described as something that was made by humans, or made by nature. In faith-based classrooms, nature-made objects were described as "God-given." Students at a Muslim school in Boston took the discussion even deeper, examining how, because humans were created by Allah, even human-made items were, in a way, gifts from Allah. Unbeknownst to these students, they were beginning to actively engage with the engineering design process: asking big questions.

In a post-implementation workshop, one of the participating teachers described her students' grasping of the nature vs. manmade concept as "empowering." She had explained to her students, "Without humans, the technology would just be sitting there." By exploring the concept through the lens of her classroom's culture of autonomy, students could clearly see the role they had to play in making KIBO work. If the students were in control, what different things could they program robots to do?

The students' ideas about what robots could do revealed connections they were already making between the technological tools and their environment. The following example shows students engaging with the next stage of the engineering design process — imagining solutions.

Students are gathered on a colorful carpet, sitting at their teacher's feet. Many small hands wave eagerly in the air, waiting to be called on to share their ideas. The teacher has just asked a most intriguing question: "If there is a bug on the ceiling, how can I reach the bug and catch it?" More than one head turns to the ceiling to see if there is a bug there. The teacher calls on the boy nearest to her, who offers that perhaps they could use a robot with an arm that "goes really high." And if the arm couldn't reach it it? Make the robot arms longer. And if that still doesn't work? He considers this a moment, looking to the ceiling, then replies confidently, "Someone could climb up the arm."

In Buenos Aires, a similar conversation took place at the secular school. Many of the students proposed ways to squash the bug. One imaginative girl did not care for this line of thinking. "Instead of killing him," she said, "I would scare him to escape." Her response was strongly influenced by the culture of her school, which focuses on environmentalism.

The idea that robots exist to help people, paired with faith or school-based contextual influences, drove many of the students' early designs. One student designed a composting robot, while others discussed ways that the robots could help stop people from cutting down trees. At the Catholic school in Boston, one child considered the ways a robot could assist in a charitable food drive.

The values displayed in these early designs were not just reflective of school and faith-based cultures. Some were influenced by the cultures of their homes.

A student at the Catholic school in Buenos Aires used his first design journal entry to create a robot that would help his mom cook. When a researcher prompted him, "Why don't you make a robot to help you store your toys?" he replied simply that his mom could help him because the robot would be cooking. In addition to the "helpful" culture of this child's home environment, here we also see the multifaceted social-emotional potential of robotics design. This little boy wanted to help his mother, but he also wanted her to be more available to spend time with him. In these examples, children envision ways that robots can support empathy, environmentalism, and family connection.

Multi-Subject Integration

A convenient and impactful way that teachers used contexts to add relevance to children's exploration of computational thinking is by incorporating the robotics lesson into other subjects being learned in the classroom. At the Jewish school in Boston, KIBO lessons were regularly integrated with Hebrew class. The secular classroom in Boston programmed their KIBOs to reenact a story they'd read about a hermit crab. The Catholic school in Buenos Aires used their KIBO project to explore environmental science, discussing things like energy and ocean pollution. The secular school in Buenos Aires strove to bring institutionalism to the project by collaborating with the gym teacher, the librarian, and others, all working together to integrate taking care of the environment with technology.

Even if the connections weren't explicitly facilitated by the teachers, students still found them. One student at the Muslim school in Argentina compared engineers to beavers because they work together to achieve a goal. Another student, while putting together a KIBO, noted, "he has electricity like some toys" and "it's like putting together a puzzle." Thinking about what robots can or cannot do, one insightful child at the Catholic school in Argentina pointed out, "robots cannot eat a tomato," while another explained that our brains give instructions to our bodies so we don't crash, and that a robot does not have a brain or heart equal to that of a human. This child, methodical and surprisingly advanced in his interests for a kindergartener, later drew a connection between the unfamiliar language of KIBO robotics and hieroglyphics, stating that the robot "reads drawings like the Egyptians" (referring to the bar codes imprinted on KIBO's wooden programming blocks).

In conceptualizing programming languages, it is helpful for students to draw connections to images they've seen in environments outside of coding. In our study, students recalled seeing barcodes in grocery stores or on the backs of books, and connected the red and green of the begin and end blocks to traffic lights. By making these sorts of connections, the students were internalizing the meaning of a coding language as familiar, attainable. Students were able to build upon these associations as they continued to master the KIBO coding language. As with tackling a math problem or a new word, students found creative ways to approach the challenges they faced with their programs.

A hunger for answers drove the students' connections with computational concepts and powerful ideas. Through their desire to master the KIBOs, they were subtly mastering the concepts of control structures and modularity and engaging directly with hardware and software.

In a final reflection, one teacher said, "The kids really grasped the concept of repeat and using brackets, and the sandwich analogy," referring to a method of teaching repeat loops in which repeating actions are placed between brackets, just like the ingredients on a sandwich go between two slices of bread. Metaphors were a useful tool for connecting coding concepts to other areas of learning. This teacher also compared the symbolic numbers of math or letters of a language to those symbols that make up a coding language, saying, "This is how coding and literacy are so connected."

Other teachers found similar connections. The teacher from the Jewish school in Boston had selected a specific song for their final project because it featured sequences, repetition, relevant Hebrew vocabulary, and an overarching message about community. A teacher from the Muslim school in Buenos Aires commented that her students found a way to mention KIBO in every class. Her students drew connections between left-to-right writing systems and how the program must be arranged for KIBO to read it.

Throughout the design process, the students grew increasingly comfortable with the process of experimentation, from asking initial questions to testing for results. They learned to be flexible when experiments did not go as planned, to debug by looking for alternative approaches to their goal.

Coding and Character Development

Regardless of cultural differences, early childhood classrooms share many of the same rules - be kind to each other, help your friends, take care of objects in the classroom. Throughout the study, we found that the classroom environments shaped the way that students interacted with KIBO and with one another, and that as their confidence in their own abilities developed, students quickly transitioned to helping others.

The children understood that KIBO was a tool that could break (one child compared it to her family's broken computer), and through discussions about caring for materials, we noticed that values emblematic of their classroom environments emerged. At the Catholic school in Boston, one young boy offered that we take care of KIBO because if it breaks, the students next year can't use it. This forward thinking shows a level of generosity that seems surprising for a kindergartener, but he was not the only one.

At each school, students were working with the KIBOs in groups of two or more. While many students naturally understood that they should treat their peers with respect, others struggled with collaboration. We observed that in classrooms where students sat at the same assigned tables every day, they seemed more used to working in pairs or small groups.

Classrooms with more flexible layouts had successful collaborations when pairs stayed the same for the whole curriculum, whereas classrooms where partners changed regularly showed fewer generous moments. In most classrooms, students were helping one another grow, not only as programmers, but as people.

"You have to be patient," a boy at the Buenos Aires secular school advised some peers who were struggling to scan. "I can help you scan!" a boy at the Boston Catholic school said, eagerly leaving his own project to assist his neighbors. When a student at the Jewish school in Buenos Aires returned from vacation, his classmate offered to show him what he had missed. A student at the Muslim school in Buenos Aires resolved a conflict between two peers who were fighting over KIBO wheels, saying "There are two wheels, one for you and one for her."

The sheer act of collaboration incites some conflict, but these students were learning that to accomplish their goals, they needed to work together. At the start of the pilot study, students were more eager to engage with the KIBO than with one another. They hoarded blocks, grabbed robots, and showed little interest in what other students were doing.

Towards the end of the twelve weeks, however, there was a shift towards more patient listening, observing one another's projects, and asking for help from peers. They shared materials freely. Some students even went out of their way to give compliments on another student's successes. They had begun to internalize a sense of community, and recognize the benefits of collaboration, of other perspectives, to reaching positive outcomes.

EXPLORING CULTURE AND COMMUNITY

Throughout the curriculum, teachers were encouraged to adapt the lessons to better meet the needs of their classrooms, their school culture, and their faith. Nowhere was the success of these adaptations more evident than in the planning and execution of the final project - "Our Treasure." The theme of treasure was used to emphasize the value and specialness of whatever they chose to express.

The final projects varied greatly (as shown in the next section), ranging from treasure hunts to parades to theatrical performances. Whatever approach a teacher decided, the goal was for students to "think deeply about what makes their school a unique community" (DevTech, 2018). The lesson that outlines the final project explains:

This [...] will give children the creative agency to choose how to represent the things that their school finds important through the treasure that they choose [...] Children will be pushed to think about their personal and community identities. Their ideas of representation will be explored as they find different ways to portray these parts of their identities in creative and abstract ways.

The final projects of the robotics curriculum served as an opportunity for the explicit examination of each classroom/school's culture. Through lengthy discussion about what made their school unique, students were able to assign meaning to abstract concepts, reflect on the symbols that represented their environments, and find creative ways to use their KIBOs to get their messages across.

This section will consist of four vignettes that capture the students' final projects, the computational thinking skills that children were engaging with, and the ways that the projects were representative of the cultures and values of their respective schools.

Vignette 1: Computational Thinking to Overcome Challenges

At the secular school in Boston there is frenzy in the classroom. The previous day, the students discussed at length the things they valued most within each other and within the school, calling them "little lights" after the school song - "This Little Light of Mine." Topics included kindness and everyone being worthy of respect. Following the discussion, students wrote letters, drew pictures, and placed them in a makeshift treasure chest. Today, to their dismay, their "treasures" have been "stolen" and hidden all around the classroom! The students must program their KIBOs to navigate through winding mazes all around their classroom until their individual treasures have been found and the chest is filled again.

One girl sets her KIBO down on the floor in front of a table covered in a blanket, giving the appearance of a tunnel. The task is for the KIBO to move through the tunnel, then turn right. She presses go. The KIBO zooms forward, but when it turns right, it hits a table leg and is sent off course into the blanket. Its blue light turns on, and it shakes as if stuck. She picks KIBO up, and this time places it further into the tunnel. The program runs again, and this time the KIBO clips the leg of the table on the other side. The girl thinks aloud. "Maybe we should try four times again?" referring to the number of "move forward" functions she programmed. "Let's see," she says, and goes back to the blocks to debug. When her KIBO finally runs without crashing into anything, she is pleased, but not yet satisfied. It moved over the treasure, when she wanted it to stop, shake, and shine a blue light right on top. She picks her KIBO back up, matter-of-factly says what she will try next, and moves to do so.

In another part of the classroom, another girl runs to the box of blocks. "We just need a 'turn left' to get my treasure," she says as she rifles around the box, looking for the block she needs. She adds the block, rescans, and groans in frustration when it turns the wrong way. Another rescanning and - "Why did it do light first?!" Back to the drawing board.

This treasure hunt created ample opportunities for the students to utilize their programming and debugging skills to complete the task, but also encapsulated the classroom cultures of divergent thinking and conflict resolution. Frustration gave way to perseverance, and perseverance gave way to pride as one by one they successfully reclaimed their treasure and returned it safely to the chest so every light could shine.

Vignette 2: Classroom Culture and Perspective-Taking

The classroom of the secular school in Argentina looks more like a stage, decorated with colorful props at every turn. In one corner sits a mouth made of poster board, its teeth made of disposable material, so large a child could sit in it. In another corner sit three potted plants decorated with paper butterflies. Another section of the classroom features a cityscape built from painted cardboard. The bottom of one of the structures reads "Xul Solar," the name of an Argentinian painter. A projection playing clips from physical education class hangs above a poster that represents their digital library. The students have been working hard for this moment. Each prop represents a classroom project they developed throughout the year — their treasures. This final project is intended to capture their ability to intertwine the KIBO curriculum with other aspects of their education (i.e. art, reading, physical education), values, and with Argentinian culture. Each team of children programs a section of the route KIBO will follow to each display. They discuss their classroom treasures, comprised not only of the projects they created, but the friendship and love that were present as they did so.

At the center of the circle of children sits a single KIBO, a GoPro camera affixed to its body. In small groups, the students select their blocks from a box and work together to create their program, scan the blocks, and test to see if it works. Later, after all the blocks have been scanned and the programs run, the students will be able to look at the GoPro footage and see what the world looks like through KIBO's "eyes" — a small but meaningful lesson in different points of view.

Vignette 3: Community Connectedness

Across the floor of a large open room at the Jewish school in Boston, a number of poster boards are set up. The posters depict colorful cutouts of the KIBO programming blocks, arranged in different orders. On the floor in front of each poster lies a "stage" that depicts different aspects of a community (houses, trees, farms). This is the landscape through which the KIBOs will move. For their final project, they are embracing the community aspect of the 6 C's. The students have invited parents and friends of the community to an expo-style presentation of their KIBO projects. Students move to their posters and await the visitors who want to see the KIBOs in action. Throughout the room, students explain their program to visitors, and carefully show them how to put the robots together. They reference the program laid out on the poster board, explain what should happen next, and at times even get the visitors involved. One child has his parent shine a flashlight on the eye attachment of KIBO while another encourages observers to clap. The KIBOs move from location to location on their respective maps, and the room buzzes with discussion as visitors ask questions and students explain their design process, decision points, successes and frustration. The youngest students in the school confidently have stepped into the role of experts and active members of their community, sharing all they've learned throughout the curriculum.

Following this portion of the presentation, the students line up and perform the song that inspired their programs: a popular Israeli children's song called 'Eretz Israel Sheli' (My Israel). Their teacher explained later at the post-study workshop that the song "teaches about building community in Israel while reinforcing some basic Hebrew vocabulary. The repetition and sequencing make it a catchy song and also support computational thinking," particularly the skill of algorithmic logic. This is the culmination of an emphasis throughout the curriculum on integrating their KIBO lessons with Hebrew language education, a subject at the heart of the school's mission.

Vignette 4: Celebration of Differences

A large blue sheet of paper lies on the floor of a classroom in the Muslim school in Buenos Aires. Painted atop the blue "sea" of paper are continents, and in the left-hand corner, the shape of South America. The students, kindergarteners, gather around the "map," clustering near the parts of the map they are originally from. They place their KIBO robots, topped with drawings students made of themselves, on their respective countries of origin. Then, one by one, the students press the "go" button on their robot, and watch how their programs run. The first group's KIBO spins once, then runs off the blue paper in an unintended direction and stops. The children pick up the robot and rescan their blocks, before trying again. This time, the KIBO spins, then zooms across the blue paper to South America, and shakes, almost as if joyful to have arrived in its intended location. Another KIBO zooms forward, and sings upon arrival to its spot on the map.

The next pair of students to try their program run into a different issue. Their KIBO moves too far forward, across the sheet and into the leg of another student. One of the programmers shakes her head - this isn't right. She picks it up and they reconfigure their program. After a few attempts at a working program, running and rerunning their tests, the KIBO finally lands on South America. The students are satisfied. Now that the dress rehearsal is over, the students and their KIBOs get in place, and press the "go" buttons once more. One by one, the KIBOs join in Argentina. The students applaud and cheer. At the bottom of the map, a large white caption reads: "Somos un montón de gente todos diferentes y nos encontramos para aprender jugar juntos." In English, this means "We are a lot of people, all different, and we meet to learn and play together."

Perspective-taking is a theme for this classroom. Their use of treasure as a metaphor for their students' immigrant experiences was not only a creative approach to storytelling - it was a means of celebrating multiple perspectives, respecting differences, and reflecting more explicitly on how they each were an important part of their school community and of Argentina, no matter where they'd come from.

CROSS-CULTURAL COLLABORATIONS

In addition to the way computational thinking can provide a context for the exploration and expression of one's own cultures, ample opportunity exists for cross-cultural collaborations. In the professional development workshops, we saw the benefits of people from different cultures and communities coming together and sharing their perspectives. It was important that the outcomes of the individual schools' projects should not exist in a vacuum — they had all pursued the same curriculum, after all.

We created a website where teachers could share videos and pictures from their students' final projects. They were encouraged to share the projects from other schools within their own classrooms and leave comments. This created an opportunity for students not only to glimpse other children's work, but to have conversations and ask questions about cultural differences, using their conception of multiple perspectives to enrich their understanding of the world outside their classroom, home, and faith-based environments.

Teachers were invited to a final workshop to talk about the challenges, successes, and interesting moments that arose throughout the curriculum. It was striking how similar many of their experiences were, and how readily teachers offered suggestions, resources, and techniques that had worked for them.

The teachers discussed the perspective-taking abilities their students developed as a more nuanced form of collaboration — working with partners, coordinating, and compromising, but also, as the teacher from the Boston Jewish school described it, standing back and saying, "Wow, we all programmed the same song using the same equipment, but look how different they all are." Through the development of computational thinking skills, implicit values were made explicit and multiple opportunities to exercise these values emerged, all while internalizing programming concepts.

Each school benefited from seeing the other schools' approaches, which inspired conversations among the teachers about the different capacities of their institutions. Some teachers approached the curriculum from an integrated standpoint and felt that their students had a well-rounded experience. Others felt that by not integrating KIBO into other subjects they had missed the opportunity to dig deep into their values and ultimately their projects were limited. Teachers from schools without a designated computer teacher reflected on how that might have impacted the lessons. Teachers who were nervous about implementing the curriculum and had followed the guidelines strictly expressed that, now they'd grown comfortable with the materials, next time they would approach the lessons with more creativity and flexibility.

For the students, however, these insecurities went unnoticed. The Boston secular school's teacher put it well, stating, "In [children's] minds there's no math, there's no technology. Everything is part of the learning experience."

CONCLUSION

This chapter set out to examine the ways that socio-cultural environments inform the teaching and learning of computational concepts in diverse kindergarten classrooms. The schools that participated in our study explored how a robotics curriculum could be taught in an integrated way with their own values, faith and culture, while supporting perspective-taking and learning about others. The experience highlighted a microcosm of the challenges that face education, where words like integration, social-emotional growth and STEM get thrown around without much context, and teachers are constantly asked to find ways to put them together through project-based experiences. This project attempted to reverse course by saying to teachers: start with your classroom, your school, your faith tradition, your community, and your values, and you will find ways to integrate the teaching of robotics.

At the beginning of the project, teachers and researchers alike were uncertain about what we would find. Over the twelve weeks, however, it was clear that the salient moments we were looking for were everywhere. We watched students struggle, argue, ask, and overcome. We watched them help unprompted and share challenges they faced openly and unashamed. We watched students push their creative boundaries, question the limits of their programming abilities, and glow with pride when their program proved successful.

The play that is facilitated by the KIBO curriculum inherently requires problem solving, collaboration, perspective shifting, and creative approaches. The students were able to use these computational thinking skills beyond coding, to engage with the abstract concepts that define the cultures of their classrooms, schools, and religions, and use their KIBOs to turn those abstract concepts into tangible, creative expressions.

In this chapter, we saw how culture can be harnessed as a tool to help young children conceptualize computational concepts, but also, how the culture of the learning environment influences the problems children see fit to be solved. On a deeper level, KIBO was a vehicle for imagining other technological tools the students could build to serve a greater purpose.

ACKNOWLEDGMENT

This research was supported by the Templeton World Charity Foundation.

REFERENCES

Bers, M. (2008). *Blocks to robots: Learning with technology in the early childhood classroom*. Teachers College Press.

Bers, M. (2018). *Coding as a playground: Programming and computational thinking in the early childhood classroom*. Routledge Press.

Bers, M. U. (2012). *Designing digital experiences for positive youth development: From playpen to playground.* Oxford University Press. doi:10.1093/acprof:oso/9780199757022.001.0001

Bers, M. U. (2019). Coding as another language: A pedagogical approach for teaching computer science in early childhood. *Journal of Computers in Education*, *6*(4), 499–528. doi:10.100740692-019-00147-3

DevTech Research Group. (2018). *Our treasure: A KIBO coding curriculum for emergent readers.* Tufts University.

Govind, M., & Bers, M. U. (2019). *Parents Don't Need to Be Coding Experts, Just Willing to Learn With Their Children.* EdSurge. https://www.edsurge.com/news/2019-12-11-parents-don-t-need-to-be-coding-experts-just-willing-to-learn-with-their-children

K-12 Computer Science Framework Steering Committee. (2016). *K-12 computer science framework.* https://k12cs.org/

Lee, K., Sullivan, A., & Bers, M. U. (2013). Collaboration by design: Using robotics to foster social interaction in kindergarten. *Computers in the Schools*, *30*(3), 271–281.

Modan, N. (2019, September 11). *33 states adopted 57 computer science ed policies since 2018.* K-12 Dive. https://www.educationdive.com/news/33-states-adopted-57-computer-science-ed-policies-since-2018/562530/

Sano, A. (2019, March 27). *Coding will be mandatory in Japan's primary schools from 2020.* Nikkei Asia. https://asia.nikkei.com/Economy/Coding-will-be-mandatory-in-Japan-s-primary-schools-from-2020#:~:text=TOKYO%20%2D%2D%20Computer%20programming%20will,highly%20sought%20information%20technology%20skills

Seow, P., Looi, C. K., How, M. L., Wadhwa, B., & Wu, L. K. (2019). Educational policy and implementation of computational thinking and programming: Case study of Singapore. In *Computational thinking education* (pp. 345–361). Springer.

Toikkanen, T., & Leinonen, T. (2017). The coding ABC MOOC: Experiences from a coding and computational thinking MOOC for Finnish primary school teachers. In *Emerging research, practice, and policy on computational thinking* (pp. 239–248). Springer International Publishing.

Chapter 11
Supporting Girls' Computational Thinking Skillsets:
Why Early Exposure Is Critical to Success

Amanda Sullivan
Tufts University, USA

ABSTRACT

The representation of women in technical fields such as computer science and engineering continues to be an issue in the United States, despite decades of research and interventions. According to the most recent Bureau of Labor Statistics reports, only 21.1% of computer programmers are women, and only 16.5% of engineering and architecture positions are filled by women. This chapter discusses the long-term importance of exposing girls to computational thinking during their formative early childhood years (Kindergarten through second grade) in order to set them up for equal opportunities in technical fields throughout their later educational and career years. This chapter presents a case example of a K-2nd grade robotics and coding curriculum in order to highlight examples of developmentally appropriate technologies, activities, and strategies that educators can implement to foster young girls' computational thinking skills. Best practices and instructional strategies to support girls—as well as young children of any gender identity—are discussed.

DOI: 10.4018/978-1-7998-7308-2.ch011

INTRODUCTION

From smart home devices to cell phone applications, the fields of technology, engineering, and computer science drive the development of everyday innovations we all rely on. So what happens when female voices are not represented in these fields? We see "innovations" like cell phones that fit men's hands better than women's (Tufecki, 2013; Ryan, 2013) and health tracking apps that ignore women's menstrual cycles (Duhaime-Ross, 2014). We see virtual assistants that have a harder time answering women's questions than men's – that can suggest help for a heart attack but not for domestic violence or rape (Chemaly, 2016; Miner et al, 2016). In short, we see masculine biases in almost all of the technology we engage with on a daily basis. Although most of the issues in these particular examples have been addressed by developers since coming to light, they are issues that would have been unlikely to occur to begin with if female voices had been equally involved and valued during the development of these technologies.

The problem is that female voices[1] are not – and historically, have never been– well represented in the fields driving innovation. While this issue of female under-representation in technical STEM (Science, Technology, Engineering, and Mathematics) fields is not a new one, it is a persistent one. Despite decades of research and interventions, the disparity between the representation of men and women in technical fields in the United States continues to persist. According to the most recent Bureau of Labor Statistics numbers, only 21.1% of computer programmers are women and only 16.5% of engineering and architecture positions are filled by women (Bureau of Labor Statistics, 2020).

Although there is a major spotlight on workforce representation of women in technology, it is important to note that the issue of female representation in engineering and computing fields begins long before the career years. Beginning in early childhood and throughout their middle and high school years, girls and young women are exposed to stereotypes that inform ideas about their identity, abilities, and interest in STEM fields (e.g., McKown & Weinstein, 2003; Kuhn, Nash, & Brucken, 1978; Signorella, Bigler, & Liben, 1993; Metz, 2007; Steele, 1997; Sullivan, 2019). By high school, research has shown that male students are more likely than female students to take the standardized exams closely associated with the fields of engineering and computing (Corbett & Hill, 2015).

With the rise of the K-12 coding and computational thinking education movement in recent years, educators have a renewed opportunity to begin addressing this divide from an early age – and potentially address the gender and STEM gap before it becomes pronounced. This chapter will explore best practices for gender-inclusive computational thinking curriculum that can be implemented during the foundational early childhood years (kindergarten through second grade). While the focus of this chapter is on increasing STEM access to girls and women (and as such the research and language used will focus on those identifying as girls) the ultimate goal of this line of work is to encourage researchers and educators to develop curriculum and technologies that are bias-free, gender-neutral, and equally appealing to all young children.

Through the lens of a curriculum unit called "Helpful Robots," this chapter will highlight suggestions for choosing appropriate tools, curricular themes, and adult role-modeling practices that can positively engage girls – and all students, regardless of gender identity– in playful computational learning from an early age, with the ultimate goal of ensuring all young children are afforded equal opportunities to succeed in STEM and beyond.

LITERATURE REVIEW

Women in STEM

In the United States, men have consistently outnumbered women in numerous technical STEM fields, particularly fields like computer science and engineering (Hill, Corbett, & St. Rose, 2010). The representation of women in science and engineering education and employment is substantially lower than their representation in the U.S. population. According to the National Science Foundation (2017), the fields of computer science, physics, and engineering were all overwhelmingly male.

Today, the Bureau of Labor Statistics (2020) confirms that men continue to dominate these fields. Women make up only 21.1% of computer programmers, and only 16.5% of combined engineering and architecture positions are filled by women. Looking at computer and mathematical sciences combined, women fill only around a quarter of these positions (25.2%). (Bureau of Labor Statistics, 2020).

Stereotypes & the Importance of Early Exposure

It has been long theorized that a social psychological phenomenon known as "stereotype threat" may influence the participation of girls and women in STEM. Stereotype threat refers to the anxiety that one's performance on a task or activity will be seen through the lens of a negative stereotype (e.g., Steele, 1997; Steele, 1999; Steele & Aronson, 1995; Spencer, Steele, & Quinn 1999). For example, Spencer, Steele, & Quinn (1999) found that women performed significantly worse on a math test if they were first shown information indicating that women do not perform as highly as men on math tasks (to induce the negative stereotype). If the negative stereotype was not triggered (i.e. participants were told that there were no gender differences associated with the math test) women and men performed similarly on the test.

While most research on the influence of stereotype threat has focused on adolescent and adult research participants, we know that basic stereotypes do begin to develop in children around two to three years of age (Kuhn, Nash, & Brucken, 1978; Signorella, Bigler, & Liben, 1993). Children learn to make sense of the vast and confusing world around them by putting things into neat categories (often stereotypes), based on their observations of their peers, families, books, and other media they are exposed to. As children grow older, stereotypes about sports, occupations, and adult roles expand, and their gender associations become more sophisticated (Sinno & Killen, 2009). Adults should be aware of these newly forming stereotypes in order to expand on them (or, disprove them) by providing children with new role-models, experiences, and media that can help shift children's belief system.

Early experiences have the potential to play an ongoing role in children's sense of belonging and confidence in different STEM activities and their own developing identity as they grow up. Forming a positive "STEM Identity" (Aschbacher, Li, & Roth, 2010) during this time can be pivotal to maintaining girls' interest in these fields. Prior research has shown that early childhood experiences with technology and engineering – or lack thereof – can continue to impact young women during middle school and high school, even those on competitive robotics and programming teams (Sullivan & Bers, under review; Sullivan & Bers, 2019). Taken together with the past body of work on stereotypes, it is critical to begin reaching girls (and all children) with positive, developmentally-appropriate experiences with technical STEM content from an early age.

Supporting Young Children's Computational Thinking Skills

We have been focusing on the "technical" STEM fields (i.e., coding, engineering, etc.) because these are the fields in which women continue to be sorely under-represented at the professional level. There is great variability across the fields in these domains, but one thing they have in common is their reliance on computational thinking (hereafter, CT) skillsets. While "coding" can be considered a technical skill and "computer science" is typically thought of as an academic discipline, "CT" can be thought of as a problem-solving process central to computer science that can be applied more broadly to problem solving and learning in any discipline (Avengine et al, 2017).

CT was brought into the public discourse by Wing (2006), who asserted that "computational thinking represents a universally applicable attitude and skillset everyone, not just computer scientists, would be eager to learn and use" (p. 33). Wing (2006) went on to define CT as a way of solving problems, designing systems, and understanding human behavior that draws on concepts fundamental to computer science. Since Wing's pivotal work defining CT, research has shown that CT is, in fact, applicable across disciplines from the arts to the sciences. Chapter 7 in this book for example, depicts how CT relates and supports the dramatic arts (Strawhacker & Sullivan, 2021).

But what does CT look like at the early childhood level? As we will explore in the following case study, it can look a lot like hands-on play, independent and collaborative problem-solving, and tinkering with developmentally-appropriate computing technologies for children. Bers (2020) describes 7 "powerful ideas" from CT that young children begin to master including: algorithms, modularity, control structures, representation, hardware/software, the design process, and debugging. Later in this chapter, we will look at how each of these concepts described by Bers can be introduced to young girls beginning in pre-kindergarten.

Case Example: "Helpful Robots" Curriculum

This case example will walk through the design and implementation of the Helpful Robots curriculum which introduced coding, engineering, and CT concepts to children in K-2nd grade with the goal of increasing girls' interest in engineering. It will highlight research on the curriculum's efficacy and strategies from the unit that educators can employ in their own early learning settings.

Overview of the Curriculum

The "Helpful Robots" theme was developed in collaboration between the lead researcher for this project and the participating K-2nd grade teachers at a public elementary school located in Somerville, Massachusetts. Teachers were interested in a theme that would foster community, helping, and caring. These behaviors were also aligned with the Positive Technological Development (PTD) Framework developed by Bers (2012). The PTD framework was developed as an extension of the computer literacy and the technological fluency movements to guide the development, implementation, and evaluation of educational programs that use new technologies (Bers, 2012). The PTD framework proposes six positive behaviors (commonly referred to as the "six C's") that should be supported by educational programs that use new educational technologies including: content creation, creativity, communication, collaboration, community building and choices of conduct (Bers, 2012). These six C's were used as a theoretical guide when developing the "Helpful Robots" curriculum.

The teachers and the lead researcher developing the curriculum worked together to decide upon a theme that they believed would be equally appealing to all young children, regardless of gender. Throughout the curriculum, children learned about robots that perform helpful jobs in the real world (such as hospital robots, robots that clean like the Roomba, etc.). As a final project, children worked in groups to create their own "Helpful Robots" robots using the KIBO robotics kit (described in the next section) to do helpful classroom jobs, teach important ideas, and demonstrate respectful behaviors and school rules.

Each week, children spent one hour learning and practicing new CT, engineering, and robotics concepts such as sequencing, repeat loops, sensors, and conditional statements. While the same curricular structure was used across classes and grades, modifications were designed to make the curriculum developmentally appropriate. For example, younger grades spent more time on new concepts while older grades moved through the same concepts more quickly (See Table 1).

The exploration of new concepts continued for the first five weeks of the curriculum. Children always worked in groups ranging from 2-4 children, based on class size and factors decided upon by the classroom teacher. Children also gathered as a full group for discussions, games, read-alouds, and/or showcases during each session. The final two weeks of the curriculum were spent working on final "helpful robot" creations and culminated in a final showcase of projects.

Technology Used

The Helpful Robots curriculum unit utilized an early prototype version of the KIBO robotics kit, developed by the DevTech Research Group at Tufts University and now manufactured by KinderLab Robotics. KIBO is a screen-free robotics construction kit that children assemble, decorate, and then program using wooden programming blocks to make the robot move and react to stimuli (Sullivan and Bers 2015). The kit contains wheels, motors, a light output, and a variety of sensors that are easy for children to attach to the robot . KIBO is programmed to move using interlocking wooden programming blocks that each have a unique barcode. KIBO uses an embedded scanner in the robot body to scan the barcodes one at a time, sending the program to the robot (see Figure 1).

Technological learning tools like KIBO are perfect for engaging young girls (and all young children) in CT for a few key reasons. First off, KIBO is designed for open-ended play that allows girls to make almost anything they want based on their own personal interests. KIBO can be used to act out a scene from a story or movie, it can be decorated to look like an animal, it can be a carousel or a fire truck (Sullivan, 2020). Therefore, it can be used to help explore almost any interest that a young girl has (see figures 2 and 3 for examples of final Helpful Robot projects).

Table 1. Scope and Sequence of the Helpful Robots Curriculum as described by Sullivan (2016) and Sullivan & Bers (2018)

Lesson	Kindergarten	First Grade	Second Grade
1	**What is a robot? Who are engineers?** Children learn the engineering design process and build sturdy robots that can carry a ball of paper to the recycling bin.	**What is a robot? Who are engineers?** Children learn the engineering design process and build sturdy robots that can carry a ball of paper to the recycling bin.	**What is a robot? Who are engineers?** Children learn the engineering design process and build sturdy robots that can carry a ball of paper to the recycling bin.
2	**What is a program (pt. 1)?** Children learn sequencing & program their robots to dance the Hokey Pokey.	**What is a program?** Children learn sequencing & program their robots to dance the Hokey Pokey.	**What is a program?** Children learn sequencing & program their robots to dance the Hokey Pokey.
3	**What is a program (pt. 2)?** Children continue to practice sequencing a program by navigating masking tape maps on the floor.	**What are sensors?** Children add sound sensors to their robots and program them to wait for their clap.	**What are sensors?** Children add sound sensors to their robots and program them to wait for their clap.
4	**What are sensors?** Children add sound sensors to their robots and program them to wait for their clap.	**What are repeat loops with number parameters?** Children practice estimation while using repeat loops and number parameters to make their robots navigate floor maps.	**What are repeats loops with number parameters and sensor parameters?** Children practice estimation while using repeat loops and number parameters to make their robots navigate floor maps. Next, children navigate the same maps using distance and light parameters.
5	**What are repeats loops with number parameters?** Children practice estimation while using repeat loops and number parameters to make their robots navigate floor maps.	**What are repeat loops with sensor parameters?** Children learn about the distance and light sensors and program them to work with their robots using repeat loops.	**What are conditional statements?** Children learn about conditional "if" blocks. They program their robots to respond to light and distance sensor input in order to "decide" what to do.
6	**Final Project-** Children plan, build, and begin to program their Helpful Robots.	**Final Project-** Children plan, build, and begin to program their Helpful Robots.	**Final Project-** Children plan, build, and begin to program their Helpful Robots.
7	**Final Project-** Children finish their projects. In a final exhibition, they share their final projects.	**Final Project-** Children finish their projects. In a final exhibition, they share their final projects.	**Final Project-** Children finish their projects. In a final exhibition, they share their final projects.

Figure 1. Example KIBO robot and interlocking wooden programming blocks

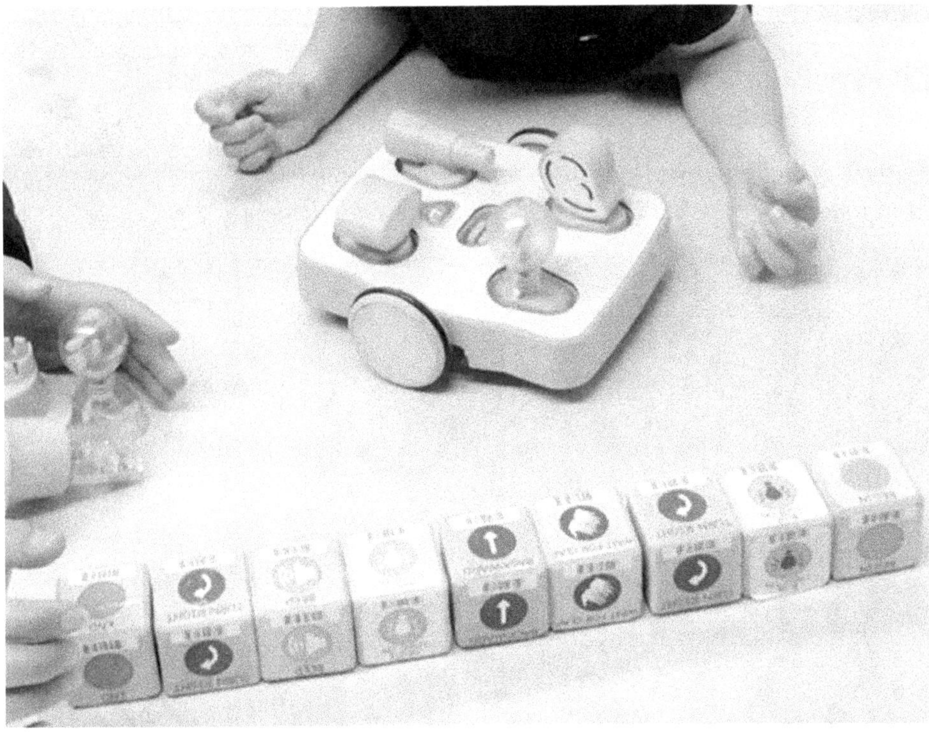

KIBO was explicitly designed to have neutral aesthetic, making it equally appealing to children of any gender. While many early childhood robots are designed to portray "character" (think of Beebot's friendly "bee" aesthetic or Dash's blue, one-eyed, robot appearance) KIBO takes the opposite approach, with neutral colored wooden parts that do not attempt to look like anything until a child builds and decorates it.

With so few truly gender-neutral toys out there (even LEGO has long legacy of being deemed a "boy's toy"), neutrally designed kits like KIBO can be useful to equally reach children in mixed gendered classrooms. Finally, KIBO engages girls in hands-on building and tinkering as well as CT problem-solving and coding.

Supporting CT in Young Children

Throughout the curriculum, a variety of problem-solving, and specifically CT related, skills and concepts were taught through the use of KIBO as well as unplugged games and activities. Although the Helpful Robots curriculum was first designed in 2016, before the Bers (2020) approach to CT in early childhood was released, it is still useful to map the curriculum onto each of the 7 powerful ideas of CT presented by Bers.

This helps to illustrate how programmable robotics kits like KIBO can support the learning of a range of CT skills and concepts that are developmentally appropriate for young children. Table 2 connects each of the seven CT concepts presented by Bers (2020) with activities that from in the Helpful Robots unit.

Figure 2. KIBO prototype decorated for the final to remind children to listen. This was a final project created for the Helpful Robots curriculum

Figure 3. A final project designed to help carry school supplies for the Helpful Robots curriculum

Table 2. Exploring CT concepts through the Helpful Robots Unit

Bers (2020) CT Concepts	"Helpful Robots" Curricular Activities
Algorithms – A series of ordered steps taken in sequence.	Creating algorithms for the KIBO robot. Playing unplugged sequencing focused games such as "Coder Says" (a version of Simon Says).
Modularity – Breaking down tasks and procedures into simpler, manageable units.	Breaking down Final Project tasks into smaller jobs (e.g., planning code, creating program in small pieces, decorating robot, etc.).
Control Structures – Controlling the sequence in which a program is executed. Making decisions based on conditions.	Exploring Repeat Loops and Conditional Statements with KIBO's block language.
Representation – Concepts can be represented by symbols.	Learning that each block represents a different action for KIBO.
Hardware/Software - Computing systems need both hardware & software to operate.	Learning that KIBO works because of hardware (i.e. the robot chassis, motors, sensors) and software (block programming language).
Design Process – An iterative process used to develop programs & artifacts with multiple steps.	Children worked iteratively to revise, edit, and improve their Helpful Robot projects over multiple class sessions.
Debugging – Fixing problems in our programs.	Children worked to solve problems with their robot's hardware as well as their syntactical problems with their code.

Strategies Employed

From the first glance, this may seem like a typical robotics and computer science curricular unit. And in many ways, it was. However, in its *implementation* it differed in a few ways that may have made it especially effective at increasing girls' interest in STEM and reducing gender stereotypes of both boys and girls. These include:

- **Focusing on female role models** – Prior research has shown that role-modeling may be an important piece of engaging girls and women in STEM (Amelink & Creamer, 2010). This unit was originally taught by an all-female teaching team from Tufts University.
- **Choosing a gender-neutral tech** – The unit used the neutrally designed KIBO robotics kit, as opposed to stereotypically "girly" materials or a traditionally masculine STEM products like LEGO, programmable cars, or drones.
- **Focusing on collaboration over competition** – Prior research has shown that girls and women may respond more to collaboration than competition (e.g., Sullivan & Bers, under review; Rusk, Berg, & Resnick, 2005). Still, many educational robotics initiatives that exist in schools are centralized around competitive goals. This unit differed by focusing on collaboration in

all aspects of the unit, from the theme itself (creating robots that help the school community), to content taught (e.g., learning about robots that help in hospitals and other settings), to the set-up activities (all children worked collaboratively in partners or small groups).

These classroom practices, as well as other best practices for engaging girls in CT and STEM in general, are explored further later in this chapter.

Efficacy of this Curricular Approach

The efficacy of the *Helpful Robots* curriculum on increasing girls' interest in engineering was published by Sullivan & Bers (2018a) in the *International Journal of Technology & Design Education*. In this study by Sullivan & Bers (2018a), findings from a sample of 105 children in K-2nd grade demonstrated that after completing the curriculum, students identifying as girls had a statistically significant positive change in their desire to be an engineer. This change was not present in a control group from the school that did not receive the curriculum. Prior to the curriculum intervention, pretest findings also showed that student identifying as boys were significantly more interested in being an engineer than girls. After completing the curriculum, there was no longer a significant difference between boys' and girls' interest in engineering. Additionally, there were no significant effects for gender on mastery of CT concepts measured– indicating boys and girls mastered the computational concepts equally well (Sullivan & Bers, 2018a). This, ultimately, should be the goal: to create equitable learning situations where children of any gender are able to demonstrate equal mastery.

It is important to note that in this study, the child participants themselves were asked to share how they identified (not a teacher or parent). They were given the opportunity to provide non-binary responses or to say they "didn't know" however, in this sample, all children chose to identify as either boys or girls. Future research should aim to find samples of children that captures non-binary identities as well.

Other work by Sullivan & Bers found that the female teaching team may have been especially important to the success of the program (Sullivan & Bers, 2018b). This study demonstrated preliminary evidence that having a female instructor may positively impact girls' performance on certain programming tasks and reduce the number of gender differences between boys and girls in their mastery of programming concepts.

Taken together, findings suggest that a combination of technology choice, female role-modeling of instructors, and the actual curricular content itself led to positive outcomes for increasing girls' interest in engineering, while supporting children in mastering computational concepts equally well, regardless of gender.

Table 3. Best practices for engaging girls in CT

Best Practices	What it Might Look Like in the Classroom...
Choosing the Right Tools	• Choosing open ended tools or applications that engage girls as *creators* rather than *consumers* of their digital experiences. • Choosing tools that support CT learning, engineering, and design.
Integrative STEAM Approach	• Rather than teaching CT concepts or coding as a "standalone," integrating CT with the arts, music, culture, history and more to reach a wider range of students.
Fostering Collaboration	• Providing girls with opportunities to work in partnerships, small groups, and large groups. • Work toward collaborative STEM events. (i.e., showcases) as opposed to competitive ones (i.e., final competitions or contests).
Breaking/Preventing/Disrupting Stereotypes	• Talking about stereotypes that are presented in the shows, books, and movies they are exposed to. • Proving examples that contradict negative stereotypes.
Female STEM Role Models	• Expose all children (including boys and children of any gender identity) to female STEM role models through books, media, and real-world introductions. • Female educators should be sure to model their own positive attitudes toward CT and STEM in general.
Language Matters: Choosing Words Wisely	• Choose gender-neutral terms to talk about robots and technology (e.g., "it" rather than "he"). • Carefully choosing how you talk about STEM concepts, fields, and professionals.

Practices to Support Girls' CT Learning

In the previous case study, we saw that certain practices were successful at increasing girls' interest in engineering and teaching computational thinking to all young children. Table 3 highlights key practices from the case study, as well as other best practices for positively engaging girls in CT from an early age. The suggestions presented in Table 3, while rooted in research on girls in STEM, are applicable for young children of any gender identity.

The case study focused on the use of the KIBO robotics kit, but similar curricular interventions could be implemented using a range of materials. Sullivan (2019) explains that when choosing technologies and apps for young girls, one of the most important factors educators should consider choosing applications that engage girls as creators of digital content rather than consumers of digital content. Choosing tools that prompt girls not just to watch but to do. These might include programming applications like ScratchJr, robotics kits like KIBO or Code-a-Pillar, or DIY kits from companies like Goldie Blox.

One of the key features of the Helpful Robots curriculum was the focus (explicit and implicit) on collaboration, teamwork, and support. Over the past decade, robotics competitions have become an increasingly popular way for K-12 students to become involved with computer science and engineering in a way that has been thought to increase student interest in math and science and (Hendricks, Alemdar, & Olgletree, 2012; Petre & Price 2004). However, while robotics competitions may motivate students to learn more about fields such as computer science, research also demonstrates that gender gaps persist in these competitive environments and appear to widen as students grow older and enter more advanced competitions (Witherspoon, Schunn, Higashi, & Baehr, 2016). By offering opportunities for children to explore coding, CT, and engineering in collaborative contexts, it may be possible to increase the number of girls who master CT skills and decide to pursue computational subjects and fields as they grow up.

The final key component of Helpful Robots curriculum worth highlighting is the early age of the age of the curricular intervention: Kindergarten through second grade. There is a growing body of work on the importance of early exposure to coding and engineering content in order to help pique girls' interest in technical STEM fields from an early age (e.g., Sullivan, 2019; Sullivan & Bers, 2018a, Sullivan & Bers, 2018b; Sullivan, 2016). Other work has shown that even beginning in preschool, children can successfully learn basic CT concepts (e.g., Elkin, Sullivan, & Bers, 2016). So why wait until later in life when stereotypes and beliefs about abilities are more firmly ingrained? Instead, we should consider early childhood a pivotal time to begin reaching girls with engaging STEM and CT content, and continue to support them – and all children – throughout their educational journeys.

CONCLUSION

Five years later and we still have a ways to go in order to truly make CT and computer science equitable and available for all students. This chapter has highlighted strategies and approaches that educators can use to begin to address issues of STEM equity early on. Ensuring that girls are exposed to developmentally appropriate tools that encourage engineering, computational thinking, and creating from an early age through collaborative contexts is key for setting the stage for success later on. Additionally, it is important to expose all young children to female role models from STEM fields and to carefully choosing gender-neutral terms to talk about technology to children. Finally, explicitly talking to young children about stereotypes and addressing biases from books and media will help to tackle the issue of stereotypes head on.

By following these guidelines and encouraging CT exposure early on, with a focus on consciously choosing teaching tools and themes that will be equally appealing to all students, we can take an important first step toward realizing the vision of "computer science for all."

ACKNOWLEDGMENT

The research presented in this chapter was supported by the National Science Foundation [grant number DRL-1118897].

REFERENCES

Amelink, C. T., & Creamer, E. G. (2010). Gender differences in elements of the undergraduate experience that influence satisfaction with the engineering major and the intent to pursue engineering as a career. *Journal of Engineering Education*, *99*(1), 81–92. doi:10.1002/j.2168-9830.2010.tb01044.x

American Psychological Association. (2015). Guidelines for psychological practice with transgender and gender nonconforming people. *The American Psychologist*, *70*(9), 832–864. doi:10.1037/a0039906 PMID:26653312

American Psychological Association. (2021). *APA Resolution on Gender Identity Change Efforts*. American Psychological Association. Retrieved from: https://www.apa.org/about/policy/resolution-gender-identity-change-efforts.pdf

Angevine, C., Cator, K., Roschelle, J., Thomas, S. A., Waite, C., & Weisgrau, J. (2017). *Computational Thinking for a Computational World*. Academic Press.

Bers, M. (2020). *Coding as a Playground: Programming and Computational Thinking in the Early Childhood Classroom* (2nd ed.). Routledge Press. doi:10.4324/9781003022602

Bers, M. U. (2012). Designing Digital Experiences for Positive Youth Development: From Playpen to Playground. Cary, NC: Oxford. doi:10.1093/acprof:oso/9780199757022.001.0001

Bureau of Labor Statistics. (2020). *Labor Force Statistics from the Current Population Survey*. Retrieved from: https://www.bls.gov/cps/cpsaat11.htm

Chemaly, S. (2016, March 16). The problem with a technology revolution designed primarily for men. *Quartz*. Retrieved from https://qz.com/640302/why-is-so-much-of-our-new-technology-designed-primarily-for-men/

Corbett, C., & Hill, C. (2015). *Solving the equation: the variables for women's success in engineering and computing*. The American Association of University Women.

Doerschuk, P., Liu, J., & Mann, J. (2007). Pilot summer camps in computing for middle school girls. *ACM SIGCSE Bulletin*, *39*(3), 4–8. doi:10.1145/1269900.1268789

Duhaime-Ross, A. (2014, September 25). Apple promised an expansive health app, so why can't I track menstruation? *The Verge*. Retrieved from https://www.theverge.com/2014/9/25/6844021/apple-promised-an-expansive-health-app-so-why-cant-i-track

Elkin, M., Sullivan, A., & Bers, M. U. (2016). Programming with the KIBO Robotics Kit in Preschool Classrooms. *Computers in the Schools*, *33*(3), 169–186. doi:10.1080/07380569.2016.1216251

Gal-Ezer, J., & Stephenson, C. (2009). *The current state of computer science in US high schools: a report from two national surveys.* Retrieved from Computer Science Teachers Association website, https://csta.acm.org/Research/sub/ Projects/ResearchFiles/StateofCSEDHighSchool.pdf

Hendricks, C. C., Alemdar, M., & Ogletree, T. W. (2012). *The impact of participation in VEX robotics competition on middle and high school students' interest in pursuing STEM studies and STEM-related careers.* Paper presented at the ASEE Annual Conference, San Antonio, TX. Retrieved from https://peer.asee.org/22069

Hill, C., Corbett, C., & St Rose, A. (2010). Why so few? Women in science, technology, engineering, and mathematics. American Association of University Women.

Kafai, Y., & Margolis, J. (2014, October 7). Why the 'coding for all' movement is more than a boutique reform. *Washington Post.* Retrieved from https:// www.washingtonpost.com/news/answer-sheet/wp/2014/10/17/whythe-coding-for-all-movement-is-more-than-a-boutique-reform

Kuhn, D., Nash, S. C., & Brucken, L. (1978). Sex role concepts of two- and three-year-olds. *Child Development*, *49*(2), 445–451. doi:10.2307/1128709 PMID:679779

McKown, C., & Weinstein, R. S. (2003). The development and consequences of stereotype-consciousness in middle childhood. *Child Development*, *74*(2), 498–515. doi:10.1111/1467-8624.7402012 PMID:12705569

Metz, S. S. (2007). Attracting the engineering of 2020 today. In R. Burke & M. Mattis (Eds.), *Women and Minorities in Science, Technology, Engineering and Mathematics: Upping the Numbers* (pp. 184–209). Edward Elgar Publishing. doi:10.4337/9781847206879.00018

Miner, A. S., Milstein, A., Schueller, S., Hegde, R., Mangurian, C., & Linos, E. (2016). Smartphone-based conversational agents and responses to questions about mental health, interpersonal violence, and physical health. *JAMA Internal Medicine, 176*(5), 619–625. doi:10.1001/jamainternmed.2016.0400 PMID:26974260

National Science Foundation. (2017). *Women, Minorities, and Persons with Disabilities in Science and Engineering: 2017*. Special Report NSF 17-310. Available at www.nsf.gov/statistics/wmpd/

Petre, M., & Price, B. (2004). Using robotics to motivate 'back door' learning. *Education and Information Technologies, 9*(2), 147–158. doi:.0000027927.78380.60 doi:10.1023/B:EAIT

Rusk, N., Berg, R., & Resnick, M. (2005). *Rethinking robotics: Engaging girls in creative engineering*. Proposal to the National Science Foundation, Cambridge. Retrieved from https://www.media.mit.edu/publications/rethinking-robotics-engaging-girls-in-creative-engineering-2/

Ryan, E. G. (2013, November 8). Smartphones are made for giant man-hands. *Jezebel*. Retrieved from https://jezebel.com/smartphones-are-made-for-giant-man-hands-1461122433

Signorella, M. L., Bigler, R. S., & Liben, L. S. (1993). Developmental differences in children's gender schemata about others: A meta-analytic review. *Developmental Review, 13*(2), 147–183. doi:10.1006/drev.1993.1007

Spencer, S. J., Steele, C. M., & Quinn, D. M. (1999). Stereotype threat and women's math performance. *Journal of Experimental Social Psychology, 35*(1), 4–28. doi:10.1006/jesp.1998.1373

Steele, C. M. (1997). A threat in the air: How stereotypes shape intellectual identity and performance. *The American Psychologist, 52*(6), 613–629. doi:10.1037/0003-066X.52.6.613 PMID:9174398

Steele, C. M. (1999). Thin ice: "Stereotype threat" and black college students. *Atlantic Monthly, 284*(2), 44–47, 50–54.

Steele, C. M., & Aronson, J. (1995). Stereotype threat and the intellectual test performance of African-Americans. *Journal of Personality and Social Psychology, 69*(5), 797–811. doi:10.1037/0022-3514.69.5.797 PMID:7473032

Strawhacker, A., & Sullivan, A. (2021). Computational Expression: How dramatic arts support computational thinking in young children. In M. U. Bers (Ed.), *Computational thinking and coding in early childhood*. IGI Global.

Sullivan, A. (2016). *Breaking the STEM Stereotype: Investigating the Use of Robotics to Change Young Children's Gender Stereotypes About Technology and Engineering* (Doctoral Dissertation). Tufts University, Medford, MA.

Sullivan, A. (2019). *Breaking the STEM Stereotype: Reaching Girls in Early Childhood.* Rowman & Littlefield.

Sullivan, A. (2020). *STEM Tools, Games, and Products to Engage Girls in Pre-K through Early Elementary School. Technological Horizons in Education.*

Sullivan, A., & Bers, M. U. (2018a). Investigating the use of robotics to increase girls' interest in engineering during early elementary school. *International Journal of Technology and Design Education, 29*(5), 1033–1051. doi:10.100710798-018-9483-y

Sullivan, A., & Bers, M. U. (2018b). The Impact of Teacher Gender on Girls' Performance on Programming Tasks in Early Elementary School. *Journal of Information Technology Education: Innovations in Practice, 17*, 153–162. doi:10.28945/4082

Sullivan, A., & Bers, M. U. (2019). VEX Robotics Competitions: Gender differences in student attitudes and experiences. *Journal of Information Technology Education, 18*, 97–112. doi:10.28945/4193

Sullivan, A., & Bers, M. U. (Manuscript submitted for publication). Increasing female representation on VEX robotics competition teams: Results from a three-year study. *International Journal of Technology and Design Education.*

Sullivan, A., Elkin, M., & Bers, M. U. (2015). KIBO Robot Demo: Engaging young children in programming and engineering. In *Proceedings of the 14th International Conference on Interaction Design and Children (IDC '15)*. ACM.

Tufekci, Z. (2013, November 4). It's a man's phone. *Medium.* Retrieved from https://medium.com/technology-and-society/its-a-mans-phone-a26c6bee1b69

Wing, J. M. (2006). Computational thinking. *Communications of the ACM, 49*(3), 33–35. doi:10.1145/1118178.1118215

Witherspoon, E. B., Schunn, C. D., Higashi, R. M., & Baehr, E. C. (2016). Gender, interest, and prior experience shape opportunities to learn programming in robotics competitions. *International Journal of STEM Education, 3*(1), 18. doi:10.118640594-016-0052-1

Zweben, S., & Bizrot, B. (2015). *2014 Taulbee survey.* Retrieved from the Computing Research Association website, https://cra.org/wp-content/uploads/2015/06/2014-Taulbee-Surv

ADDITIONAL READING

Bers, M. U. (2017). *Coding as a playground: Programming and computational thinking in the early childhood classroom*. Routledge. doi:10.4324/9781315398945

Corbett, C., & Hill, C. (2015). *Solving the equation: the variables for women's success in engineering and computing*. The American Association of University Women.

Hill, C., Corbett, C., & St Rose, A. (2010). *Why so few? Women in science, technology, engineering, and mathematics*. American Association of University Women.

Sullivan, A. (2017). Breaking Gender Stereotypes Through Early Exposure to Robotics. *Education Week: Education Futures: Emerging Trends in K-12.*

Sullivan, A. (2019). *Breaking the STEM Stereotype: Reaching Girls in Early Childhood*. Rowman & Littlefield.

Sullivan, A. (2019). *Supporting Girls' STEM Confidence & Competence: 7 Tips for Early Childhood Educators*. EdTech Review.

Sullivan, A., & Bers, M. U. (2019). VEX Robotics Competitions: Gender differences in student attitudes and experiences. *Journal of Information Technology Education, 18*, 97–112. doi:10.28945/4193

Sullivan, A., Bers, M. U., & Mihm, C. (2017). Imagining, Playing, & Coding with KIBO: Using KIBO Robotics to Foster Computational Thinking in Young Children: *Proceedings of the International Conference on Computational Thinking Education*. Wanchai, Hong Kong.

KEY TERMS AND DEFINITIONS

KIBO: A screen-free programmable robotics kit for young children with blocks, sensors, modules, and art platforms.

Stereotype: A widely held but fixed and oversimplified image or idea of a particular type of person, group, or thing.

Stereotype Threat: A socially premised psychological threat that arises when one is in a situation for which a negative stereotype about one's group applies.

ENDNOTE

1 Use of the words and phrases "female voices," "women," "girls," and "female" throughout this chapter refers to anyone self-identifying as female. The American Psychological Association (APA) defines "gender identity" as "a person's deep felt, inherent sense of being a girl, woman, or female; a boy, a man, or male; a blend of male or female; [or another] gender" as well as nonbinary individuals (APA, 2015, p. 862; APA, 2021). While this chapter focuses on female-identifying individuals, the best practices and suggestions presented are applicable to working with young children of any gender identity.

Chapter 12
Including Students With Disabilities in the Coding Classroom

Tess Levinson
Tufts University, USA

Libby Hunt
Tufts University, USA

Ziva Hassenfeld
Brandeis University, USA

ABSTRACT

This chapter discusses understandings of coding and computational thinking education for students with disabilities. The chapter describes the special education system in the United States, including limitations in how computer science education is made available to students receiving special education services. The chapter then provides a summary of research in computer science education for students with disabilities, including both high-incidence and low-incidence disabilities. A case study of a young student with a mild disability learning in a general education computational thinking program is then presented, and the implications of the case study for future research directions are discussed.

DOI: 10.4018/978-1-7998-7308-2.ch012

INTRODUCTION

Sophie[1], a 5-year-old girl enters Ms. Locke's kindergarten class smiling. She sits on her spot on the carpet and waits for class to begin. The girls around her argue over who will sit next to who. She seems intentionally oblivious. At some point, she is drawn into this seating dance as another student asks if she will move so that said girl can sit next to another student. She obliges, undisturbed. Soon Ms. Locke begins class and sings the robot part song. Sophie stands delighted and dances along, moving side to side: "The body is connected to the motor; the motor is connected to the... so move robot move." Today is the day the teacher informs the class they will finally get to play with the KIBO robot. The teacher has organized and sorted all the KIBO parts into different storage bins in the "materials" part of her classroom. The students are broken into pairs and called into a line to collect their materials. Sophie and her partner, Pete, wait patiently as students mull over the KIBO bins. Finally, it is their turn. Sophie and Pete take turns filling their tray with all the KIBO parts.

Soon they find a quiet place on the rug and begin building. They work collaboratively, taking turns, co-constructing a path for the KIBO robot to travel and the corresponding program that will allow KIBO to travel. Sophie plans her project in her Design Journal and references that plan as she and Pete create their program. Instead of becoming discouraged when the scanning of the coding blocks does not work, they work together to problem-solve, and Sophie scans the coding blocks with the robot. One would not know from this short snapshot of the classroom that Sophie does not talk in school. She doesn't speak out loud to Pete as they build their program, and her design plan does not include the voice recorder and associated blocks. Still, she and Pete work together, excited by the possibilities KIBO offers for creativity and expression. In this chapter we will explore what Sophie's teacher did to accommodate her disability so that she could access KIBO learning alongside her peers. More broadly, we will discuss how computer science education can be used towards inclusive classrooms and pedagogy.

STUDENTS WITH DISABILITIES AND COMPUTER SCIENCE

Fourteen percent of public-school students in the United States ages 3-21 receive special education services under the Individuals with Disabilities Education Act (Congress, 1975) for some form of disability, which can range from specific learning disorder, to speech or language impairment, to autism spectrum disorder (*Students with Disabilities*, 2020). As each student's individual needs vary, so do the special education services provided. A student with a *high-incidence disability*, a category including but not limited to learning disabilities, emotional and/or behavioral disorders, and speech or language impairments, may spend most of their day with their peers in the general education classroom and only receive an hour or so of special education services for domain-specific instruction (Gage et al., 2012). Students with disabilities of this nature comprise the majority of students with disabilities (Gage et al., 2012; National Center for Education Statistics, 2021). In contrast, students with *low-incidence disabilities* have disabilities that affect learning across domains, such as significant sensory or cognitive impairments (Congress, 1975). Depending on the nature of their disability and needs, students with more-significant intellectual disabilities or other domain-general disabilities may spend most of the day receiving special education services, meaning much of their education is provided by the special education teacher. As suggested by the term, the minority of students with disabilities have disabilities that are classified as low-incidence.

Over 60% of students with disabilities spend more than 80% of their day in the general education classroom (National Center for Education Statistics, 2021). However, students with disabilities do not have equal access to computer science and computational thinking education as their nondisabled peers, which ultimately leads to knowledge gaps for students with disabilities in increasingly important 21st century skills. For example, while approximately 10% of students without disabilities scored below proficient for the National Assessment of Educational Progress technology and engineering literacy content area, nearly half of students with disabilities scored below proficient (National Center for Education Statistics, 2021). Groups and initiatives such as AccessCSForAll and Deaf Kids Code are increasing access to computer science and computational thinking programming for kids with disabilities (deafkidscode.org, n.d.; Ladner & Israel, 2016). Additionally, researchers are developing dedicated educational programs for students with disabilities, as well as best practices for accommodation, in order to improve the quality of computer science education for these students.

Many of the specific programs and interventions relating to computer science and coding instruction for students with disabilities have focused on developing educational programs for students with low-incidence disabilities and autism (Taylor, 2018). Much of this research focuses on educational pedagogies based around explicit instruction. In a curriculum based on explicit instruction, a student might learn, for example two control structures, and then practice them by programming a specific game. Using evidence-based explicit instruction, computer programming has been taught to students with Down syndrome, autism, and intellectual disability (Pivetti et al., 2020). For example, Knight, Wright, and DeFreese (2019) used an explicit instruction pedagogy to teach an elementary student with autism and significant behaviors to code using the Ozobot robot. Following the instruction period, the student was able to generalize the coding skills to new coding challenge (Knight et al., 2019). However, skills taught through explicit instruction do not necessarily generalize. This means a child may be able to use a skill within a specific setting but cannot use the skill in a new setting or to create an unknown program. For example, Taylor (2018) used explicit instruction to teach preschool, kindergarten, and first grade students with intellectual disabilities to use the Dash robot, and although all the students learned to code the robot, no student was able to generalize the skills to complete a novel coding challenge (Taylor, 2018).

These evidence-based explicit instructionist pedagogies used by special educators are in tension with the constructionist pedagogies for computational thinking (Bers, 2020). Constructionist models allow for student-driven play to drive learning, whereas explicit instruction provides a structure for learning. For example, Munoz et al (2018) taught students with autism to create video games using an instructionist pedagogy that provided students with the prompt, characters, and code (Munoz et al., 2018). Through this instructionist video-game learning curriculum, students with autism learned computational thinking skills such as abstraction, problem decomposition, and data representation. In contrast, in a constructionist robotics curriculum focused on cause and effect, students participated in guided free-play involving coding and sensors (Albo-Canals et al., 2018). The primary goal of Albo-Canals et al.'s (2018) research was understanding student engagement with educational robots, rather than computational thinking learning, but the findings suggest that the students gained some computational thinking knowledge, including sequencing and cause-effect. There has not yet been research specifically on computational thinking learning through constructionist curricula for students with disabilities.

Most research on computer science education for students with disabilities has focused on students with low-incidence disabilities and autism who may receive more significant accommodations or modifications to their educational materials. However, the majority of students with disabilities have high-incidence disabilities, and as mentioned above, most of them receive education at least partially within the general education setting (Gage et al., 2012; *Students with Disabilities*, 2020). Services for students with high-incidence disabilities, which include specific learning disabilities (e.g., reading disabilities, math disabilities), speech and language impairments, and emotional and behavioral disorders, are often targeted to a student's specific area of need. For example, a student with a specific learning disability in reading may receive special education services in literacy and language arts but might not receive individualized attention or accommodations in computer science. Bouck and Yadav (2020) showed that students with high-incidence disabilities in an upper elementary school resource room learned computational thinking concepts such as algorithms through a combination of explicit instruction and unplugged activities. They also suggest use of instructional methods such as pre-teaching vocabulary and providing information in multiple formats (Bouck & Yadav, 2020). Israel et al. (2015) reinforce the use of multiple instructional methods and emphasize the use of Universal Design for Learning practices, which uses multiple means of representation, action and expression, and engagement to create an inclusive and accessible curriculum (Israel et al., 2015).

Here, we describe a case study of a student with a disability served primarily in the general education classroom, selective mutism. Selective mutism is defined as "a complex childhood anxiety disorder characterized by a child's inability to speak and communicate effectively in select social settings, such as school" (American Speech-Language-Hearing Association, n.d.). The condition must cause impairment either academically or socially and must not be explained by another communication or developmental disorder (Viana et al., 2009). Although speech or language impairment is classified as high-incidence with regard to special education services, selective mutism is thought to be a relatively rare diagnosis, with prevalence estimated to be between 0.47% and 0.76% (Viana et al., 2009). There is no known single cause of selective mutism, and while there is evidence suggesting an association with anxiety disorders, some students also express externalizing behaviors or ADHD (Viana et al., 2009). The complexity and variations of the disorder create further challenges for a teacher of a student with selective mutism, as there is no single approach to accommodate a student with this diagnosis. The curriculum presented in this case study was not intended as a program or intervention to teach computer science or coding to students with disabilities. Rather, by accommodating the needs of a student with a disability, the teacher was able to create an inclusive and accessible constructionist, coding environment. As such, the case study we present explores exciting new possibilities for using constructionist pedagogies in teaching computer science with students with high-incidence disabilities.

CASE STUDY: CODING AS ACCESSIBLE COMMUNICATION

At first or even second glance, Sophie's classroom participation was similar to that of any other child in her kindergarten class. She sat amid her peers during carpet circle times, raised her hand during participatory questions, and turned her head to anyone who addressed her. Sophie has selective mutism and does not speak, but she was fully included in her class's computer science programming. Sophie and her kindergarten class took part in a larger research project investigating how religious and secular elementary schools used tangible robotics as an opportunity to foster character development (see Chapter 10 in this book). As a research team, we were interested in the different ways that kindergarten-age children would interact with one another in the context of robotics, and how their classroom environment would influence those interactions.

Ms. Locke's classroom was a place where Sophie's disability was accommodated and accepted. Ms. Locke explained Sophie's disability to the researcher's when explaining an accommodation made to the curriculum, and throughout the implementation of the KIBO tangible robotics curriculum, Ms. Locke made notes about how she modified discussion-based activities to allow non-verbal participation. Ms. Locke made turned open-ended questions into "raise your hand if you agree" questions, allowing her to contribute non-verbally without standing out among her peers. Ms. Locke also seemed to have an eye out for Sophie. In one classroom activity we observed, we watched as Sophie began to look a little despondent while her peers shouted their ideas. Ms. Locke noticed and turned to Sophie, saying, "Tell me, should we do a dog?" Sophie smiled and nodded. The acceptance and accommodation of Sophie's disability modeled by Ms. Locke appeared to translate to the other students' acceptance and inclusion of Sophie. In another activity, while creating underwater scenes with crayons, Sophie's classmate leaned over the table to look at her drawing. "I love yours! Look how Sophie did hers!" her classmate remarked, drawing everyone's attention to Sophie. "So pretty," another classmate said. Sophie did not look up but smiled slightly and continued coloring.

Throughout the tangible robotics curriculum, Sophie had the same partner, her classmate Pete. In her notes, Ms. Locke writes that Pete "continues to show kindness and patience towards his partner. Sophie is very quiet, and Pete takes time to explain/ talk with Sophie about KIBO." In the hands-on robotics activity, Pete and Sophie worked to build the KIBO robotics kit together. Sophie poked Pete to get his attention. He never denied her the chance to touch the KIBO robot even when she was having difficulty scanning the tangible programming blocks. Ms. Locke wrote in her lesson notes: "Pete didn't take KIBO away and didn't do the scanning himself, he just held his friend's hands from above and controlled her hand movements." At another point in the curriculum, Ms. Locke used Sophie and Pete's program as an example for the whole class. Although Pete and Ms. Locke did the verbal presentation, they consistently used the plural pronouns "them and their" to give ownership to Pete and Sophie, not just Pete. In another class discussion about who helped other students work with their KIBOs, Pete raised his hand. Sophie noticed and raised her hand. Ms. Locke called on Pete. He announced to the class that Sophie had helped him because she scanned the barcodes of the tangible block program for him. Sophie smiled big and looked down, but the smile lingered for moments after.

We found through analysis of our ethnographic data and video observations that Sophie demonstrated more communicative acts during KIBO robotics activities than during discussion-based activities. She ran from spot to spot with her classmates during "Robot Corners," a game about differentiating between items that are or are not robots, but during sharing circles, she appeared distracted and uninterested. While this finding may feel intuitive, this serves as a reminder of the role of tactile and kinesthetic learning tools for students with communication-related disabilities. The fact that this particular robotics kit, KIBO, centers on student expression and the teaching of coding as a language for communication, makes this finding even more promising for future applications of KIBO as a tool for students with communication-related disabilities to learn computational thinking.

While Sophie offered consistent communicative gestures whenever she was engaged with the KIBO robotics kit, no activity in the curriculum showed her engagement with the tangible tool more than her final project. Figure 1 below shows her planning sheet in her Design Journal for her final project. The assignment asked the students to create "Gratitude Floats" celebrating things special to the students and their community: Because this lesson took place close to Thanksgiving, this was an opportunity for the students to examine the tenets of their school and reflect on what they were grateful for.

GRATITUDE FLOATS (15 min) Ask students to think about what makes their school special. Often, things that are special to you have some sort of meaning that signifies who you are or where you come from. Tell students that today, they will be making "Gratitude Floats," similar to a Thanksgiving Parade, celebrating their school and what makes it special. Ask students: What's important to you? Is it important to other people in the school too? What is different about our school than other schools? Students then should draw images of the things they felt made their school special. These images will later be used to decorate their Gratitude Floats.

The project continued:

PLAN THE PARADE (15 min) Before giving the students their KIBO, have the students plan out their parade. They should imagine if they could take their parade around the school, where they will go (e.g., other classrooms, the cafeteria, the chapel) and why. If time allows, children could even draw their route in the form of a map in their Curiosity Journals.

The blocks Sophie circled in her project plan suggest that she had a developing technical understanding of the KIBO programming language. First, she circled that she would use both a begin block and end block, both necessary for any KIBO program. This is significant because it ties into a basic understanding of the foundations of programming and connects to the powerful idea of algorithms and sequencing. Second, she circled movement blocks in her program, suggesting an expanded vocabulary of programming functions. Third, she circled both the light bulb and the light block, suggesting an emergent understanding of hardware-software correspondence. Although she did not yet show a mastery of this concept, for example, selecting the light and distance sensors without the corresponding blocks, this is significant as it connects to the powerful ideas of representation and multiple tools of communication.

Figure 1. Sophie's final project plan (IGI, 2021)
Source: IGI, 2021

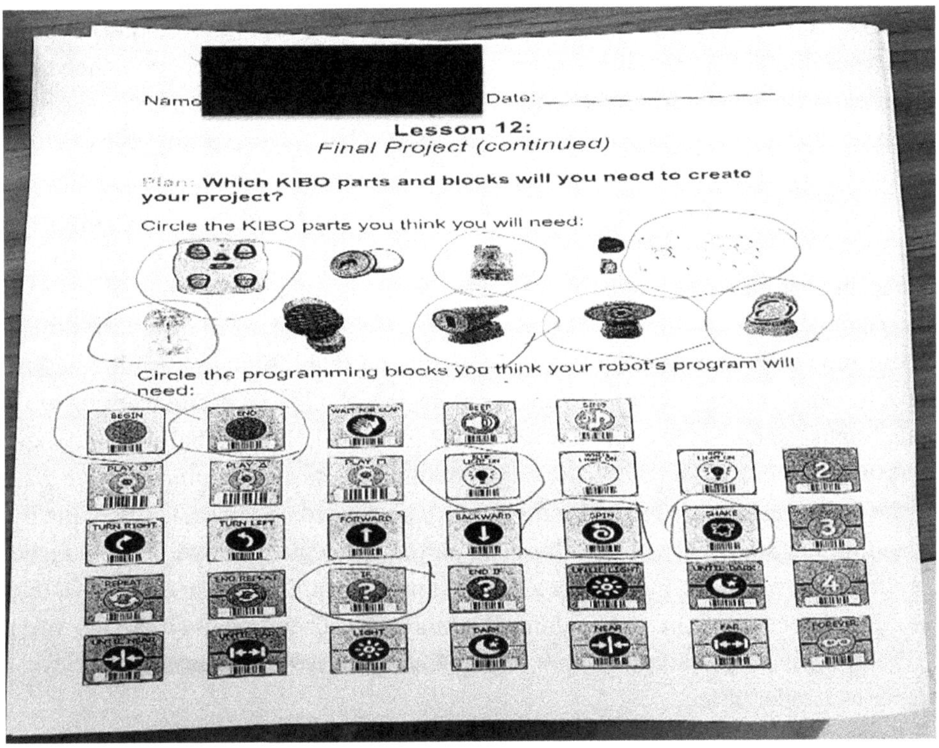

While Sophie's project planning sheet showed her technical understanding of the KIBO robotics kit and block programming language, the sheet also revealed that Sophie saw the KIBO programming language as a language that she could use, express in, and communicate with. Particularly noteworthy was that Sophie felt empowered to circle every sensor except the voice recorder. Ms. Locke created a classroom culture in which Sophie was included and her disability was accommodated, and Sophie was comfortable in this classroom to express herself using every accommodating aspect of the KIBO language while rejecting the unaccommodating aspects. Within the KIBO language, she was able to advocate for and accommodate her own needs, making the language work for her.

CONCLUSION

Sophie's classroom experience suggests that even young children with disabilities can access mainstream, constructionist computer science learning environments with classroom accommodations, and that this opportunity to explore the coding platform leads to creative expression and classroom communication using the coding language. Her planning sheet communicates that she felt empowered to use the KIBO robotics kit to build her "Gratitude Float." She communicated in her plan that the program would require use of all the robotic sensors, except one, the voice recorder. With the classroom accommodations provided by her teacher, she could compose and self-express using the KIBO robotics language and was able to write programs expressing and accommodating her individual needs.

Sophie's successful experience reinforces previous research on students with disabilities and computer science on how to incorporate Universal Design for Learning and other accommodations into computer science instruction (Israel et al., 2015). For example, Israel et al. suggest that teachers give students with disabilities roles within project groups that allow them to focus on their strengths, while altering expectations for the student as necessary (Israel et al., 2015). While working with Pete, Sophie scanned the code (a non-verbal task), while Pete verbally shared their work with the class. The constructionist tangible robotics curriculum used in Ms. Locke's class also used many of Israel et. al.'s suggested practices, for example by providing the students with a culturally-relevant project or including unplugged activities to provide for multiple means of action and expression (Israel et al., 2015). Sophie's success with this curriculum suggests that teachers can use these practices to create an inclusive and accommodating coding classroom even for students as young as Kindergarten.

Computer science education for students with disabilities is important. These students are entitled to equally access all educational opportunities as their non-disabled peers, including computer science education (IDEA, 2004). Although most students with disabilities do not have computer-science or robotics specific accommodations, previous research suggests applying the supports already in place for other classroom subjects will lead to successful learning outcomes in computer science for students with disabilities (Snodgrass et al., 2016). We saw this with Sophie, who was included, engaged, and ultimately successful in the student-centered tangible robotics curriculum because her teacher's existing supports allowed for alternate methods of communication. For other students, existing supports might include access to assistive technology, KIBO blocks modified to include braille, or multiple modes of providing instructions.

Recently, there have been increasing opportunities for students with disabilities to learn computer science and access computer science curricula. As mentioned earlier, organizations and initiatives such as AccessCSForAll and Deaf Kids Code are bringing computer science opportunities to more students with disabilities (deafkidscode.org, n.d.; Ladner & Israel, 2016). Educational programs in robotics and computational thinking are being developed and assessed for students with disabilities using traditional special education practices (Knight et al., 2019; Munoz et al., 2018; Taylor et al., 2017). As computer science education becomes more available to young children, students with disabilities have the right to learn these 21st century skills alongside their nondisabled peers. Our work with Sophie suggests even young students with disabilities can learn computer science in student-centered learning environments alongside their nondisabled peers, including experiencing the benefits of the student-centered computer science pedagogy. By expanding their existing supports to new computer science curricula, teachers can offer inclusive and exciting computer science opportunities to engage students with and without disabilities in new ways of thinking and expression.

ACKNOWLEDGMENT

This research was supported by the Templeton World Charity Foundation.

REFERENCES

Albo-Canals, J., Martelo, A. B., Relkin, E., Hannon, D., Heerink, M., Heinemann, M., Leidl, K., & Bers, M. U. (2018). A Pilot Study of the KIBO Robot in Children with Severe ASD. *International Journal of Social Robotics*, *10*(3), 371–383. doi:10.100712369-018-0479-2

American Speech-Language-Hearing Association. (n.d.). *Practice Portal: Clinical Topics: Selective Mutism.* American Speech-Language-Hearing Association. Retrieved February 15, 2020, from https://www.asha.org/Practice-Portal/Clinical-Topics/Selective-Mutism/#collapse_8

Bers, M. U. (2020). *Coding as a playground: Programming and computational thinking in the early childhood classroom*. Routledge. doi:10.4324/9781003022602

Bouck, E. C., & Yadav, A. (2020). Providing Access and Opportunity for Computational Thinking and Computer Science to Support Mathematics for Students With Disabilities. *Journal of Special Education Technology*. Advance online publication. doi:10.1177/0162643420978564

Congress, U. (1975). The Individuals with Disabilities Education Act–IDEA.

deafkidscode.org. (n.d.). *Our Story*. Retrieved February 15, 2020, from https://www.deafkidscode.org/our-story

Gage, N. A., Lierheimer, K. S., & Goran, L. G. (2012). Characteristics of Students With High-Incidence Disabilities Broadly Defined. *Journal of Disability Policy Studies, 23*(3), 168–178. doi:10.1177/1044207311425385

Israel, M., Wherfel, Q. M., Pearson, J., Shehab, S., & Tapia, T. (2015). Empowering K–12 Students With Disabilities to Learn Computational Thinking and Computer Programming. *Teaching Exceptional Children, 48*(1), 45–53. doi:10.1177/0040059915594790

Knight, V. F., Wright, J., & DeFreese, A. (2019). Teaching Robotics Coding to a Student with ASD and Severe Problem Behavior. *Journal of Autism and Developmental Disorders, 49*(6), 2632–2636. doi:10.100710803-019-03888-3 PMID:30734176

Ladner, R. E., & Israel, M. (2016). For all" in" computer science for all. *Communications of the ACM, 59*(9), 26–28. doi:10.1145/2971329

Munoz, R., Villarroel, R., Barcelos, T. S., Riquelme, F., Quezada, A., & Bustos-Valenzuela, P. (2018). Developing Computational Thinking Skills in Adolescents With Autism Spectrum Disorder Through Digital Game Programming. *IEEE Access: Practical Innovations, Open Solutions, 6*, 63880–63889. doi:10.1109/ACCESS.2018.2877417

National Center for Education Statistics. (2021). *Digest of Education Statistics: 2019*. U.S. Department of Education. https://nces.ed.gov/programs/digest/d19/

Pivetti, M., Di Battista, S., Agatolio, F., Simaku, B., Moro, M., & Menegatti, E. (2020). Educational Robotics for children with neurodevelopmental disorders: A systematic review. *Heliyon, 6*(10), e05160. doi:10.1016/j.heliyon.2020.e05160 PMID:33072917

Snodgrass, M. R., Israel, M., & Reese, G. C. (2016). Instructional supports for students with disabilities in K-5 computing: Findings from a cross-case analysis. *Computers & Education, 100*, 1–17. doi:10.1016/j.compedu.2016.04.011

Taylor, M. S. (2018). Computer Programming With Pre-K Through First-Grade Students With Intellectual Disabilities. *The Journal of Special Education, 52*(2), 78–88. doi:10.1177/0022466918761120

Taylor, M. S., Vasquez, E., & Donehower, C. (2017). Computer Programming with Early Elementary Students with Down Syndrome. *Journal of Special Education Technology, 32*(3), 149–159. doi:10.1177/0162643417704439

The Condition of Education: Students with Disabilities. (2020). National Center of Education Statistics. https://nces.ed.gov/programs/coe/indicator_cgg.asp

Viana, A. G., Beidel, D. C., & Rabian, B. (2009). Selective mutism: A review and integration of the last 15 years. *Clinical Psychology Review, 29*(1), 57–67. doi:10.1016/j.cpr.2008.09.009 PMID:18986742

ADDITIONAL READING

Bargagna, S., Castro, E., Cecchi, F., Cioni, G., Dario, P., Dell'Omo, M., Di Lieto, M. C., Inguaggiato, E., Martinelli, A., Pecini, C., & Sgandurra, G. (2019). Educational Robotics in Down Syndrome: A Feasibility Study. *Technology. Knowledge and Learning, 24*(2), 315–323. doi:10.100710758-018-9366-z

González-González, C. S., Herrera-González, E., Moreno-Ruiz, L., Reyes-Alonso, N., Hernández-Morales, S., Guzmán-Franco, M. D., & Infante-Moro, A. (2019). Computational Thinking and Down Syndrome: An Exploratory Study Using the KIBO Robot. *Informatics (MDPI), 6*(2), 25. doi:10.3390/informatics6020025

Israel, M., Ray, M. J., Maa, W. C., Jeong, G. K., Lee, C. E., & Lash, T. (2018). School-Embedded and District-Wide Coaching in K-8 Computer Science: Implications for Including Students with Disabilities. *Journal of Technology and Teacher Education, 26*(3), 471–501.

Israel, M., Jeong, G., Ray, M., & Lash, T. (2020). Teaching Elementary Computer Science through Universal Design for Learning. *Proceedings of the 51st ACM Technical Symposium on Computer Science Education*, 1220–1226. 10.1145/3328778.3366823

Knight, V. F., Wright, J., Wilson, K., & Hooper, A. (2019). Teaching Digital, Block-Based Coding of Robots to High School Students with Autism Spectrum Disorder and Challenging Behavior. *Journal of Autism and Developmental Disorders, 49*(8), 3113–3126. doi:10.100710803-019-04033-w PMID:31055684

Ladner, R. E., Stefik, A., Naumann, J., & Peach, E. (2020). Computer Science Principles for Teachers of Deaf Students. *2020 Research on Equity and Sustained Participation in Engineering, Computing, and Technology (RESPECT)*, 1–4.

Stefik, A., Ladner, R. E., Allee, W., & Mealin, S. (2019). Computer Science Principles for Teachers of Blind and Visually Impaired Students. *Proceedings of the 50th ACM Technical Symposium on Computer Science Education*, 766–772. 10.1145/3287324.3287453

Wille, S., Century, J., & Pike, M. (2017). Exploratory Research to Expand Opportunities in Computer Science for Students with Learning Differences. *Computing in Science & Engineering*, *19*(3), 40–50. doi:10.1109/MCSE.2017.43

KEY TERMS AND DEFINITIONS

Constructionism: A student-directed pedagogy in which students' learning is self-directed based on individual questions and interests.

Explicit Instruction: A structured, teacher-directed pedagogy in which teachers provide direct instruction to students, provide students with a scaffolded learning environment, and assess student learning based on correctness of answers.

General Education Environment: The learning environment (including curriculum, teachers, standards, social environment, and physical environment) provided to children without disabilities.

High-Incidence Disability: A category of disabilities that includes specific learning disorders, speech or language impairments, ADHD, and emotional and behavioral disabilities.

Individuals With Disabilities in Education Act: The law that mandates special education services be provided to students with disabilities, and that students with disabilities are entitled to a free appropriate public education in the least restrictive learning environment.

Low-Incidence Disability: A category of disabilities that affect learning across domains, such as significant sensory impairments or significant cognitive impairments.

Special Education Services: Services provided by the school or school district to support students with disabilities, including special education teachers, paraprofessionals, and specialized curricula.

ENDNOTE

1 All names are pseudonyms.

Section 4
Evaluation

Chapter 13
TechCheck:
Creation of an Unplugged Computational Thinking Assessment for Young Children

Emily Relkin
Tufts University, USA

ABSTRACT

This chapter describes the development and validation of TechCheck, a novel instrument for rapidly assessing computational thinking (CT) skills in 5-9 years old children. TechCheck assessments can be administered in classroom or online settings regardless of whether students have prior knowledge of coding. This assessment probes six domains of CT described by Bers (2018) as developmentally appropriate for young children including algorithms, modularity, control structures, representation, hardware/software, and debugging. TechCheck demonstrates good psychometric properties and can readily distinguish among young children with different CT abilities.

INTRODUCTION

Numerous studies have demonstrated that children as young as four years of age are capable of learning to code (Kazakoff & Bers, 2014; Bers, 2018; Clements & Gullo, 1984; Strawhacker & Bers, 2018). In the process of acquiring coding skills, children often simultaneously develop a set of thought processes known as computational thinking (CT) that are useful for framing and solving problems using computers and other technologies (Wing, 2006; Wing 2011). CT is valuable for coding but also applicable to other disciplines including problem-solving in everyday

DOI: 10.4018/978-1-7998-7308-2.ch013

life. Promoting the acquisition of CT is accordingly one of the goals of computer science (CS) education (Grover & Pea, 2013; Lye & Koh, 2014; NRC, 2011). One of the challenges to achieving this goal has been the lack of availability of suitable instruments for assessing CT skills in young children (Lockwood & Mooney, 2018; Grover & Pea, 2013; Lee et al., 2011; Román-González et al., 2019). A reliable and validated CT assessment tool can be used to monitor young students' CT progress and allow educators to gauge the effectiveness of CS lessons. CT assessment can also be used to identify students in need of extra support as well as those with exceptional talents. CT assessment can provide new insights into how children's CT abilities develop and can assist in the development of new curricula and best practices for CS education.

Challenges of CT Assessment in Young Children

Assessing CT in young children requires taking into account stages of cognitive development. A young child's literacy, numeracy, and abstract reasoning undergo gradual development (Piaget, 1971). Their developmental stage can impact their ability to understand certain CS concepts and their readiness for CT assessment (Chen et al., 2017). A kindergarten student may not be able to fully understand CT principles such as "if-then" conditionals (Barrouillet & Lecas, 1999; Janveau-Brennan & Markovits, 1999; Muller et al., 2001). Aspects of abstract representations such as programming variables may be inaccessible to them. They may express magical thinking as an explanation for the action of computers and other technology (Flavell et al., 1993; Mioduser et al., 2009). These and other constraints may affect the design and implementation of CT assessments for young children.

Instruments for assessing CT in older students and adults have existed for some time (Fraillon at al., 2018; Werner at al., 2012; Chen at al., 2017). A common approach involves the use of coding exercises that are designed to elicit the same type of logic and reasoning that is involved in programming. However, coding-based assessments require prior knowledge of a coding language and can conflate coding ability with CT skills (Yadav et al., 2017). Assessments that require knowledge of coding cannot readily be used to assess baseline CT abilities in coding-naive students. In addition, research with older children has indicated that coding can become automatic. Therefore, coding exercises may not be the most effective way to probe CT (Werner at al., 2014).

It is advantageous to be able to measure CT skills in children regardless of whether they have past knowledge or experience with computer programming (Grover et al., 2014). With this in mind, our group began exploring the use of code-free instruments to assess CT skills in children. Our basis for creating a coding-free CT assessment was the realization that CT is exercised in the context of many "unplugged" activities

(Bell & Vahrenhold, 2018; Zapata-Cáceres et al., 2020). Unplugged activities typically involve puzzles, games and exercises that exemplify CS concepts without requiring explicit knowledge of coding or the use of computers (see Relkin & Strawhacker, Chapter 3). Unplugged activities have been used to teach CS concepts for over two decades (e.g., CSUnplugged.com; code.org) and in recent years have started to be used for the purposes of assessment. It has been argued that unplugged assessments offer advantages because they do rely on a particular computer language or curricula and are therefore purer reflections of CT abilities (Dagiene & Futschek, 2008).

Conceptual Foundations for Unplugged CT Assessment

Unplugged CT activities involve the participation in tasks that exercise analogous thinking processes to those involved in CT. However, there is a lack of consensus about the precise definition of CT and its subdomains. A number of definitions have been put forth, most of which place CT outside of the context of early childhood. (Aho, 2012; Barr & Stephenson, 2011; Cuny at al., 2010; Grover & Pea, 2013; Kalelioğlu at al., 2016; Lu & Fletcher, 2009; Shute at al., 2017; Wing, 2006; Wing, 2008; Tang at al., 2020). To operationalize unplugged CT assessment, it is important to use a conceptual framework that is developmentally appropriate for young children.

To identify CT's cognitive subdomains in young children, Bers (2018) drew on Papert's definition of "powerful ideas" as skills within a domain or discipline that are individually meaningful and change how we think or perceive the world and problem solve (Papert, 1980). This led to the formation of the "Seven Powerful Ideas" that operationalize CT in a developmentally appropriate way that can be taught to young children through a CS or robotics curriculum (Bers, 2018). These powerful ideas include algorithms, modularity, control structures, representation, hardware/software, design process, and debugging (see table 1). To create a coding-free assessment of CT, we identified unplugged activities that were associated with six of these domains and confirmed the associations through the consensus of a panel of child development and computer science professionals.

Table 1. This table describes each of the developmentally appropriate seven powerful ideas of Computer Science that were selected by Bers (2018) and used as domains of CT in TechCheck

Bers' Seven Powerful Ideas	Description
Algorithms	A step-by-step sequential process that helps achieve a goal/task
Modularity	Breaking up large tasks into smaller parts; reusing modules
Control Structures	Recognizing patterns and repetition, cause and effect
Representation	Symbolic representation, forming models
Hardware/Software	Understanding that smart objects are not magical but are human-engineered
Debugging	Finding and fixing problems, troubleshooting
Design Process	An iterative and creative process often involving perseverance

What Features Should a CT Assessment for Young Children Ideally Possess?

Assessment of young children requires careful attention to certain elements of design and content selection that differ from those involved in creating comparable instruments for adults (Goldstein & Flake, 2016). Assessment instruments for young children must use developmentally appropriate language, symbols and tasks to assure that factors such as literacy, numeracy and fine motor skills are not limiting (Chen et al., 2017; Sattler, 2014). Activities and artifacts employed must be familiar and non-threatening to young children and as free as possible from cultural biases (Tang et al., 2020; McMillan, 2013; Mullis & Martin, 2019). In light of the shorter attention span of young children (Moyer & Gilmer, 1953) and the likelihood of test fatigue, the duration of the assessment must be kept sufficiently brief to allow routine use in educational settings (Basu et al., 2016; Werner et al., 2014; Chen et al., 2017). Educational assessments should not be so lengthy and or complex that teachers are unable to administer them in routine classroom settings. Likewise, it may be impractical to require that assessment be carried out one-on-one because of the constraints of class time. Ideally, teachers who are not particularly skilled at coding themselves should be able to administer CT assessments. As emphasized above, the administration should be possible regardless of the student's past programming experience. The rating system employed should use simple outcome categories and/or numeric scores that are straightforward to calculate and interpret (Koretz et al., 1992). A CT assessment for children should demonstrate good construct and face validity (essentially, confirmation that the assessment measures what it was designed to measure) as well as acceptable inter-rater reliability and sensitivity to change. These and other considerations can make the creation of assessments for early childhood a particularly challenging enterprise (Snow et al., 2008).

Features of the *TechCheck* CT Assessment

TechCheck is a 15 item, multiple-choice, unplugged CT assessment that was created at Tufts University in 2018 for children ages 5-9. The assessment is easy to administer to students regardless of their familiarity with coding. The concepts behind *TechCheck* followed from experience with a platform-specific CT assessment instrument called TACTIC-KIBO that we previously designed. TACTIC-KIBO required one-on-one administration as well as familiarity with the KIBO robotics platform (Relkin, 2018; Relkin & Bers, 2019). In contrast, *TechCheck* is platform-independent, meaning it does not require knowledge of programming or the use of a computer to be completed. Six of the "Seven Powerful Ideas" of computer science (Bers, 2018) that are developmentally appropriate for early childhood are probed by this instrument. The remaining powerful idea, Design Process, was not included in the assessment because it is an iterative and open-ended process that does not lend itself to a multiple-choice format. A multiple-choice format was used because it simplifies data collection, helps make scoring more objective and facilitates administration to large numbers of students simultaneously.

TechCheck can be administered one-on-one, to whole classrooms, or even to geographically distributed groups. It can be printed out or displayed on a device (the mode of presentation does not alter the "unplugged" nature of the assessment content). A proctor reads out loud the stem (question/challenge) for each of the 15 items. Children respond by checking off or clicking on one of the answer options. The children can understand answer options as they have little to no text in them (see Table 2). Each correct response is awarded one point, with a maximum total score of 15 points. Two practice questions are included at the beginning of the assessment to familiarize students with the format but are not included in the scoring. This is a "forced choice" assessment, meaning all questions must be answered. Students are instructed to guess if they do not know the answer and informed that there is no penalty for guessing. The administration time varies by grade but in most cases, *TechCheck* can be completed in under 20 minutes. The use of an answer key makes scoring straightforward. Mean scores and distribution by grade have were determined in nearly 800 students first and second grade.

Figure 1. A Density plot of scores on original TechCheck for first and second grade. A slight ceiling effect is evident second graders

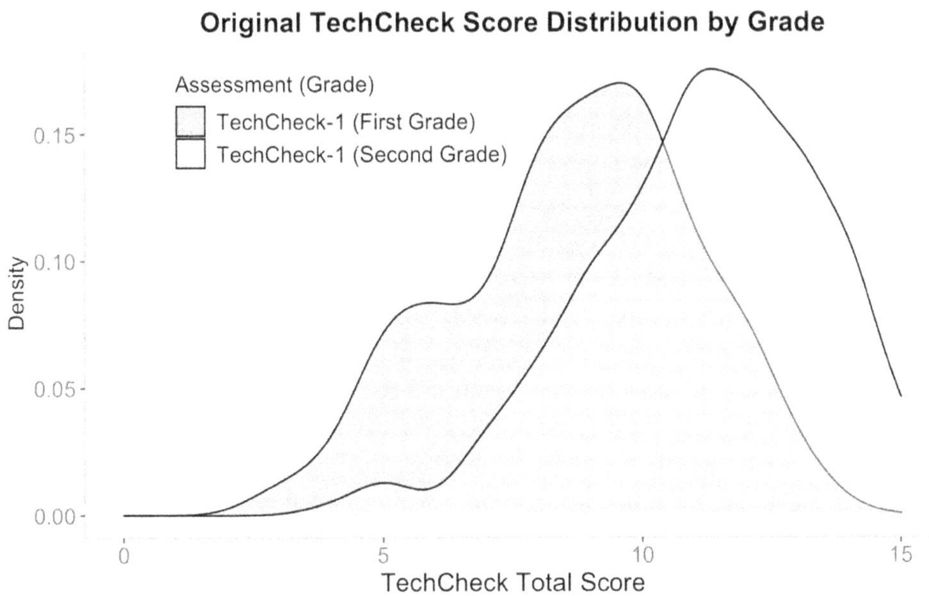

Design and Validation

After developing prototypes of *TechCheck* questions, we assembled a group of nineteen evaluators (CS researchers, educators and students) with various levels of expertise in CT to judge whether or not the questions embodied the domains of Bers' Seven Powerful Ideas. Inter-rater agreement was then assessed. There was an average agreement of 81% among raters. Fleiss' Kappa indicated consensus among evaluators about the CT domain most associated with each question $\kappa = 0.63$ (95% CI) $p < 0.001$. Although all prototypes were judged to probe the intended CT domains, some questions were rejected because their content was judged to fall outside the common knowledge base of typical 5-to-9-year-olds (Relkin et al., 2020).

We initially validated *TechCheck* in a cohort of first and second graders participating in a research study involving the CAL-KIBO curriculum. *TechCheck* showed good reliability and validity according to measures of classical test theory (CTT) and item response theory (IRT). CTT and IRT are models that are commonly used to examine items on an assessment and individual responses to better understand their relationship to the underlying concept being measured (Kingsbury & Weiss 1983). *TechCheck's* discrimination between skill levels was found to be adequate. The difficulty was suitable for first graders and low for second graders. *TechCheck* scores correlated moderately with a previously validated CT assessment tool (TACTIC-KIBO).

In a prospective longitudinal study, Relkin et al., (2021) used *TechCheck* to compare children receiving the CAL-KIBO programming curriculum (*N*=667*)* to a control group (*N*=181) who participated in typical classroom activities without coding (No-CAL). A sequential regression showed TACTIC-KIBO scores and a child's baseline *TechCheck* score predicted the endpoint *TechCheck* score (Relkin & Bers, 2020). Over the course of the study, children who received CAL-KIBO improved on *TechCheck* (M_{change}= 0.94, p<0.001) whereas the No-CAL group did not change significantly (M_{change}=0.27, *p*=.07). This change equated to the change in *TechCheck* scores estimated to occur over approximately 6 months of typical development. Generalized Linear Mixed Model (GLMM) and Bayesian analyses revealed that exposure to the CAL-KIBO curriculum predicted the *TechCheck* outcome score, taking into account differences in baseline *TechCheck* performance and other demographic and environmental effects. Children who received CAL-KIBO showed the most improvement in the CT domains algorithms, modularity, representation. The significant changes observed in *TechCheck* scores in this study demonstrates the utility of this measure for longitudinal assessment.

Although results using *TechCheck* in first and second graders were encouraging, there was a noticeable ceiling effect in the second-grade cohort manifesting in a smaller window to observe change compared to the first-grade cohort. The assessment for second grade children ages 7-9 was subsequently modified to increase item difficulty. We conducted an item analysis of all the questions and modified those that had low difficulty, discrimination, and/or point biserial correlations. In the initial pilot test of the assessment, five experimental questions were also added to test whether any of the newer questions performed better than the previous ones. All modifications to original questions and three of the five experimental questions were included in the final version of *TechCheck-2.*

When we began administering *TechCheck* to kindergarten students, we realized that further modifications were required. Previous research has shown that the working memory of children of kindergarten age (~5 years old) limits them to hold an average of three items in immediate memory, compared to children in first and second grade (~6-9 years old) who can hold an average of four items (Cowan, 2016; Simmering, 2012). This limit can potentially impact kindergartener's performance on multiple-choice assessments. Consequently, we reduced the number of response options from four to three in *TechCheck-K* (the Kindergarten version). We accomplished this by systematically eliminating one of a pair of distractors for each item that had close to the same response probabilities. We followed this procedure in an effort to maintain the overall difficulty and discrimination levels on a par with the original version of *TechCheck*. An example of a *TechCheck* question for each of the three grade levels is shown in Table 2. The impact of *TechCheck-K* and *TechCheck-2* on the score distributions by grade are shown in Figure 2.

Table 2. Sample Debugging Symmetry Problem and Algorithms Missing Symbol Series questions for kindergarteners, first graders, and second graders

TechCheck-K Debugging	*TechCheck-K Algorithms*
This seesaw isn't going up and down. How can it be changed so it works? 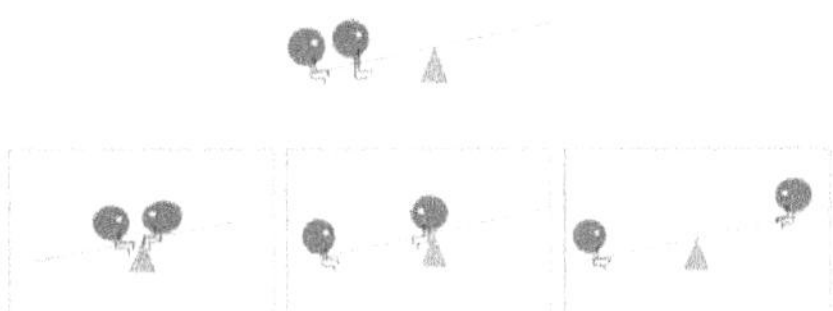	What comes next? 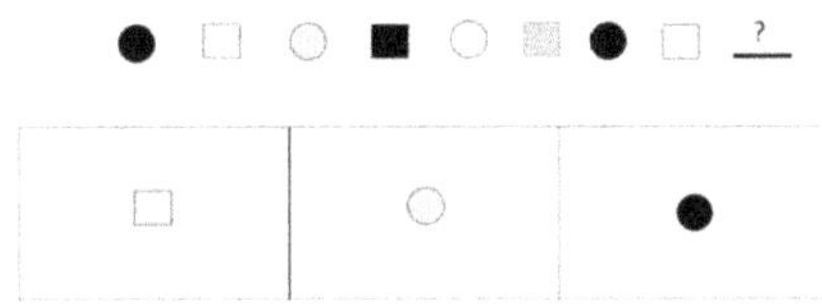

TechCheck-1 Debugging	*TechCheck-1 Algorithms*
This seesaw isn't going up and down. How can it be changed so it works? 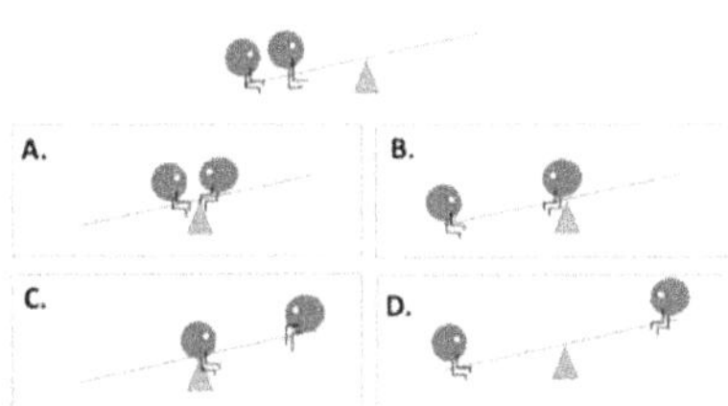	What comes next? 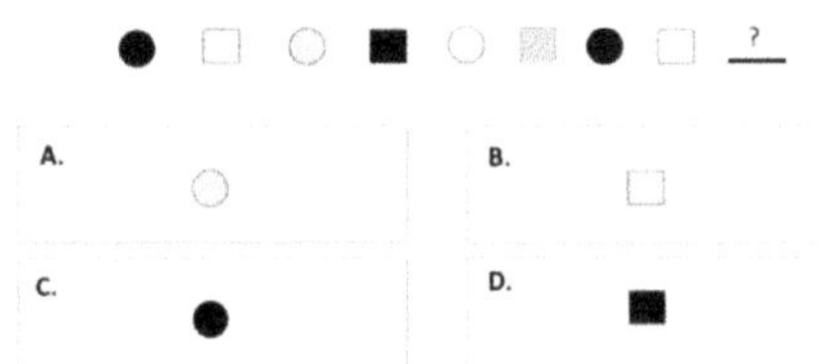

TechCheck-2 Debugging	*TechCheck-2 Algorithms*
This seesaw isn't going up and down. How can it be changed so it works? 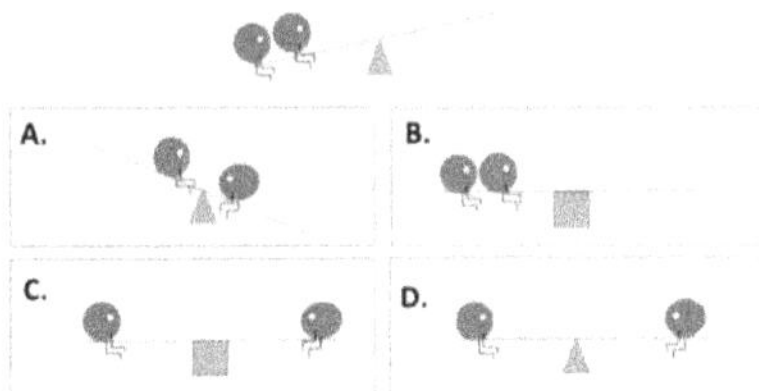	What comes next? 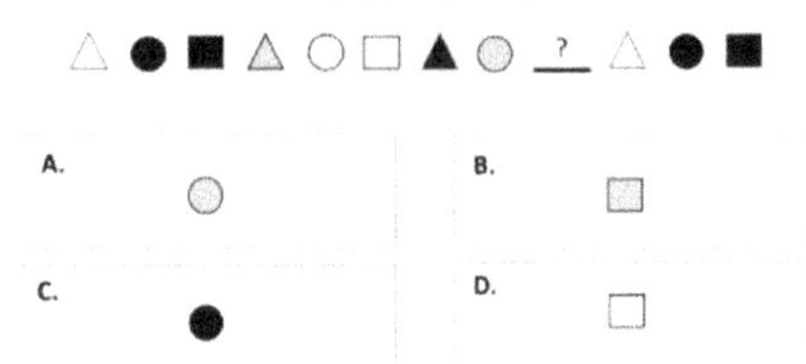

Figure 2. This figure shows a density plot with the distribution of TechCheck scores by grade for the revised and updated formats of the assessment

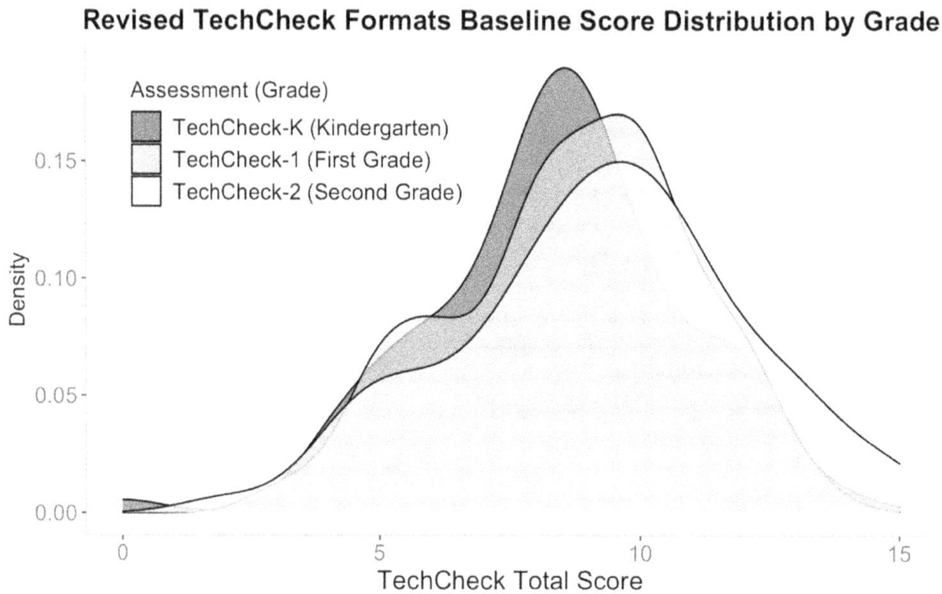

REFLECTIONS AND FUTURE DIRECTIONS

Our experience to date in the development, validation and research application of *TechCheck* has been gratifying. *TechCheck* has been successfully administered in a variety of formats including in-person or remotely, online and on paper, to groups of students and individuals in many countries. The instrument has been translated into several languages in addition to English (e.g., Spanish, Turkish, Chinese) and is currently being used in a variety of educational and research settings around the world.

Despite these successes, *TechCheck* can be further improved. The multiple-choice format of the instrument does not lend itself to creative self-expression and open-ended problem solving which is a significant part of CT. We are currently examining other testing formats including more naturalistic and open-ended formats that may better address this shortcoming. Román-González at al., (2019) pointed out that CT assessments often focus on concepts rather than "practices and perspectives", and as a consequence become "static and decontextualized." To address this concern, strategies such as game-based assessment may offer a window on CT applied to real-time problem-solving. Currently, *TechCheck* is designed for K-2nd grades. There is potential to extend this to other grades, including preschool and higher elementary grade levels. Future studies should also explore whether the assessment can be used with neuro-diverse children and other contexts.

ACKNOWLEDGMENT

I would like to thank the principal investigator of this project, Marina Bers as well as our Project Coordinators, Angela de Mik and Megan Bennie for their many contributions to this work. I would like to thank my many colleagues at the DevTech research group at Tufts University who assisted in the development and field testing of *TechCheck* . Lastly, I would like to thank all of the educators, parents, and students who participated in this study. This work was funded by the Department of Defense Education Activity (DoDEA) grant entitled "Operation: Break the Code for College and Career Readiness." Unique Entity Identifier: "WORLDCL10" and the Department of Education PR/Award Number: U411C190006

REFERENCES

Barrouillet, P., & Lecas, J. (1999). Mental models in conditional reasoning and working memory. *Thinking & Reasoning*, *5*(4), 289–302. doi:10.1080/135467899393940

Basu, S., Biswas, G., Sengupta, P., Dickes, A., Kinnebrew, J. S., & Clark, D. (2016). Identifying middle school students' challenges in computational thinking-based science learning. *Research and Practice in Technology Enhanced Learning*, *11*(1), 13. doi:10.118641039-016-0036-2 PMID:30613246

Bell, T., & Vahrenhold, J. (2018). CS Unplugged—How Is It Used, and Does It Work? In H.-J. Böckenhauer, D. Komm, & W. Unger (Eds.), Adventures Between Lower Bounds and Higher Altitudes: Essays Dedicated to Juraj Hromkovič on the Occasion of His 60th Birthday. doi:10.1007/978-3-319-98355-4_29

Bers, M. U. (2018). *Coding as a playground: Programming and computational thinking in the early childhood classroom*. Routledge.

Chen, G., Shen, J., Barth-Cohen, L., Jiang, S., Huang, X., & Eltoukhy, M. (2017). Assessing elementary students' computational thinking in everyday reasoning and robotics programming. *Computers & Education*, *109*, 162–175. doi:10.1016/j.compedu.2017.03.001

Clements, D. H., & Gullo, D. F. (1984). Effects of Computer Programming on Young Children's Cognition. *Journal of Educational Psychology, 76*(6), 1051–1058. doi:10.1037/0022-0663.76.6.1051

Code.org. (2019). Retrieved from https://code.org/

Dagiene, V., & Stupurienė, G. (2016). Bebras–a sustainable community building model for the concept based learning of informatics and computational thinking. *Informatics in Education, 15*(1), 25–44. . doi:10.15388/infedu.2016.02

Flavell, J. H., Miller, P. H., & Miller, S. A. (1993). *Cognitive development* (3rd ed.). Prentice Hall.

Fraillon, J., Ainley, J., Schulz, W., Duckworth, D., & Friedman, T. (2018). *International Computer and Information Literacy Study*. ICILS 2018: Technical Report.

Goldstein, J., & Flake, J. K. (2016). Towards a framework for the validation of early childhood assessment systems. *Educational Assessment, Evaluation and Accountability, 28*(3), 273–293. doi:10.100711092-015-9231-8

Grover, S., Cooper, S., & Pea, R. (2014). Assessing computational learning in K-12. In *Proceedings of the 2014 conference on Innovation & technology in computer science education* (pp. 57-62). ACM. 10.1145/2591708.2591713

Grover, S., & Pea, R. (2013). Computational thinking in K–12: A review of the state of the field. *Educational Researcher, 42*(1), 38–43. doi:10.3102/0013189X12463051

Janveau-Brennan, G., & Markovits, H. (1999). The development of reasoning with causal conditionals. *Developmental Psychology, 35*(4), 904–911. doi:10.1037/0012-1649.35.4.904 PMID:10442860

Kazakoff, E. R., & Bers, M. U. (2014). Put your robot in, Put your robot out: Sequencing through programming robots in early childhood. *Journal of Educational Computing Research, 50*(4), 553–573. doi:10.2190/EC.50.4.f

Kingsbury, G. G., & Weiss, D. J. (1983). A comparison of IRT-based adaptive mastery testing and a sequential mastery testing procedure. In *New horizons in testing* (pp. 257–283). Academic Press. doi:10.1016/B978-0-12-742780-5.50024-X

Koretz, D., McCaffrey, D. F., Klein, S. P., Bell, R. M., & Stecher, B. M. (1992). *The Reliability of Scores from the 1992 Vermont Portfolio Assessment Program*. Academic Press.

Lee, I., Martin, F., Denner, J., Coulter, B., Allan, W., Erickson, J., Malyn-Smith, J., & Werner, L. (2011). Computational thinking for youth in practice. *ACM Inroads*, *2*(1), 32–37. doi:10.1145/1929887.1929902

Lockwood, J., & Mooney, A. (2018). Computational Thinking in education: Where does it fit? A systematic literary review. *International Journal of Computer Science Education in Schools*, *2*(1), 41–60. doi:10.21585/ijcses.v2i1.26

Lye, S. Y., & Koh, J. H. L. (2014). Review on teaching and learning of computational thinking through programming: What is next for K-12? *Computers in Human Behavior*, *41*, 51–61. doi:10.1016/j.chb.2014.09.012

McMillan, J. H. (2013). *Classroom assessment: Principles and practice for effective instruction* (6th ed.). Pearson/Allyn and Bacon.

Mioduser, D., Levy, S. T., & Talis, V. (2009). Episodes to scripts to rules: Concrete-abstractions in kindergarten children's explanations of a robot's behavior. *International Journal of Technology and Design Education*, *19*(1), 15–36. doi:10.100710798-007-9040-6

Moyer, K., & Gilmer, B. V. H. (1953). The Concept of Attention Spans in Children. *The Elementary School Journal*, *54*(1), 464–466. doi:10.1086/458623

Mullis, I. V., & Martin, M. O. (2019). *PIRLS 2021 Assessment Frameworks*. International Association for the Evaluation of Educational Achievement. Retrieved from https://eric.ed.gov/?id=ED606056

National Research Council. (2011). *Report of a workshop on the pedagogical aspects of computational thinking*. National Academies Press.

Piaget, J. (1971). Developmental stages and developmental processes. In D. R. Green, M. P. Ford, & G. B. Flamer (Eds.), *Measurement and Piaget* (pp. 172–188). McGraw-Hill.

Relkin, E. (2018). *Assessing young children's computational thinking abilities* (Master's thesis). Retrieved from ProQuest Dissertations and Theses database. (UMI No. 10813994)

Relkin, E., & Bers, M. (2021). *TechCheck-K: A Measure of Computational Thinking for Kindergarten Children. In 2021 IEEE Global Engineering Education Conference (EDUCON)*. IEEE. Retrieved from https://sites.tufts.edu/devtech/files/2021/05/1487.pdf

Relkin, E., & Bers, M. U. (2019). Designing an Assessment of Computational Thinking Abilities for Young Children. In L. E. Cohen & S. Waite-Stupiansky (Eds.), *STEM for Early Childhood Learners: How Science, Technology, Engineering and Mathematics Strengthen Learning* (pp. 85–98). Routledge. doi:10.4324/9780429453755-5

Relkin, E., & Bers, M. U. (2020). *Exploring the Relationship Among Coding, Computational Thinking, and Problem Solving in Early Elementary School Students* [Symposium]. Annual Meeting of the American Educational Research Association (AERA), San Francisco, CA.

Relkin, E., de Ruiter, L., & Bers, M. U. (2020). TechCheck: Development and Validation of an Unplugged Assessment of Computational Thinking in Early Childhood Education. *Journal of Science Education and Technology*, *29*(4), 482–498. Advance online publication. doi:10.100710956-020-09831-x

Relkin, E., de Ruiter, L., & Bers, M. U. (2021). Learning to Code and the Acquisition of Computational Thinking by Young Children. *Computers & Education*, *169*, 104222. Advance online publication. doi:10.1016/j.compedu.2021.104222

Román-González, M., Moreno-León, J., & Robles, G. (2019). Combining Assessment Tools for a Comprehensive Evaluation of Computational Thinking Interventions. In *Computational Thinking Education* (pp. 79–98). Springer. doi:10.1007/978-981-13-6528-7_6

Sattler, J. M. (2014). *Foundations of behavioral, social and clinical assessment of children*. Jerome M. Sattler, Publisher, Incorporated.

Snow, C. E., Van Hemel, S. B., & Committee on Developmental Outcomes Assessments for Young Children. (2008). *Early childhood assessment: Why, what, and how*. Washington, DC: National Academies Press.

Strawhacker, A., & Bers, M. U. (2018). What they learn when they learn coding: Investigating cognitive domains and computer programming knowledge in young children. *Educational Technology Research and Development*. Advance online publication. doi:10.100711423-018-9622-x

Tang, X., Yin, Y., Lin, Q., Hadad, R., & Zhai, X. (2020). Assessing computational thinking: A systematic review of empirical studies. *Computers & Education*, *148*, 103798. doi:10.1016/j.compedu.2019.103798

Werner, L., Denner, J., & Campe, S. (2014). Using computer game programming to teach computational thinking skills. *Learning, Education And Games*, *37*. Retrieved from https://dl.acm.org/citation.cfm?id=2811150

Werner, L., Denner, J., Campe, S., & Kawamoto, D. C. (2012). The fairy performance assessment: measuring computational thinking in middle school. *Proceedings of the 43rd ACM Technical Symposium on Computer Science Education*, 215–220. 10.1145/2157136.2157200

Wing, J. (2011). Research notebook: Computational thinking—What and why? *The Link Magazine*. Retrieved from https://www.cs.cmu.edu/link/research-notebookcomputational-thinking-what-and-why

Wing, J. M. (2006). Computational thinking. *Communications of the ACM*, *49*(3), 33–35. doi:10.1145/1118178.1118215

Yadav, A., Good, J., Voogt, J., & Fisser, P. (2017). Computational thinking as an emerging competence domain. In Technical and vocational education and training (Vol. 23, pp. 1051–1067). doi:10.1007/978-3-319-41713-4_49

Zapata-Cáceres, M., Martín-Barroso, E., & Román-González, M. (2020). Computational Thinking Test for Beginners: Design and Content Validation. In *2020 IEEE Global Engineering Education Conference (EDUCON)* (pp. 1905-1914). IEEE. 10.1109/EDUCON45650.2020.9125368

ADDITIONAL READING

Brennan, K., Haduong, P., & Veno, E. (2020). *Assessing Creativity in Computing Classrooms*. Creative Computing Lab.

Clarke-Midura, J., Silvis, D., Shumway, J. F., Lee, V. R., & Kozlowski, J. S. (2021). Developing a kindergarten computational thinking assessment using evidence-centered design: The case of algorithmic thinking. *Computer Science Education*, *31*(2), 1–24. doi:10.1080/08993408.2021.1877988

Hogenboom, S. A., Hermans, F. F., & Van der Maas, H. L. (2021). Computerized adaptive assessment of understanding of programming concepts in primary school children. *Computer Science Education*, 1–30. doi:10.1080/08993408.2021.1914461

Relkin, E., & Bers, M. U. (2021). Factors Influencing Learning of Computational Thinking Skills in Young Children. Virtual Annual Meeting of the American Educational Research Association (AERA).

Román-González, M., Pérez-González, J. C., & Jiménez-Fernández, C. (2017). Which cognitive abilities underlie computational thinking? Criterion validity of the Computational Thinking Test. *Computers in Human Behavior*, *72*, 678–691. doi:10.1016/j.chb.2016.08.047

Shute, V. J., Sun, C., & Asbell-Clarke, J. (2017). Demystifying computational thinking. *Educational Research Review*, *22*, 142–158. doi:10.1016/j.edurev.2017.09.003

Zhang, L., & Nouri, J. (2019). A systematic review of learning computational thinking through Scratch in K-9. *Computers & Education*, *141*, 103607. doi:10.1016/j.compedu.2019.103607

KEY TERMS AND DEFINITIONS

CAL-KIBO: A KIBO robotics coding curriculum appropriate for children in preschool through second grade that combines teaching programming skills, self-expression, and literacy.

KIBO: A screen-free programmable robotics kit for young children with blocks, sensors, modules, and art platforms.

Powerful Ideas: Skills within a domain or discipline that are individually meaningful and change how we think or perceive the world and problem solve. Bers' seven powerful operationalize domains of CT that are developmentally appropriate for young children.

ScratchJr: A free block-based programming application for young children.

TACTIC-KIBO: A CT measure that requires knowledge of coding with the KIBO robot. The assessment classifies children into one of four programming proficiency levels.

TechCheck: A "unplugged" assessment of Computational Thinking (CT) designed for children in kindergarten, first and second grades. *TechCheck- K* is for kindergarteners, *TechCheck-1* is for first graders, and *TechCheck-2* is for second graders The *TechCheck* assessments can be used to assess CT regardless of a child's familiarity with coding.

Unplugged: Describes activities such as games and puzzles that aid the teaching and learning of computer science but without requiring the use of computers and other technologies.

Chapter 14
Examining Young Children's Computational Artifacts

Apittha Unahalekhaka
Tufts University, USA

Madhu Govind
Tufts University, USA

ABSTRACT

Computational thinking (CT), in line with the constructionist perspective, is often best displayed when children have the opportunity to demonstrate their skills by producing creative coding artifacts. Performance-based or project portfolio assessments of young children's coding artifacts are a rich and useful approach to explore how children develop and apply CT abilities. In this chapter, the authors examine various rubrics and assessment tools used to measure the levels of programming competency, creativity, and purposefulness displayed in students' coding artifacts. The authors then discuss the development of ScratchJr and KIBO project rubrics for researchers and educators, including examples to illustrate how these highly diverse projects provide insight into children's CT abilities. Finally, the authors conclude with implications and practical strategies for using rubrics in both educational and research settings.

INTRODUCTION

Elisa is in first grade. She wants to create a ScratchJr project where she throws a birthday party in space and invites her classmates to eat a birthday cake with aliens. Elisa has a complex plan to include multiple scenes in her project. Using the paint

DOI: 10.4018/978-1-7998-7308-2.ch014

editor tool on ScratchJr, she customizes her characters: a girl who looks like and represents herself, a rocket, aliens, a birthday cake, and friends. The first scene is at the street in front of her house. Elisa starts her program with the green flag. A voice recording plays, "Please come to my birthday party," after which a purple rocket flies off into space. As soon as the rocket lands on the moon (second scene), character-Elisa and her friends are greeted by an alien that looks like a jellyfish. This jellyfish brings out a blue gigantic birthday cake that gets bigger every time Elisa taps on the cake. After the cake triples in size, it jumps, makes a pop noise and spins away. The project ends in a bedroom (third scene) where character-Elisa wakes up, realizes her space birthday party is all a dream, and exclaims with a text bubble, "That was a strange dream I had!"

In the kindergarten classroom down the hall, Shiro sits down with his KIBO robotics kit, excited to return to his final project. His teacher had just read aloud the book *Pete the Cat: Robo-Pete* and tasked the class with a final project to create their own KIBO robot-friend. Shiro wonders, "What will my KIBO look like? I like to play soccer, so I want KIBO to play soccer with me." He looks at the programming blocks and begins to assemble a program: Begin, Forward, Turn Left, Turn Right, Shake, End. Shiro scans each block carefully using the KIBO robot's embedded barcode scanner and then runs his program. Shiro's teacher comes to check on his progress and asks Shiro about his project idea. Shiro explains how his robot-friend is moving around on the soccer field to kick the soccer ball into the goal. Shiro's teacher comments, "That's a neat idea! Is there anything you could add to show that KIBO made the goal?" Shiro thinks for a moment and looks over at his blocks. He notices the lightbulb module, and a light goes off in his head. He responds, "I'm going to add a white light at the end to show that KIBO made the goal!" He places the White Light On block between the Shake and End blocks, inserts the lightbulb module into one of KIBO's ports, and scans his revised program. He exclaims, "I love it! My KIBO scored the winning goal, hooray!"

Millions of computational artifacts like Elisa's and Shiro's have been created, remixed, and shared all around the world. Each of these projects is special in its own regard comprising unique sequences of coding blocks, but they all share one thing in common: each project reflects something about the child's ability to think computationally. In Chapter 1 of this book, Dr. Bers writes, "As children make computational media, they develop computational thinking (CT). This involves more than just problem-solving or logical thinking; it means gaining the concepts, skills, and habits of mind to express themselves through coding." In this chapter we will explore this notion further by examining how children develop and display their CT abilities through producing creative and personally meaningful projects.

We begin this chapter by presenting existing literature on the assessment of children's coding projects. We then introduce two rubrics we developed for ScratchJr and KIBO that assess the coding concepts and project design elements displayed in children's computational artifacts. Next, we discuss some of the similarities and differences between the two rubrics, highlighting the CT opportunities afforded by different interfaces. Finally, we end the chapter with implications and practical strategies for using project rubrics in educational and research settings.

Background

Questionnaires and task-based assessments are the two most common ways to assess children's programming mastery (Shute, Sun & Asbell-Clarke, 2017). These forms of assessments involve children answering a set of pre-made questions or performing tasks that typically have one correct solution. However, some studies have used performance-based assessments to evaluate young children's programming competencies from their projects (Basu, 2019; Denner, Werner, & Ortiz, 2012; Wangenheim et al., 2018; Wilson, Hainey, & Connolly, 2013). Performance-based assessments allow children to demonstrate their knowledge and skills by making authentic products that are driven by their individual interests and identities (Chen & Martin, 2000). These assessments provide insights into how children apply their acquired cross-disciplined skills into their coding projects.

Performance-based (or often called project portfolio) analysis on children's coding projects is a rich and concrete approach to explore how children develop and apply their understanding of programming concepts. For example, one study conducted by Brennan and Resnick (2012) compared Scratch (a block-based programming language for children ages eight and up) project portfolios between a novice child programmer and an expert child programmer. The expert child experimented with a wider variety and number of coding blocks across projects compared to the novice child. The duration of experience with Scratch seemed to affect how these two children displayed their understanding of programming concepts. The expert programmer child in the study had been coding with Scratch for three years, whereas the novice programmer child had just started for one week. Through examining each child's Scratch program, researchers could better identify the computational skills and concepts each child had acquired and displayed through their projects.

One drawback of a performance-based assessment is that it only captures the *product* and fails to acknowledge the *process* of creation. The process of creation, or design process, is a component of CT needed for young children to create solutions to problems (Bers, 2020). Brennan & Resnick (2012) acknowledged that after interviewing the child, the project portfolio analysis did not accurately reflect the child's CT. In other words, the use of particular coding blocks in the child's project did not necessarily mean that the child could explain how those programming concepts actually worked. Thus, instead of analyzing final projects exclusively, it may be beneficial to include incomplete coding projects in children's portfolios (Brennan & Resnick, 2012) or utilize a "system of assessments" (Grover, 2017) to provide a more holistic view of children's understanding.

It is important to note that researchers and educators may be interested in understanding different aspects of students' coding artifacts, or products created through the activity of computational making. This next section explores coding project rubrics that were specifically developed for researchers versus educators.

Project Rubrics for Researchers

A few studies have explored the development and use of coding project rubrics to assess students' coding performances. Although each rubric differs in the categories of criteria, the two most common categories are programming concepts and project design. Whereas programming concepts refer to concrete skills and computational thinking practices necessary for students to plan and construct their coding artifacts, project design refers to the range of aesthetic elements used in the project. The project design criterion generally includes character and stage customization (Funke & Geldreich, 2017). For example, Denner et al. (2012) developed a rubric to grade coding projects on Stagecast Creator software, a visual programming language for children and adults to create games. In this rubric, the three main criteria are 1) programming, 2) code organization, and 3) design. Although most rubrics examine both the programming and design components of projects, some studies focus only on the programming (Moreno-León & Robles, 2015).

In addition to rubric criteria, another essential component of rubrics is calculating and interpreting projects' final scores or mastery levels. Different researchers have used different approaches to determine final scores. For example, Wangenheim et al. (2018) calculated the final project scores by adding raw scores from all criteria, dividing them by the maximum possible score, and categorizing the total scores into ten competency levels. Other studies, instead of having one final project score, used percentages to report how many times each concept occurs, which helps researchers understand the most frequently used concepts (Denner et al., 2012; Wilson et al., 2013). For example, Wilson et al. (2013) reported that the most common competencies found in 8-11-year-old children's projects were sequencing and events (if statements). The least frequently displayed competencies were random numbers and keyboard input. Another study by Funke & Geldreich (2017) had an additional rubric criterion used to determine children's overall level of understanding, in addition to calculating the frequency of each competency.

One lacking element from researchers' rubrics is the design process, which is more commonly incorporated in educators' coding rubrics. Why might this be the case? Although both researchers and educators have similar interests in examining children's computational artifacts, they may have differing aims for how to use the project scores. For instance, researchers might probe into specific characteristics of projects or use rubric scores to understand the kinds of programming competencies exhibited in a project. On the other hand, educators may use rubrics to track students' design thinking or as an instructional tool to understand how to better support student learning. In this next section, we discuss coding project rubrics developed by and for educators.

Project Rubrics for Educators

As computer science education becomes increasingly prevalent in schools, educators may also find coding project rubrics useful for their classrooms. The Creative Computing Lab at the Harvard Graduate School of Education shared a report synthesizing 50 teachers' highly diverse rubrics used to assess their students' coding projects (Brennan, Haduong, & Veno, 2020). Similar to the coding project rubrics used in research, the rubrics used in classrooms also have programming and project design components. However, some of the classroom rubrics also have extra elements that capture the working process—for example, students' iterative process, time management, and collaboration. Because one key takeaway from this report was that students should receive feedback on their coding projects from multiple perspectives (e.g., self, teachers, peers, and family members), additional rubrics were designed for these various evaluators.

The Creative Computing Lab also developed a rubric to assess students' development of computational practices through artifact-based interviews (Creative Computing Lab, n.d.). Computational practices differ from computational thinking, as computational practices focus more on the process of thinking than on the concept learned (Brennan & Resnick, 2012). This rubric consists of four sets of questions related to children's projects—Experimenting and Iterating, Testing and Debugging, Reusing and Remixing, Abstracting and Modularizing. For example, one of the questions under the computational practices of Testing and Debugging is "Describe a time when your project didn't run as you wanted." Each response the child provides to a question is given a proficiency rating of low, medium, or high.

Connecting Children's Projects to Computational Thinking

Programming concepts are a component of project rubrics that are most directly related to computational thinking (CT; Moreno-LeÓn et al., 2020; Seiter & Foreman, 2013). Different researchers define CT differently, but the seven most common concepts across studies according to Rose et al. (2017) are: 1) using sequences of steps (algorithms); 2) using if-then statements (conditionals); 3) having more than one code running concurrently (parallelism); 4) breaking a big problem into smaller parts (decomposition); 5) solving problems (debugging); 6) formulating a problem in an understandable manner (abstraction and generalization); 7) and analyzing data to solve problems (data collection). Most of these seven common CT concepts can be found in coding project rubrics, especially under the programming criteria. However, concepts related to thinking processes, such as debugging, may be harder to assess from projects alone.

The components of CT mentioned by Rose et al. (2017) apply generally across all ages of coders. However, some CT concepts in the way they are defined for adults may not be developmentally suitable for young children. In Chapter 1, Bers outlined the seven powerful ideas of CT that are relevant and appropriate for young children, which include algorithms, modularity, control structures, representation, hardware/software, design process, and debugging. Bers (2020) also connected the seven powerful ideas to the ScratchJr app and the KIBO robotics kit, two coding platforms for children ages 4-7 that were developed by the DevTech Research Group. Both of these coding tools enable children to acquire and display CT abilities, which we will explore in this chapter by describing the ScratchJr and KIBO Project Rubrics.

Although programming concepts may be most directly related to CT, the design elements of children's coding projects also invite children to think in computational ways. CT is a way of thinking that children can apply to various life situations not limited to programming (Relkin & Bers, 2019; Wing, 2006). The depth of children's project design may also reflect CT. For example, if a child wants to differentiate the main character in her ScratchJr project, she may experiment with different solutions, such as enlarging the character or changing the color of the main character. The process of designing the main character, although not involving programming, still enables the child to engage with computational practices and concepts. Thus, the ScratchJr and KIBO Project Rubrics presented in these next sections take into account both the programming concepts and design elements of coding projects.

SCRATCHJR PROJECT RUBRIC

The ScratchJr Project Rubric was developed by the DevTech Research Group with the primary goal to assess young children's purposeful creation with ScratchJr projects. The rubric, which is outlined in Figure 1, specifically assesses for the comprehension of coding concepts and the project design ability, while also getting at the purposefulness of their creations.

Figure 1. ScratchJr Project Rubric domains (Coding Concepts and Project Design) and their sub-categories
Source: IGI, 2021

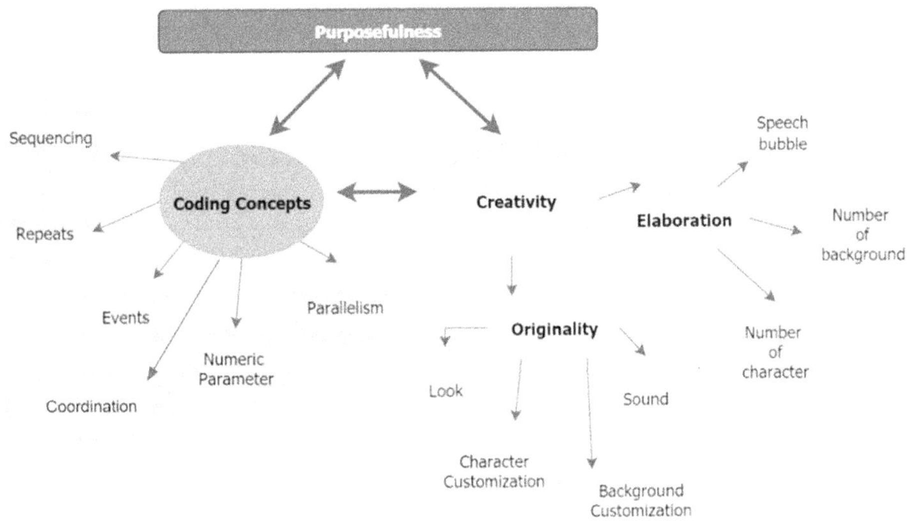

The ScratchJr Project Rubric went through multiple iterations until reaching a high level of agreement between graders. Four research assistants were involved in testing and revising the rubric over the span of three months. Each grader assessed 245 projects created by children or adults: 176 practice projects and 69 test projects with a finalized rubric.

Coding Concepts

Coding concepts are the first set of criteria as shown in Table 1 and are based on previous studies that reported various coding concepts from ScratchJr (Flannery et al., 2013; Strawhacker & Bers, 2019). This work was also inspired from studies that used coding rubrics to examine elementary and middle school students' Scratch coding projects (Basu, 2019; Moreno-León & Robles, 2015). For example, "Dr. Scratch" is an example of a project rubric for Scratch that assesses CT development (Moreno-León & Robles, 2015). The seven programming concepts for Dr. Scratch are abstraction, parallelism, logical thinking, synchronization, flow control, user interactivity, and data representation.

Table 1. ScratchJr Project Rubric criteria

Scoring Criteria	Sub-Categories	Description
Programming Concepts (A)	A1. Sequencing	Is the program functional? How many coding blocks were included (repeated blocks were counted once)?
	A2. Repeats	Does the program utilize any repeat blocks? If so, to what complexity are they used?
	A3. Events	Does the program utilize any start on tap, bump, message, or go to page? If so, to what complexity are they used?
	A4. Parallelism	Is there more than one program being executed simultaneously? If so, to what complexity are they used?
	A5. Coordination	Does the program utilize any wait, speeding, or stop blocks? If so, to what complexity and intentionality are they used?
	A6. Number Parameter	Is the number bubble being used to execute the action to a certain number of times?
Project Design Elements (B)	B1. Character Customization	To what extent did the child customize the character? How many different approaches did the child used?
	B2. Background Customization	To what extent did the child customize the background? How many different approaches did the child used?
	B3. Look	Are any of the look blocks used to change the appearance of the character?
	B4. Sound	Does the child use any pop block or record sound? Does the added sound make logical connection to the project?
	B5. Number of Characters	How many characters with at least one coding block was included in the project?
	B6. Number of Settings	Out of the 4 possible pages to add background, how many different settings did the child choose?
	B7. Speech bubbles	Is the speech bubble included? Is there only one word, a few words, a sentence, or an entire conversation?

Source: IGI, 2021

The six sub-categories of the ScratchJr Coding Project's coding concepts criterion include (A1) Sequencing, (A2) Repeat, (A3) Events, (A4) Parallelism, (A5) Coordination, and (A6) Number parameters. These six sub-categories can be examined in projects and are connected to the definition of CT by Rose et al. (2017) and Bers (2020). The rubric explicitly examines sequencing and algorithms, repeat loops and control structures, events and conditionals.

The ScratchJr project shown in Figure 2 displays all six coding concepts in the rubric. This project shows programming sequences (A1) of the bird's movement and appearance. The bird flies one block down then 15 blocks (A6) to the right. Simultaneously, the bird repeatedly (A2) gets bigger eight times as it is flying, making it seem that it flies closer to the screen. Parallel sequences (A4) make it possible for the bird to move and change in appearance at the same time. The bird then touches, which triggers (A3) the flower to wait one second (A5) before it enlarges 10 times.

To expand on the coding concepts, this project has sequencing, which is the ability to create ordered steps to achieve a goal. Repeat block tells a command to replay, shown in Figure 2 when the bird gets bigger repeatedly. Parallelism is when there is more than one order executing at the same time, when the bird grows when moves right. Events is the ability to trigger an if-then command to start, shown in Figure 2 when a bumping block that tells the flower's syntax to start playing after the bird touches the flower. Coordination is when a child programs two or more characters to interact with one another, shown by the flower that is waiting for the bird to fly by and starts growing after a few seconds. Lastly, number parameter is the ability for children to specify the number of times they want each coding block to play. Number parameters are the white bubbles at the bottom of each coding block.

Figure 2. ScratchJr project example that shows parallel coding sequences for a bird character and a single coding sequence for a flower character
Source: IGI, 2021

Project Design

In Table 1, the project design category of the project was inspired from creativity applications for education and creativity assessment literature (O'Quin & Besemer, 1989; Plucker, Beghetto, & Dow, 2004). The literature describes creativity, in this context, as having originality, elaboration, and purposefulness through the process of meaningful creation (O'Quinn & Besemer, 1989; Plucker et al., 2004). Therefore, we developed a final project rubric for design based on three dimensions:

originality, elaboration, and purposefulness. Furthermore, we adapted the rubric sub-categories to be applicable with the ScratchJr functions. Our rubric not only captures purposefulness in the project design criterion, but it also measures whether children use coding blocks purposefully.

The project design in the ScratchJr Project Rubric has seven sub-categories in total, four sub-categories are under originality including (B1) character customization, (B2) background customization, (B3) look blocks, and (B4) sound. The three remaining subcategories fall under elaboration, which includes (B5) number of characters, (B6) number of settings, and (B7) speech bubble. There are limitless ways children can display their creativity and imagination with ScratchJr. Figure 3 displays how children can insert photos of their faces, record their voices, insert shapes, write, draw, paint, and more. Thus, there are a wide variety of project design elements assessed in the rubric.

Figure 3. ScratchJr project example that has customized characters created using the draw, paint, and insert shape tools. One cat character is painted red, and the other cat has long hand drawn hair.
Source: IGI, 2021

Projects that contain the same coding concepts and project design sub-categories may not receive the same final project score. This is because the score of each subcategory ranges from 0 to 4 points, depending on the mastery level in each area. For example, a syntax that only uses one coding block in multiple repeat loops shown on the top of Figure 4 is less complex than a syntax that has a nested loop shown on the bottom of Figure 4. Therefore, in a repeat coding concept, a project that has one coding block inside a repeat loop would get a lower score, whereas a project that has a nested loop would the highest score.

Figure 4. Both sequences command a character to make the same movement. However, the bottom sequence is more efficient and displays the advanced concept of nested loops.
Source: IGI, 2021

From the final 69 projects that were created by children and adults, sequencing was the sub-category with the highest mean score, following by number parameter. All projects had 4-5 coding blocks on average. The lowest mean scores went to coordination and parallelism, which are the two undeniably most difficult concepts, both involving high levels of purposefulness and coding mastery. To get the highest score in coordination, there must be a clear intention that two or more characters interact by using a certain set of coding blocks (e.g., wait and change speed).

Figure 5 is an example of a project that has a score of 4 for coordination. In this example, there is intentional coordination between the two cars that are racing across the city. Both cars' sequences have an orange speed block; however, the top sequence has a speed up block, whereas the bottom sequence has a speed down block. Therefore, the top car ends up winning the race. From the seven powerful ideas (Bers, 2020), this project displays the CT concepts of algorithms, modularity, and control structures. The creator of this project formulated coding sequences in two different chucks for the two cars. Additionally, the speeding blocks in this project are examples of control structures.

Figure 5. A car racing project which receives the highest score under the coordination sub-category in the ScratchJr Project Rubric
Source: IGI, 2021

Figure 6. A music composition project that will play a different instrumental sound after tapping on each character
Source: IGI, 2021

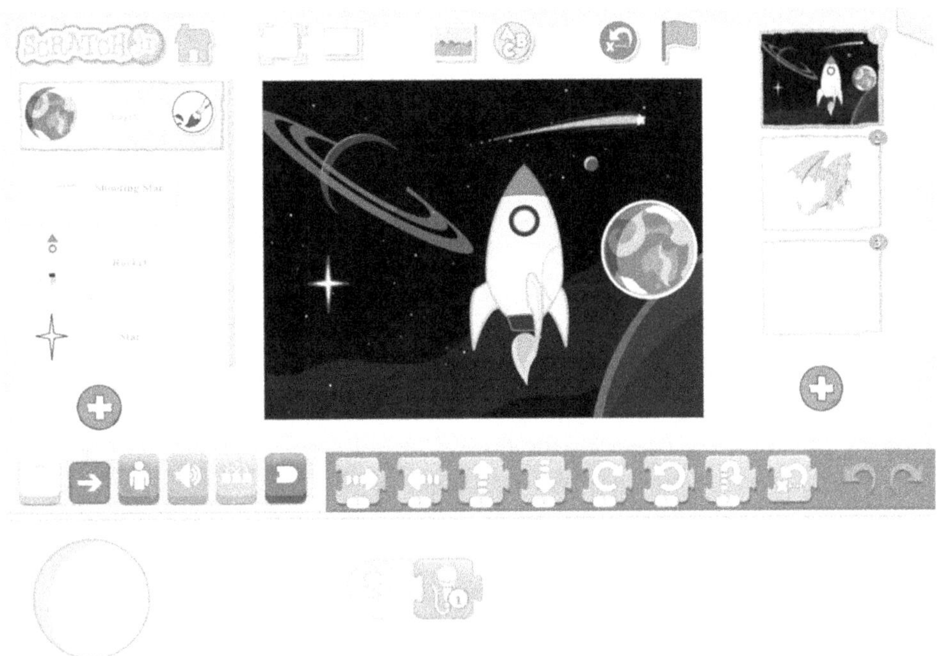

Different Types of ScratchJr Projects

During the rubric development, we came across various project types. We originally tried to group ScratchJr projects into three types: stories, games, and collages. However, we soon realized that it was impossible to fit the projects into only three types. We were surprised by the remarkable versatility of ScratchJr projects and describe several of these creative projects in the following section.

Space Musical Jam

The project in Figure 6 features four characters: Earth, Shooting Star, Rocket, Star. Using only two coding blocks per character, each has a unique recorded sound that will start playing when the character is tapped. The rocket has a drum sound, the star has a piano melody, the earth has a high humming voice, and the shooting star has a low humming voice. The combination of all four characters' sounds makes a complete musical piece. The project exemplifies how the project design aspect (sound recording) can be highlighted through programming (conditional statements).

Lighting Candles

For the project shown in Figure 7, if the user taps on the candle stick, a message is sent from the candle stick to the flames, which then receive the message and begin appearing and disappearing. This repeated action mimics a flickering candle. This project uses sending messages (conditional statement), which is one of the hardest and thus less frequently used coding blocks for young children.

Figure 7. A candle lighting projects that uses tapping and sending messages blocks for the flames to flicker
Source: IGI, 2021

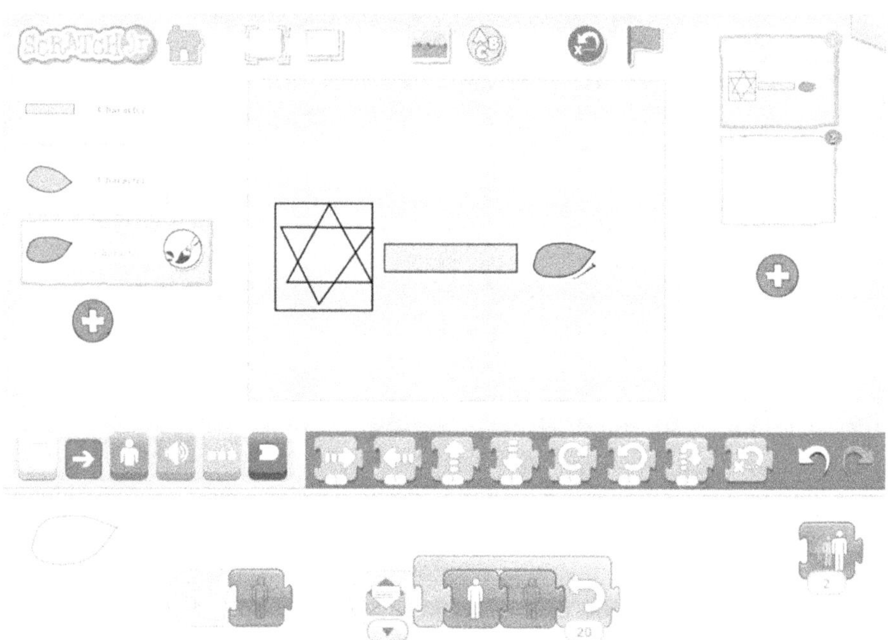

Holiday E-Card

For the project shown in Figure 8, the creator first took a picture of a snowy Christmas tree background from the internet and set it as the project background. The author then painted a Santa Claus character from an existing ScratchJr character. The last step was to add a speech bubble for Santa Claus to say Merry Christmas. This project presents the combination of modularity and design process. In the design process of this project, the creator customized the background and characters using various paint editor tools.

Figure 8. This project was created as a holiday e-card with speech bubbles and a winter Christmas tree background
Source: IGI, 2021

Score Tracker

The ScratchJr project in Figure 9 functioned as a score tracker board for three players as shown through the three different color cards. The project was intended to be used while the children were playing a separate game off-screen, such as an outdoor playground game. When a player earns a point in the game, the player can tap on their assigned color card. The score increases by one point with each tap. Once one of the cards reaches a certain point, six in this case, a congratulating message pops up. The programming concepts behind this project are fairly complex. To create the score cards, the child had created multiple card number characters and stacked them on top of each other, with the lowest score on top. Each card number character is programmed similarly: start on tap and then become invisible. When the top card disappears, the next number appears.

Figure 9. This project works as a score tracker for three players (red, green, pink) and involves multiple start on tap and invisible blocks
Source: IGI, 2021

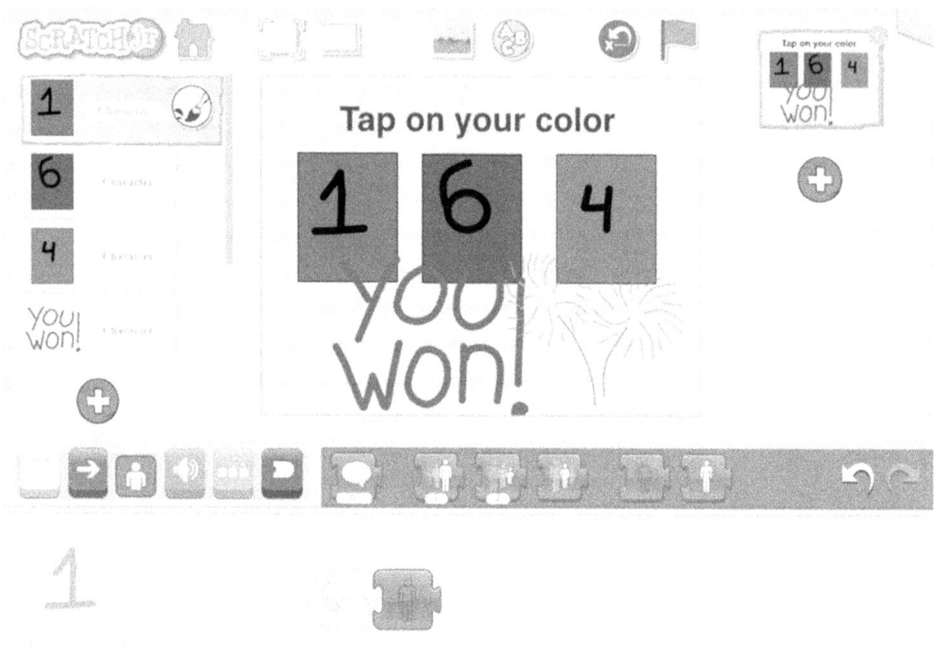

KIBO PROJECT RUBRIC

The primary purpose of the KIBO Project Rubric is to assess the programming concepts and design elements exhibited in a KIBO robotics project. The rubric enables educators and practitioners to assess children's mastery of the KIBO programming language when they are presented with the opportunity to apply their knowledge and skills to their own personalized project.

A project rubric for the KIBO robotics kit was first developed in 2018 and was aligned with existing KIBO robotics curricula (DevTech Research Group, 2018). The rubric outlined criteria such as correct usage of repeat loops and conditionals, proper sequencing of blocks to accomplish the intended task, and appropriate placement of sensors and modules, among other general project characteristics. Over the years, with feedback from researchers and educators, the rubric was revised to include specific scoring criteria, usability across various curricular activities, and step-by-step instructions for identifying children's overall level of programming mastery exhibited in the project. Versions of the rubric were tested and validated with over a hundred KIBO projects.

Similar to the ScratchJr Project Rubric, the KIBO rubric consists of two sets of scoring criteria: programming concepts and project design elements. Solely learning how to program the KIBO robot using the blocks does not necessarily constitute a high display of computational thinking abilities. Therefore, the rubric also takes into account how children choose to sequence the programming blocks, customize their robots using arts and crafts materials, and utilize music, dance, or other creative media to make their robotic creations come alive. These aspects, though not always related to coding and computer science, are activities that invite children to utilize general computational thinking abilities and to produce personally meaningful and purposeful projects. As displayed in Table 2, the two categories of scoring criteria are each split into five specific sub-categories, which are described in the following sections.

Programming Concepts

There are five sub-categories of programming concepts, which are (A1) syntactical accuracy, (A2) repeats, (A3) conditionals, (A4) module use, and (A5) data. When children display their understanding of programming concepts through their robotic creations, the thought processes required to carry out the project task invoke the powerful ideas of computational thinking introduced by Bers in Chapter 1. Syntactical accuracy, for instance, requires children to assemble KIBO programs in order from left to right, thereby engaging their use of algorithmic thinking. Repeats and conditionals are examples of control structures. When children use repeat and if blocks in their projects, they display an understanding of what control structures are and how these blocks impact the way their KIBO programs run. The use of appropriate modules and sensors necessitate children's understanding of hardware and software and how both are required to make the robot function. Storing data in the form of recorded sounds also involves hardware and software. For example, a child presses different buttons on the sound recorder module to record different sounds (hardware) and subsequently scans the corresponding sound blocks (software) to hear their sounds play aloud. Data are also present in the form of subroutines, or single blocks that are used to represent a whole sequence of actions. The use of subroutines not only invokes children's understanding of symbolic representation, but these blocks also engage children in modular thinking. The subroutine blocks can be treated like a module that can be used in multiple places throughout the program.

Table 2. KIBO Project Rubric criteria

Scoring Criteria	Sub-Categories	Description
Programming Concepts (A)	A1. Syntactical Accuracy	Is the program functional? When the blocks are scanned in order from left to right, the robot will be able to perform the programmed actions without beeping an error message.
	A2. Repeats	Does the program utilize any repeat blocks? If so, to what complexity are they used?
	A3. Conditionals	Does the program utilize any if blocks? If so, to what complexity are they used?
	A4. Module Use	What kinds of modules were attached to the ports on the KIBO body? Do these attached modules have any correspondence to the actual program? For instance, a sound sensor used without a Wait for Clap block does not display correspondence.
	A5. Data	Does the child exhibit an understanding of information storage in their programming? For instance, the child records their own sounds using the sound recorder blocks/module or makes use of subroutines, which are blocks used to substitute a set of other blocks.
Project Design Elements (B)	B1. Sequencing	How many blocks does the child use to construct their program?
	B2. Block Variety	What kinds of blocks does the child use to construct their program?
	B3. Robot Customization	How is the robot decorated and customized with arts and crafts and/or building materials? The child will be able to test their creations so that decorations are securely attached to the robot.
	B4. Setting	What (if any) additional project elements are included as part of the final project? Examples of project elements might include singing or dancing along, playing background music, or displaying some type of visual poster alongside their robotic creations.
	B5. Coordination	How are project elements used purposefully to enhance synchronization and coordination? For example, the child uses a Wait for Clap block strategically to align the robot's actions with a particular song or dance.

Source: IGI, 2021

To further examine the programming concepts assessed in the KIBO Project Rubric, let's take a look at the KIBO project displayed in Figure 10. A child scans her block sequence with her KIBO robot, which she has adorned with a handmade puppet. The first scoring criterion is syntactical accuracy (A1), referring to whether the constructed sequence represents a functional program. In this example, if the blocks were scanned in order, the KIBO robot would indeed perform the sequence of commands. In addition to proper syntax, the child has correctly utilized a repeat loop (A2) with the number 4 parameter. This parameter signifies that the two forward motions will repeat four times. There is no use of if/conditional blocks (A3) in this program. The child has attached wheels and motors, as well as a sound sensor to the KIBO body. However, the only action the robot is programmed to do is to move forward, so the attachment of the sound sensor is unnecessary. Only the wheels and motors have appropriate correspondence (A4) to the constructed program. Finally, there is no use of sound recorder blocks, subroutines, or any project elements that require storing information, so there is no supporting evidence for the Data subcategory (A5).

Figure 10. Child scans the sequence of KIBO blocks using the robot's embedded barcode scanner. The KIBO program contains a repeat loop with a number "4" parameter, indicating that the two forward blocks will repeat a total of four times (i.e., KIBO will move forward eight times).
Source: IGI, 2021

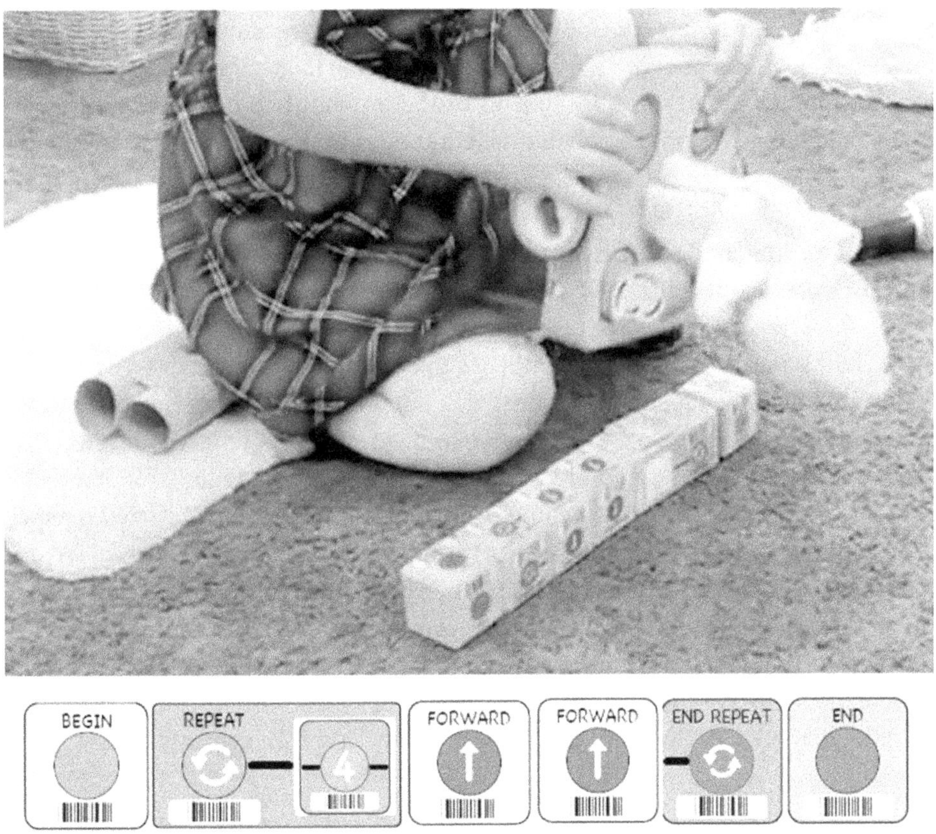

Project Design Elements

Project design elements refer to anything that the child adds to the project in order to add aesthetic appeal, display originality and creativity, or extend the complexity of their project. There are five sub-categories of project design elements, which are (B1) sequencing, (B2) block variety, (B3) robot customization, (B4) setting, and (B5) coordination.

Even the design elements of a KIBO project require that children utilize and display their CT abilities. For example, the more blocks a child uses in their KIBO project, the more opportunities the child has to plan and scan their programs, enabling the child to engage more deeply in algorithmic thinking and the design process. Using a variety of blocks also displays the use of more advanced CT abilities, particularly symbolic representation and the idea that attributes such as colors and icons signify specific types of actions. Symbolic representation also plays a role in how the child customizes their robot. What is the robot intended to be: a helper bot, an animal, a famous person? Whatever the child imagines the robot to be can be represented in the way that the child decorates and programs the robot. Even after decorating the robot, the child may later decide to extend their ideas by adding music, creating corresponding dance moves, or drawing an illustration to go along with the project. These additional project design elements further exemplify the design process and any debugging and problem-solving the child does to make their projects exactly as they intended.

Let's examine these five constructs using the KIBO project displayed in Figure 11. Students from this kindergarten classroom participated in a curriculum that integrated the KIBO robotics kit with literacy and storytelling using the theme book *Brown Bear, Brown Bear, What Do You See?* written and illustrated by Bill Martin, Jr. and Eric Carle. In this particular project, a child designed and programmed the KIBO robot to move through the taped illustrations, which represent the brown bear going to a farm and seeing a white dog. The program utilized eight programming blocks (B1) of various types: yellow Light block, gray Repeat blocks, and blue Motion blocks (B2). On top of the KIBO robot's platform, the child had securely affixed a paper plate, cup and various decorations and artwork, which personalized the project and enabled the child to showcase their imagination and creativity (B3). The illustrations taped to the floor enhance the overall project and provide a meaningful connection between the robot's actions and the theme book (B4). Finally, there is no supporting evidence that the project required specific timing or synchronization through the use of multiple robots, the Wait for Clap block, or other project elements (B5).

Figure 11. The KIBO robot is programmed to turn on its white light and then move forward and turn left for a total of three times. The robot is decorated like a brown bear and moves through different illustrations taped to the floor that represent various scenes from the children's book Brown Bear, Brown Bear.
Source: IGI, 2021

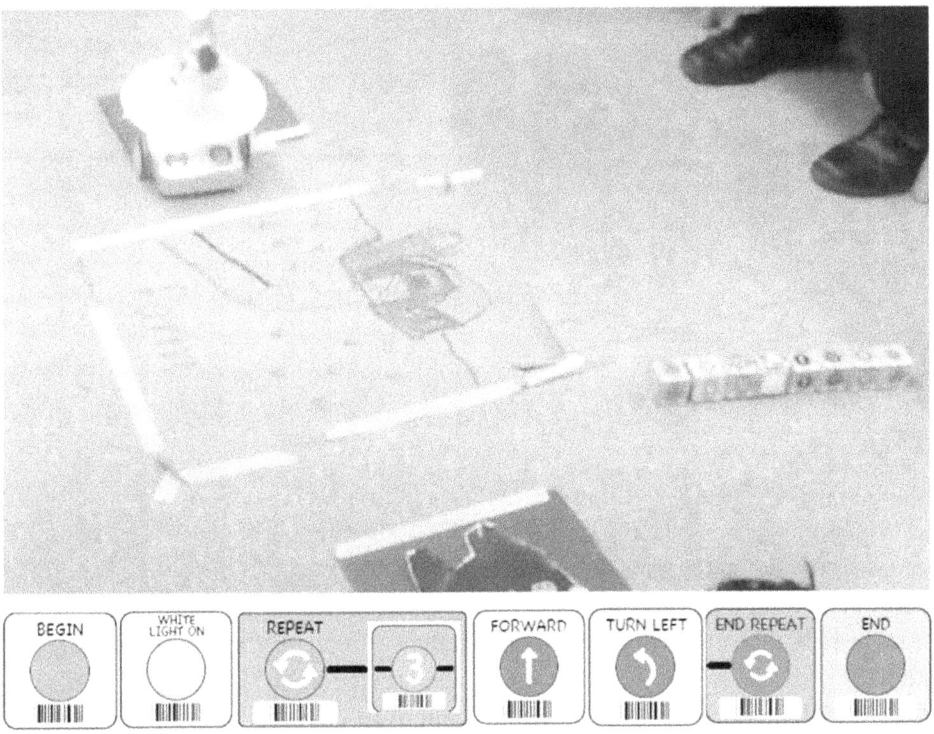

Similar to the ScratchJr Project Rubric, the score for each of the ten sub-categories ranges from 0 to 4 points, depending on the mastery level in each area. The higher the points received for a particular construct, the more advanced skill the child has exhibited. For instance, for the sub-category of Repeats, the point allocation is as follows:

0 points: No repeat blocks used
1 point: Repeat attempted but missing or misplaced the Begin/End and/or parameter or no blocks in-between Begin/End blocks
2 points: At least one successful repeat loop with number parameter
3 points: At least one successful repeat loop with sensor parameter
4 points: At least one successful repeat loop as part of nested statement

Table 3 displays sample KIBO programs that would receive these varying point values. With each successive point, it is evident that the child has demonstrated more advanced programming competency.

Table 3. KIBO repeat loops of varying point criteria

Point Value	Example of KIBO Program
0	
1	
2	
3	
4	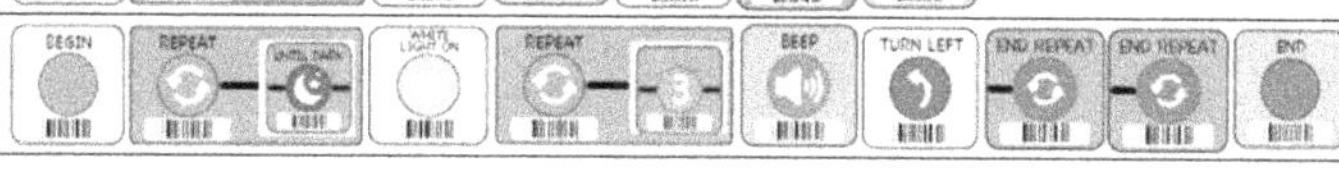

Source: IGI, 2021

With five sub-categories (each worth 0-4 points), each set of scoring criteria is worth a maximum of 20 points. However, just as coding concepts were weighted more heavily than project design in the ScratchJr Project Rubric, so is the case for the KIBO rubric. The rationale for weighting the categories differently is simple. At its core, KIBO is a tangible programming interface, not an arts and crafts activity. Although the tool offers ample integration opportunities and has aesthetic appeal, the primary educational purpose of KIBO is to introduce foundational programming concepts to young children. Thus, the KIBO Project Rubric follows the same 60-40 weighted ratio, with emphasis given to programming concepts by multiplying its summed score by 1.5. Therefore, the maximum number of points for programming concepts is 30 points, which brings the total summed score to a maximum of 50 points. This total score is then split evenly into five levels: Budding (0-9 points), Developing (10-19 points), Proficient (20-29 points), Advanced (30-39 points), and Distinguished (40-50 points).

There are several important points to note about these final project scores and levels of mastery. One is the positive framing for the names of the five level categories. Aligned with the principles of strengths-based education (Lopez & Louis, 2009), the level of mastery obtained from the KIBO Project Rubric is meant to highlight the positive aspects of children's efforts and achievement, rather than position any misconceptions in their learning as deficits. Using positive framing also serves to position children's learning and engagement with the KIBO robotics kit as a developmental activity. A child who is introduced to KIBO for the very first time, even if older by age, may not necessarily create a "proficient" KIBO project. By using the terms "budding" or "developing", we recognize that the child is growing their programming skills and with more experience and exposure, the child may have the opportunity to develop and display greater proficiency with the KIBO programming language.

Finally, it is essential to note that the final score does not indicate a child's *overall* level of programming mastery. Rather, the score provides an estimated level of mastery *as exhibited in this particular project*, which means that children's projects might be limited by factors outside of their control. For example, children who are working with KIBO-10 (an introductory kit sold by KinderLab Robotics that comes with the 10 most basic programming blocks) are likely to create projects that are less complex than children working with KIBO-21 (a more comprehensive kit that comes with additional advanced blocks and sensors). Unless children are prompted to construct a program using the most advanced KIBO parts that they have access to and know how to use, as well as provided with unlimited time and resources for building and decorating their robots, children are not necessarily expected to display the entire extent of their knowledge in a single project. As discussed in other parts of this chapter, scoring children's project artifacts is one useful way of assessing what they know, but it is certainly not the only way.

COMPARING AND CONTRASTING SCRATCHJR AND KIBO PROJECTS

The developers designed ScratchJr's and KIBO's interfaces according to children's developmental ranges. KIBO has a tangible interface that caters to younger learners (ages three and up), whereas ScratchJr is a screen-based app that targets children ages five and up. Despite the different interfaces, children generally enjoy playing with both KIBO and ScratchJr and can use either coding tool to produce creative projects. As ScratchJr and KIBO are both block-based programming languages, children are not required to be able to read text because they can differentiate blocks by their symbols and colors. ScratchJr and KIBO have similar programming concepts such as sequencing, repeats, conditionals, data, and number parameters. The main difference is that programming concepts in ScratchJr are more advanced (e.g., Parallelism). KIBO does not have parallelism because the robot can only read and run one sequence at a time. Furthermore, each KIBO robot represents a single character, whereas ScratchJr can easily have multiple characters with coordination happening between characters. However, KIBO has module use, which ScratchJr does not have. The ability to use modules is a function that makes KIBO very appealing to young children. KIBO's modules include wheels, motors, lightbulbs, and sound, light, and distance sensors, all of which enable children to engage more deeply with the computational idea of hardware versus software. In addition, children can also engage more physically with the KIBO robot, for example, by clapping to trigger the sound sensor or using their hands or other materials to trigger the distance sensor.

Although children can develop programming skills from ScratchJr and KIBO effectively (Pugnali, Sullivan, & Bers, 2017), the different interfaces inevitably lead to different levels of project creativity. As KIBO is a tangible tool, many components of project design also happen physically. For example, children can be creative with decorating the KIBO robot and creating backgrounds using different art tools, such as crayons, papers, and tapes as shown in Figure 12. These customizations differ from ScratchJr, where creativity can happen within the app through drawing or coloring. Additionally, ScratchJr has more than 50 characters with different themes (e.g., animals, nature, and people) and more than 20 backgrounds. In each ScratchJr project, children can add as many characters as they want with up to four backgrounds (pages). ScratchJr has a photo taking function, which enables children to add custom images to their characters and backgrounds. Due to these interface differences, children display different kinds of aesthetic design elements in their projects.

Finally, project sharing is another factor that differs between ScratchJr and KIBO. Project sharing is an important and beneficial approach for children to collaborate and learn from one another and from each other's projects. ScratchJr has a project sharing function through email or airdrop. According to the DevTech website (ScratchJr), as of October 2020, 600 thousand projects have been shared. The percentage of project sharing increased by 200% during the COVID-19 pandemic as illustrated in Chapter 15. In contrast, the only way to share KIBO projects is to take pictures or videorecord KIBO's actions, which enables viewers to see projects in their entirety (e.g., movements, arts and crafts, light, and music).

IMPLICATIONS FOR RESEARCH AND PRACTICE

Coding projects enable children to exhibit their CT abilities while also showcasing their creativity and imagination. As more states and countries adopt formal computer science and digital literacy standards and frameworks, the need for supporting educators and practitioners in how they assess students' learning becomes more critical. What criteria are appropriate to examine in young children's projects? Should children be provided with a prescribed rubric or set of project guidelines, or should the rubric merely be an educator-facing tool to examine the different elements of a project? To answer these questions, we offer the following practical strategies for using project rubrics in educational and research settings:

- **Documentation of process and outcome**: Brennan and Resnick (2012) identified strengths and limitations of three different assessment approaches: project analysis, artifact-based interviews, and design scenarios. Part of the challenge of project analysis, which is the focus of this chapter, is that we are only able to examine children's exhibition of CT concepts, rather than CT practices. In order to gather some sense of children's design processes as they are working on their projects, we suggest a combination of project analysis and artifact-based interviews. Through the documentation of both process and outcome, we can better understand the bidirectional relationship between children's computational thinking and their computational making.
- **Subjectivity and ambiguity in scoring:** Another challenge of a project portfolio assessment, identified by Grover (2020) and others, is that the process of assessment itself can be subjective and ambiguous. Even with detailed rubrics with specific scoring criteria, different individuals may prioritize different aspects of projects, depending on the learning setting, the activity prompt, and children's own interests and motivations. Furthermore, children may be working in pairs or teams or receiving assistance from their teachers and peers, which makes assessing an individual child's understanding even more challenging. For example, the presence of advanced programming blocks in a project merely indicates conceptual encounter but does not mean the child has fully understood the use and purpose of those blocks. Despite these limitations, however, coding projects can be an authentic way of examining what children know by what they can create, rather than by multiple-choice or task-based questions that they can answer.
- **Emphasis on purposefulness:** As Bers stated in the opening of this book, the goal of introducing computing tools and activities for young children is not so that they can all become future STEM professionals. Rather, in line with Wing's (2006) push that computational thinking is a universal set of skills and not something specially reserved for computer scientists, the opportunities to manipulate and create digital artifacts is something that should be afforded to all students. Furthermore, the assessment of those creations should forefront purposefulness, providing insight into the child's unique interests and identities. Scott, Sheridan and Clark (2015) propose that technological success should be considered based on "creative intent and social significance". These goals can be applied here in the context of children's coding projects.

There is an increasing prevalence and use of creative computing tools in classrooms and informal learning settings. As such, educators and researchers are seeking robust assessment methods for examining children's coding artifacts and extrapolating their understanding of foundational CT and programming concepts from these projects. In this chapter we discussed two such rubrics for ScratchJr and KIBO that take the perspective that CT concepts can be displayed not only through the act of programming, but also through the act of designing the aesthetic elements of a project. We believe that the ability for children to apply their CT concepts through project-based learning tasks is equally as important as knowing the concepts. Therefore, coding project rubrics will be a critical component in computer science education for practitioners to better understand children's coding competency levels.

ACKNOWLEDGMENT

This research was supported by the Scratch Foundation and the U.S. Department of Education.

REFERENCES

Basu, S. (2019). Using Rubrics Integrating Design and Coding to Assess Middle School Students' Open-ended Block-based Programming Projects. *Proceedings of the 50th ACM Technical Symposium on Computer Science Education*, 1211–1217. 10.1145/3287324.3287412

Bers, M. U. (2020). *Coding as a Playground: Programming and Computational Thinking in the Early Childhood Classroom* (2nd ed.). Routledge Press. doi:10.4324/9781003022602

Brennan, K., Haduong, P., & Veno, E. (2020). *Assessing Creativity in Computing Classrooms*. Creative Computing Lab.

Brennan, K., & Resnick, M. (2012). Using artifact-based interviews to study the development of computational thinking in interactive media design. *Annual Meeting of the American Educational Research Association (AERA)*.

Chen, Y. F., & Martin, M. A. (2000). Using Performance Assessment and Portfolio Assessment Together in the Elementary Classroom. *Reading Improvement*, *37*(1), 32–38.

Creative Computing Lab. (n.d.). *Assessing Development of Computational Practices.* https://scratched.gse.harvard.edu/ct/assessing.html

Denner, J., Werner, L., & Ortiz, E. (2012). Computer games created by middle school girls: Can they be used to measure understanding of computer science concepts? *Computers & Education*, *58*(1), 240–249. doi:10.1016/j.compedu.2011.08.006

DevTech Research Group. (2018). *General Assessment Templates.* https://sites.tufts.edu/devtech/files/2018/03/GeneralAssessments.pdf

Flannery, L. P., Silverman, B., Kazakoff, E. R., Bers, M. U., Bontá, P., & Resnick, M. (2013). Designing ScratchJr: Support for early childhood learning through computer programming. *Proceedings of the 12th International Conference on Interaction Design and Children - IDC '13*, 1–10. 10.1145/2485760.2485785

Funke, A., & Geldreich, K. (2017). Gender Differences in Scratch Programs of Primary School Children. *Proceedings of the 12th Workshop on Primary and Secondary Computing Education*, 57–64. 10.1145/3137065.3137067

Grover, S. (2017). Assessing Algorithmic and Computational Thinking in K-12: Lessons from a Middle School Classroom. In Emerging Research, Practice, and Policy on Computational Thinking (pp. 269-288). Springer International.

Grover, S., Basu, S., Bienkowski, M., Eagle, M., Diana, N., & Stamper, J. (2017). A Framework for Using Hypothesis-Driven Approaches to Support Data-Driven Learning Analytics in Measuring Computational Thinking in Block-Based Programming Environments. *ACM Transactions on Computing Education*, *17*(3), 1–25. doi:10.1145/3105910

Lopez, S. J., & Louis, M. C. (2009). The Principles of Strengths-Based Education. *Journal of College and Character*, *10*(4). Advance online publication. doi:10.2202/1940-1639.1041

Moreno-León, J., & Robles, G. (2015). Dr. Scratch: A Web Tool to Automatically Evaluate Scratch Projects. *Proceedings of the Workshop in Primary and Secondary Computing Education*, 132–133. 10.1145/2818314.2818338

Moreno-LeÓn, J., Robles, G., & Roman-Gonzalez, M. (2020). Towards Data-Driven Learning Paths to Develop Computational Thinking with Scratch. *IEEE Transactions on Emerging Topics in Computing*, *8*(1), 193–205. doi:10.1109/TETC.2017.2734818

O'Quin, K., & Besemer, S. P. (1989). The development, reliability, and validity of the revised creative product semantic scale. *Creativity Research Journal*, *2*(4), 267–278. doi:10.1080/10400418909534323

Plucker, J. A., Beghetto, R. A., & Dow, G. T. (2004). Why Isn't Creativity More Important to Educational Psychologists? Potentials, Pitfalls, and Future Directions in Creativity Research. *Educational Psychologist, 39*(2), 83–96. doi:10.120715326985ep3902_1

Pugnali, A., Sullivan, A., & Bers, M. U. (2017). The Impact of User Interface on Young Children's Computational Thinking. *Journal of Information Technology Education: Innovations in Practice, 16*, 172–193. doi:10.28945/3768

Relkin, E., & Bers, M. U. (2019). Designing an Assessment of Computational Thinking Abilities for Young Children. In L. E. Cohen & S. Waite-Stupiansky (Eds.), *STEM for Early Childhood Learners: How Science, Technology, Engineering and Mathematics Strengthen Learning* (pp. 85–98). Routledge. doi:10.4324/9780429453755-5

Resnick, M., Maloney, J., Monroy-Hernandez, A., Rusk, N., Eastmond, E., Brennan, K., Millner, A., Rosenbaum, E., Silver, J., Silverman, B., & Kafai, Y. (2009). Scratch: Programming for Everyone. *Communications of the ACM, 52*(11), 60–67. doi:10.1145/1592761.1592779

Rose, S. P., Habgood, M. P. J., & Jay, T. (2017). An Exploration of the Role of Visual Programming Tools in the Development of Young Children's Computational Thinking. *The Electronic Journal of e-Learning, 15*(4), 297-309.

Scott, K., Sheridan, K., & Clark, K. (2014). Culturally Responsive Computing: A theory revisited. *Learning, Media and Technology, 40*(4), 1–25.

Seiter, L., & Foreman, B. (2013). Modeling the learning progressions of computational thinking of primary grade students. *Proceedings of the Ninth Annual International ACM Conference on International Computing Education Research - ICER '13*, 59. 10.1145/2493394.2493403

Shute, V. J., Sun, C., & Asbell-Clarke, J. (2017). Demystifying computational thinking. *Educational Research Review, 22*, 142–158. doi:10.1016/j.edurev.2017.09.003

Strawhacker, A., & Bers, M. U. (2019). What they learn when they learn coding: Investigating cognitive domains and computer programming knowledge in young children. *Educational Technology Research and Development, 67*(3), 541–575. doi:10.100711423-018-9622-x

von Wangenheim, C. G., Hauck, J. C. R., Demetrio, M. F., Pelle, R., da Cruz Alves, N., Barbosa, H., & Azevedo, L. F. (2018). CodeMaster—Automatic Assessment and Grading of App Inventor and Snap! Programs. *Informatics in Education, 17*(1), 117–150. doi:10.15388/infedu.2018.08

Wilson, A., Hainey, T., & Connolly, T. M. (2013). Using Scratch with Primary School Children: An Evaluation of Games Constructed to Gauge Understanding of Programming Concepts. *International Journal of Game-Based Learning*, *3*(1), 93–109. doi:10.4018/ijgbl.2013010107

Wing, J. M. (2006). Computational Thinking. *Communications of the ACM*, *49*(3), 33–35. doi:10.1145/1118178.1118215

Wing, J. M. (2008). Computational thinking and thinking about computing. *Philosophical Transactions of the Royal Society A: Mathematical, Physical and Engineering Sciences, 366*(1881), 3717–3725.

ADDITIONAL READING

Adams, J. C., & Webster, A. R. (2012). What do students learn about programming from game, music video, and storytelling projects? *Proceedings of the 43rd ACM Technical Symposium on Computer Science Education - SIGCSE '12*, 643. 10.1145/2157136.2157319

Delacruz, S. (2020). Starting From Scratch (Jr.): Integrating Code Literacy in the Primary Grades. *The Reading Teacher*, *73*(6), 805–812. doi:10.1002/trtr.1909

Harel, I. E., & Papert, S. E. (1991). *Constructionism*. Ablex Publishing.

Johnston, K., Highfield, K., & Hadley, F. (2018). Supporting young children as digital citizens: The importance of shared understandings of technology to support integration in play-based learning. *British Journal of Educational Technology*, *49*(5), 896–910. doi:10.1111/bjet.12664

Papadakis, S., Kalogiannakis, M., & Zaranis, N. (2017). Designing and creating an educational app rubric for preschool teachers. *Education and Information Technologies*, *22*(6), 3147–3165. doi:10.100710639-017-9579-0

Papert, S. (2000). What's the big idea? Toward a pedagogy of idea power. *IBM Systems Journal, 39*(3.4), 720-729.

Robson, S. (2014). The Analysing Children's Creative Thinking framework: Development of an observation-led approach to identifying and analysing young children's creative thinking. *British Educational Research Journal*, *40*(1), 121–134. doi:10.1002/berj.3033

KEY TERMS AND DEFINITIONS

Computational Artifact: Anything created by a human using a computer.

Event: An action that causes something else to happen.

KIBO: A screen-free programmable robotics kit for young children with blocks, sensors, modules, and art platforms.

Parallelism: Multiple codes executed concurrently for a single character.

Project-Based Learning: Student-centered pedagogy in which students acquire knowledge and skills by actively exploring real-world projects and challenges.

ScratchJr: A free block-based programming application for young children.

Syntax: The set of rules, principles and processes of a language that govern the arrangement of words and phrases.

Chapter 15
Insights Into Young Children's Coding With Data Analytics

Apittha Unahalekhaka
Tufts University, USA

Jessica Blake-West
Tufts University, USA

XuanKhanh Nguyen
Tufts University, USA

ABSTRACT

Over the past decade, there has been a growing interest in learning analytics for research in education and psychology. It has been shown to support education by predicting learning performances such as school completion and test scores of students in late elementary and above. In this chapter, the authors discuss the potential of learning analytics as a computational thinking assessment in early childhood education. They first introduce learning analytics by discussing its various applications and the benefits and limitations that it offers to the educational field. They then provide examples of how learning analytics can deepen our understanding of computational thinking through observing young children's engagement with ScratchJr: a tablet coding app designed for K-2 students. Finally, they close this chapter with future directions for using learning analytics to support computer science education.

DOI: 10.4018/978-1-7998-7308-2.ch015

UNDERSTANDING LEARNING ANALYTICS

Assessing children's knowledge is a challenge, and when we isolate computational thinking as our measure, the challenge becomes even greater. Computational thinking is a thought process, rather than a right or wrong answer on a test, and therefore our question extends beyond 'how to assess children's knowledge' and becomes more 'how to measure a child's way of thinking'. In this chapter, we explore how Learning Analytics is used to try to answer this question. Learning Analytics is the process of collecting and analyzing data from learners in order to better understand and optimize their learning processes (Gašević et al., 2015). While Learning Analytics does not solve the issue of how to measure learning, it does offer another angle to look at a children's learning process. Combined with other assessments, Learning Analytics can provide a richer view of a child's learning process.

Learning Analytics collects a wide variety of data, all of which are outcomes of learners' interactions with learning platforms. Learning platforms include online games, applications, learning management systems, and Massive Open Online Courses (MOOCs), such as Khan Academy, (Fischer et al, 2020; Gašević et al., 2015). A user interaction can be many things including number of clicks, the number of logins, and the duration in completing a lesson. Learning Analytics help educators to identify where a student is in their learning process, which allows the educators to better meet the needs of students of all different types of learning styles and paces (Baker and Siemens, 2014). Learning Analytics is most commonly used in higher education to measure school completion and learning performances (Ifenthaler & Yau, 2020). By collecting this information, higher education institutions can better identify needs of students and therefore work to address those needs.

While Learning Analytics is used more commonly in higher education, it can be applied to early childhood education as well. One example is LAP: A Learning Analytics Platform prototype developed by PBS KIDS, a US public broadcasting service catered to children (Roberts et al, 2016). In this study, the main functions of LAP were to track, store, and analyze children's (ages 2-8 years) interactions with the PBS KIDS math and literacy content including broadcasts, online videos, games, and offline activities. Children's anonymous usage data were collected and packaged into custom reports for parents on how to better support their children's learning needs. LAP measured learning in multiple ways. One way in which LAP assessed math was by reporting the accuracy and speed in answering mathematical problems. It was found that LAP measures were able to predict a reliable level of children's math proficiency compared to the TEMA-3 scale, which is a standardized mathematics test for children from 3-8 years old (Ginsburg & Baroody, 2003). From this example, we see that usage data, collected from a Learning Analytics tool can successfully predict specific learning outcomes—even when the users are young children.

Another way in which Learning Analytics can be used, is to evaluate the appropriateness of the learning tools. In the case of Vatavu et al. (2015), Learning Analytics were used to assess the usability of touch screen devices. In the study, researchers asked 89 children (ages 3-6 years) to complete various tasks such as tapping, dragging, and dropping graphics on the touch-screen device. They found that children's sensorimotor abilities, as measured by a validated sensorimotor evaluation, were correlated with touch performance such as task completion and accuracy rate (Vatavu et al., 2015). This study helps us to better understand young children's ability to interact with touch screens, which in turn help us to create better age-appropriate touch interface designs. This is particularly relevant now, as many educational tools designed for young children come in the form of touchscreen apps. In this chapter, we explore one such app for children called ScratchJr, a freely downloadable tablet app that engages children ages 5-7 years in computer programming.

In order to apply Learning Analytics to evaluate computational thinking, researchers at the DevTech Research Group have been using Google Analytics to understand the kind of coding experiences and skills developed with the ScratchJr application. This chapter will first explore the opportunities and obstacles of using Learning Analytics as a computational thinking assessment tool. We will then focus on a particular Learning Analytics Tool: Google Analytics, and discuss how we have used it to analyze young children's coding abilities by examining their usage patterns in ScratchJr. Then, we will discuss how analyzing these usage patterns may be translated into an understanding of computational thinking in early childhood. Finally, we will highlight a few specific drawbacks of Google Analytics as a Learning Analytics tool and discuss future directions for this area of research.

Learning Analytics: Applying Data Analytics to Education

Learning Analytics is a subfield Data Analytics, which is the practice of collecting and analyzing raw data in order to make meaningful conclusions. Data Analytics has proven to be instrumental in the success of many industries including medicine and many areas of business. By analyzing data regarding anything from scheduling to diagnoses, analytics in the field of medicine aids care efficiency, reduces administrative burdens, and accelerates diagnoses. In business, analytics aids in product design, sales, and business efficiency. Netflix, for example, has been one of the most successful users of Data Analytics to tailor their product to the user – which greatly contributes to its success as a business. As Andrew Medal from *The Entrepeneur* points out, "big data can help make sense of the information gathered, such as retention cost, average transaction value and even customer satisfaction"

(Medal, 2017). Businesses thrive on Data Analytics to better market their products because Data Analytics helps them to understand the audience.

Education, however, is not about selling a product or improving efficiency of medical practices – it is about understanding how children are learning and using that knowledge to improve educational practices (Piaget, 1952; Vygotsky, 1978; Papert, 1980). Computational thinking, and education as a whole, should be measured by the child's thoughts and reasoning process, not just the programs they produce or the test scores they receive. Additionally, education requires a high level of flexibility and attention to individualism. Student learning patterns cannot be captured in one number, while company profits usually can. Applying Data Analytics to education could be useful, but requires a larger emphasis on how to capture *learning* through data rather than just learning outcomes. To bring Data Analytics to education, we must think about what types of variables we can examine that will indicate something about a child's learning process.

The other drawback of adapting Data Analytics to education is that in order to draw conclusions from data, it must be guaranteed that the data were collected in environments that are consistent with one another. In order to "demonstrate an overall effect, every learner in an intervention needs to have the same experience in the intervention, and the comparison group needs to be held constant in order for the difference to be consistent." (Cope, 2016) Children's learning and learning environments are never consistent. The lack of consistency is for good reason: teachers must constantly adapt to their students' needs and alter lesson plans quickly to optimize engagement. For example, one group of children may be incredibly excited about hatching caterpillars while another group is producing a movie on ScratchJr with 4 iPads side by side. Should the teacher force one group to put down their activity and engage all together on one task? Or should the teacher allow each student to pursue their project? If they allow each child to pursue what they want, one group will gain much more experience on ScratchJr than the other, skewing that classroom's ScratchJr usage data. Should that sort of adaptable teaching practice be discouraged in order to collect consistent data? Probably not. Therefore, due to the setup of early childhood classrooms and the dependence on flexibility, drawing reliable conclusions from Learning Analytics alone isn't always possible.

Countless other industries have benefitted from using Data Analytics. While Data Analytics has incredible promise, it is clear that bringing Data Analytics to education through Learning Analytics faces many challenges that need consideration. Despite these drawbacks, there are still many Learning Analytics tools in use today. In the next section, we will discuss the how the use of Learning Analytics in education can improve educational practices as a whole.

Learning Analytics to Improve General Education Practices

The reason Data Analytics has been so beneficial to large companies and other industries is because it allows researchers to observe large overarching trends of clients, patients, and customers. From seeing these trends, industries are able to adapt practices and products to meet the needs of the masses. With education, this mindset can be problematic. If teachers just adapted practices to the majority, then students with atypical learning styles would inevitably fall through the cracks. What collecting learner data can do, however, is find trends in the majority *and* the minority. Using Learning Analytics, trends could be found among students with atypical learning styles across classrooms and can provide teachers with more knowledge on how to address those atypical student needs. It's no longer just one teacher observing the learning in her classroom, but now many teachers combining observations across classrooms, revealing unseen patterns. The use of Learning Analytics has the ability to allow educators to make connections across classrooms and isolate variables that may not be apparent when looking at just one child's learning patterns.

While not specific to computational thinking, analyzing learning trends can improve assessment as a whole because it allows educators to gauge their students' progress, and in turn, create more specialized and appropriate learning standards. All forms of learning are unique to the student, but in order to ensure success of the masses, school districts almost always align their curricula and assessments with a form of learning standards. It's broadly agreed upon that measuring student progress and setting specific goals are crucial to school improvement (Schmoker, 1999). In order to create these standards, the scope of the skills, knowledge, and all other factors impacting learning must be well defined. In order to define this scope, data is a very promising solution (NECRL, 2004). A curriculum designer can easily set unrealistic goals for students if they cannot gauge student progress or starting points accurately. Data can be used to define the scope of learning styles and current knowledge among students, which then helps teachers and education policy makers to design realistic and informed learning frameworks. We argue that with the right measures, defining learning standards and improving assessment with Learning Analytics can be extended and specialized to defining computational thinking standards and assessment as well.

Additionally, looking at trends allows us to identify gaps and irregularities in student performance and isolate contributing factors. Trends give us an idea of the norm and therefore allow us to highlight learning patterns that are outside of that "norm". While children's learning never follows a "normal" path and deviations from the norm are typical and healthy, identifying and addressing the deviations early on can prevent learning and opportunity gaps from growing. With real-time feedback of student performance, teachers can be proactive in adapting their teaching styles.

By using Learning Analytics, teachers can become more self-aware of their practices and can have immediate feedback on their teaching. As cited in a study conducted by Zwieg and colleagues, research suggests that by monitoring students' learning and growth through collecting learning outcome data, educators become more knowledgeable about their own capacities, and can develop plans for improvement (Zweig et al., 2015). If specialized correctly, Learning Analytics could help identify learning gaps specific to computational thinking tasks. With the ability to identify struggling students more easily, teachers would be able to work more strategically to ensure computational thinking proficiency across the classroom.

Depending on which types of information are collected, Learning Analytics can provide insights about student demographic trends, learning patterns, geographic variables, and so on- the insights are fairly limitless if the right data is collected. One of the greatest strengths of Data Analytics in general, specifically in education, is the ability to zoom in and out. As said previously, trends across big groups are the most frequent focus of Data Analytics, but with the right tools, data can also be collected from a specific user. In their paper about Data Analytics coming to schools, Bill Cope and Mary Kalantzis pointed out that:

In the case of big data, scaling up or down, zooming in or out, offers a range of viable perspectives on a shared data source—a micro-moment of feedback in the writing process, for instance, to larger patterns of revision, to overall progress of a student or a class or cohort measured in terms of writing standards over a longer time frame. (Cope, 2016)

Not only does Learning Analytics allow researchers to zoom in and out on student performances, but it also has the potential to predict student performance. This type of prediction is possible through Machine Learning: the concept of computer algorithms learning from past data and using the learned patterns to predict future data points. Two ways in which this can be helpful to education is predicting student performance and improving retention. As Anozie et al., found, "by 'learning' about each student, the machine learning model can find out weaknesses and suggests ways to improve, such as additional lectures or study additional literature" (Anozie, N., Junker, B. W., 2006). This prediction is another way in which Learning Analytics can be used to close achievement gaps and help teachers to identify students with specific needs; "by identifying 'at risk' students, schools can reach out to those students and get them the help they need to be successful" (Đambić, G., Krajcar, M. & Bele, D., 2016). By "learning" about each student, the technology can identify weaknesses and help students to succeed.

While machine learning is one of the most promising applications of Data Analytics, it must be noted that predicting student outcomes is not always as accurate as it should be to be informative. Since technology in education is fairly new and constantly changing, there is a large margin of error and false conclusions can easily be made. Using large amounts of data without any theoretical framework may lead to inaccurate conclusions. If we were to throw a large number of random variables into a predictive statistical model, it is likely that the model would show us significant relationships between variables that are totally unrelated in reality. This scenario is known as, data fishing or data dredging. Longo and Montevil (2018) explained the danger of data analysis without any theoretical framework and provided an example that there is a false correlation between people who drowned after fishing and marriage rate in Kentucky. These types of correlations could easily appear in Data Analytics for education, possibly leading educators and policy makers to devote attention to the wrong areas of education.

Despite notable limitations, Learning Analytics has the potential to be instrumental in improving learning standards, tracking student progress, closing achievement gaps, and understanding learners on a deeper level. In the next section, we will discuss how specific aspects of Learning Analytics tools lend themselves well to assessing mental processes such as computational thinking.

Learning Analytics to Assess Computational Thinking

Assessing Computational Thinking

Computational Thinking (CT) is a highly valuable and transferrable skill for everyday life that is the byproduct of coding (Chp.8, Relkin). According to Bers (2020), CT emerges from the "Seven Powerful Ideas" of computer science for early childhood. The Powerful Ideas include algorithms, modularity, control structures, debugging, hardware/software correspondence, debugging, and the design process. These computational thinking concepts are not only fundamental to computer science, but also to understanding the world around us. For example, the concept of modularity can help children think about solving complex problems in life by breaking down a solution into multiple small steps. One example of this could be putting on clothes. To a five-year-old, the task is not trivial. For this very reason, we often break the task up into steps: start with your head, then get your arms through, and then finally pull the shirt down over your body! Then, when faced with putting on a sweatshirt, the child will be able to recall these steps and apply them to the new task—thus successfully applying modularity to their everyday challenges.

While it's clear that computational thinking is crucial in early childhood education, it is much less clear how this skill should be assessed. As we have mentioned, an ongoing challenge in the field of assessment is how to measure a thought process, rather than just an outcome. There are many ways to address this issue including type of assessment, timing of assessment, and mode of implementation. The use of Learning Analytics as an assessment does not offer a solution to assessing CT, but it does offer benefits over other types of assessments-- specifically when thinking about assessment for early childhood education. In the next section we will discuss how Learning Analytics has the potential to be an effective assessment tool and could be refined to assess computational thinking in the future.

Learning Analytics as an Ongoing Assessment

While many Learning Analytics tools currently lack the level of specificity of measures needed to examine computational thinking, the potential is there. One of the most notable benefits to using analytics as an assessment tool is that unlike tests or project showcases, analytics offers insights that come from ongoing observation rather than sporadic intervention. This benefit is especially notable when working with young pre-school and elementary-aged children. This type of information collection is much more developmentally appropriate for young children, and therefore offers more informative data points. The National Association for the Education of Young Children (NAEYC) states that assessments that collect learning data over a longer period of time, like analytics, are the most appropriate assessment approach for young children whose development is highly complex, dynamic, and often erratic and uneven (Ackerman & Coley 2012). This makes it difficult to capture their learning through one-time assessments that provide only a snapshot of a child in a particular moment (Riley-Ayers, 2018). Observing children over longer periods of time allows educators and researchers to get a more accurate picture of a child's development and thought processes, whereas a "snapshot" such as one test or one project is likely going to be inconsistent and uninformative. With this more accurate picture, comes the possibility of isolating specific thought processes, such as computational thinking. Learning Analytics tools, while not there yet, offer the possibility of observing infinite small moments rather than just the product of a test or a project. By using ongoing observation as an assessment tool, the possibility of isolating thought processes and cognitive patterns such as computational thinking becomes much more within reach than it is with traditional assessment measures.

Despite the promise that Learning Analytics tools have, the data collection methodologies that exist right now in the field of education don't yet have the capabilities to isolate individual students' learning processes. While we are able to collect student data from Learning Analytics tools, like Google Analytics, drawing conclusions from that data about computational thinking takes a fair bit of speculation. Learning Analytics offers huge amounts of information and provides many benefits to education as a whole. Although not specialized enough to give us information about something as complex as computational thinking, Learning Analytics does provide incredible amounts of information about a student. In the next section, we will discuss how the DevTech Research Group employs Google Analytics for the ScratchJr coding app as a Learning Analytics tool.

LEARNING ANALYTICS TOOL: GOOGLE ANALYTICS

Google Analytics is a web analytics service that is used to track and analyze user metrics, evaluate usage patterns, user behavior, and much more. In education research, the use of Google Analytics with learning tools can help educators in finding advanced methods for enhancing student learning. DevTech researchers use Google Analytics to track and report the traffic of the ScratchJr app. Google Analytics has many features that allow researchers to review how and when students are interacting with the ScratchJr app. The buildings blocks of Google Analytics reporting are *dimensions* and *metrics* as shown in Figure 1. *Dimensions* are attributes of the data such as the city, state, or country where the data traffic is coming from, what browser is used, or what language the user was using ScratchJr in. *Metrics* are the numerical data points collected such as amount of time spent on the app, the number of users from a specific location, or how many new users there are on a given day.

Figure 1. ScratchJr usage data from Jan. 1, 2021-Mar. 30, 2021. Dimension = Country. Metrics = Users, New Users, etc.
Source: IGI, 2021

	Country ▾	↓Users	New users	Engaged sessions	Engagement rate	Engaged sessions per user	Average engagement time
	Totals	3,553,432 100% of total	1,927,554 100% of total	18,060,194 100% of total	92.99% Avg 0%	5.082 Avg 0%	57m 25s Avg 0%
1	(other)	2,266,378	0	0	0%	0	0m 00s
2	United States	1,070,038	432,444	7,225,758	92.03%	6.753	1h 13m
3	Japan	334,307	244,504	1,747,285	94.69%	5.227	1h 00m
4	United Kingdom	258,155	111,937	974,645	92.43%	3.775	44m 56s
5	Australia	155,357	69,020	492,463	95.55%	3.17	35m 49s
6	India	143,456	105,462	701,790	93.1%	4.892	40m 55s
7	Canada	140,271	60,811	652,746	93.34%	4.653	1h 00m
8	Sweden	92,482	34,897	357,914	92.75%	3.87	42m 34s
9	Peru	91,170	71,043	245,700	94.92%	2.695	16m 59s
10	Germany	88,669	59,702	305,084	91.93%	3.441	32m 06s

The main measure in Google Analytics is number of users. As of January 2021, ScratchJr has had over 19 million users accounted for by Google Analytics. It must be noted, though, that Google Analytics does not record the usage on the oldest versions of ScratchJr, and therefore users working on older devices, such as non-touchscreen Chromebooks, are not accounted for in this user count. Within the user count, we are able to see who is a new user (someone that is opening the app on their device for the first time) or returning users (someone that has already used ScratchJr for at least one session). Additionally, user data can be parsed by geography. Not only can we see the city, state, and country a user is from, but we can also see whether that user selected "Home" or "School" for their app setting. This tells us where a student is located when using the app and allows us to make inferences about how structured their ScratchJr playtime may have been. For example, if a user is registered as "School", it is likely that they are using the app in the context of a class with instruction and supervision. If the user selected "Home", on the other hand, we might infer that the child is using ScratchJr on their own time and at their own pace. We acknowledge that these inferences are much weaker during times of remote-schooling due to the COVID-19 pandemic, but we still want to emphasize the general importance of acknowledging a child's setting and the impact it can have on their learning. Knowing the structure of users playing time could be very powerful and informative for future ScratchJr curricula development. Google Analytics also collects the time and duration that the app is used, as shown in Figure 1. The duration of sessions is recorded each time the app is opened and then closed and is often used in analysis to indicate levels of engagement and interest from the users.

In addition to knowing when and where users are on the app, Google Analytics also offers insights into how the user may be using the app. Google Analytics collects event data, meaning any type of action a user does on the app is recorded. In ScratchJr, this means that all the block types that are selected for programs, page additions, presentation mode, paint editor, new characters, etc., are all recorded and averaged across geographical areas. For example, in Figure 2, Google Analytics will record the types of blocks put in the programming area, the character used, and any design choices the user makes on the app. It is important to note that this data is never shown for a single user: the closest we come to identifying specific users is by the city they are in, but rarely is that used in our analysis. Therefore, all details about how the app is used is only informative about how the app is used across a large group, not from user to user.

Figure 2. Example of a ScratchJr project
Source: IGI, 2021

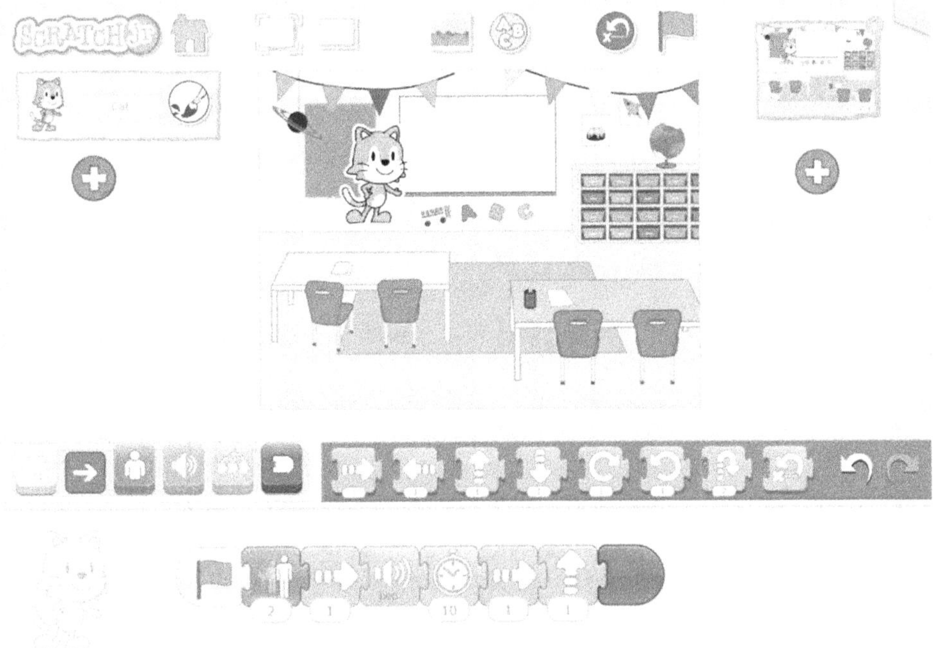

In short, Google Analytics data can be used to define the scope of research questions. It allows us to gain an overview of the app user demographics and how the app is most commonly being used. Beyond just broad demographic information, it allows us to create user personas and predict behaviors of those personas, allowing for more effective and targeted app development in the future. In order to highlight the research possibilities that Google Analytics offers, we will spend the next section discussing our newest findings on ScratchJr usage at home versus school, and how this was impacted by the COVID-19 pandemic.

Global Usage of ScratchJr Learning Analytics

ScratchJr has been using Google Analytics to track the app usage data since the early release. The purpose of exploring ScratchJr Analytics is to gain deeper insights into children's interaction with the app and to investigate how usage pattern may be related to CT (Leidl et al., 2017). Since 2015, there have been over 19 million users across 194 countries world-wide. Users have created over 52 million projects and edited projects over 70 million times. Globally, there was an increasing trend in the number of ScratchJr sessions from 2016 to 2019 as shown in Figure 3. Children commonly used ScratchJr in schools, as shown by the fact that usage dropped during the US summer break from June-August. However, the usage trends in 2020 were different due to school closures caused by the COVID-19 pandemic.

Figure 3. Number of ScratchJr users from 2016-2020. Note that in 2019, ScratchJr Analytics collected data only from January to October of 2019 due to a transition into the new Google Analytics version (Firebase)
Source: IGI, 2021

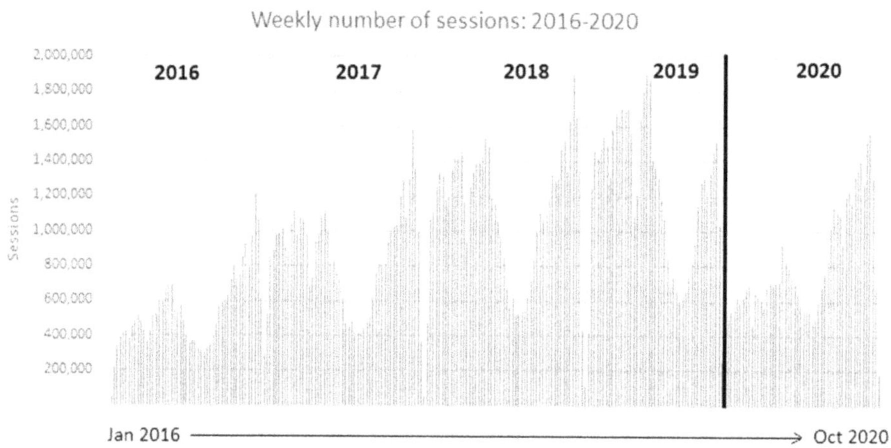

The daily ScratchJr usage pattern differed across countries, the top ten countries with the most sessions in 2020 are shown in Figure 4. Figures 5 and 6 compare the number of users from January 1, 2020 to February 30, 2020 at home and at school in United States and Japan. This particular time period was chosen because it was before the COVID-19 pandemic severely impacted the world. Figure 5 shows the similarity between the usage pattern across settings, home and school, in the U.S. In both home and school settings, ScratchJr usage by US children peaked on the weekdays and dropped on the weekends. In contrast, shown in Figure 6, Japan showed the reversed pattern of increases and drops between home and school users. When comparing between each setting in Japan, there was more usage at home on the weekend and more usage at school on the weekday.

Figure 4. The top 10 countries with most ScratchJr sessions in 2020
Source: IGI, 2021

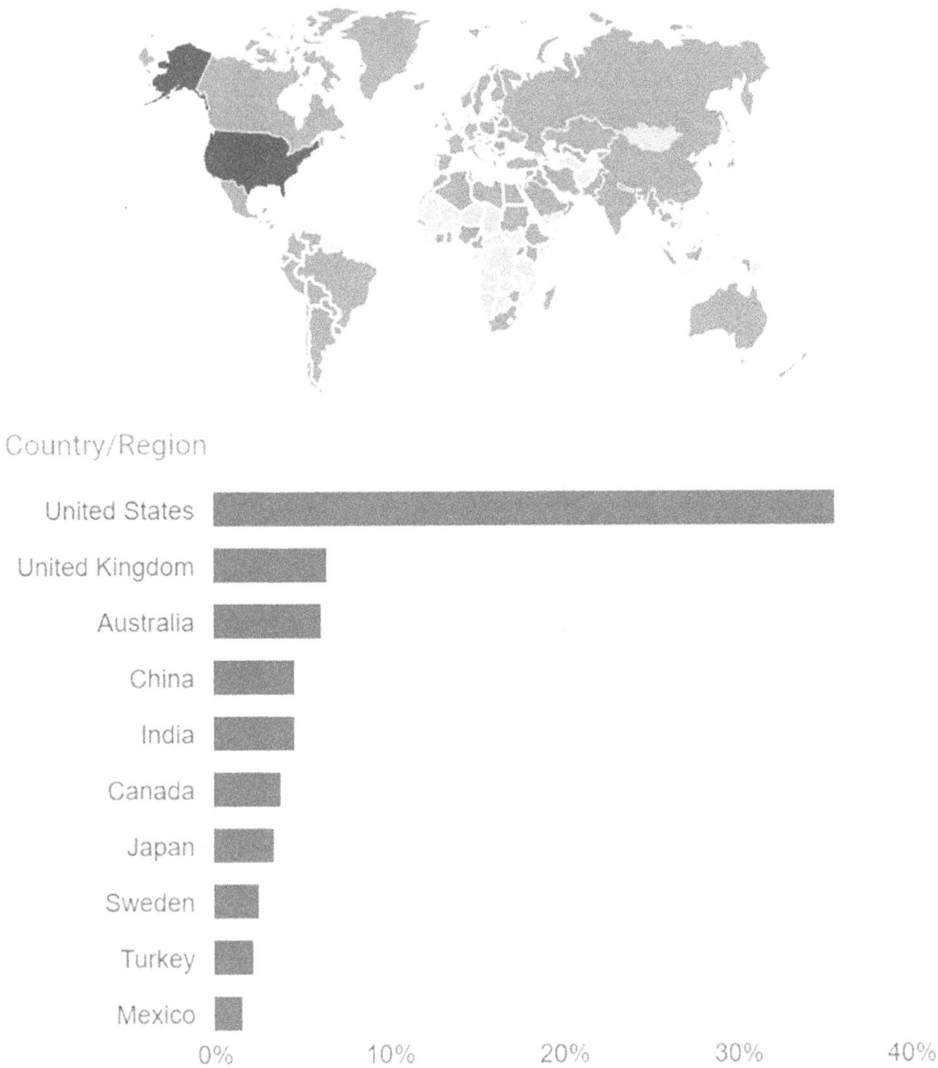

A possible explanation for the different usage pattern between USA and Japan is that ScratchJr is commonly used at school in the US, therefore the pattern follows school periods. We also saw peaks of home usage in the US when children should have been at school. This could be explained by the possibility that the app may have tracked tablets used at schools as being used at home as there were reports that children in the US also bring their home tablet to use at school (*Trends in Digital Learning: Students' Views on Innovative Classroom Models*, 2014).

Figure 5. USA ScratchJr Users: January 1, 2020 to February 30, 2020. Dotted Line= Home users, Solid Line= School users
Source: IGI, 2021

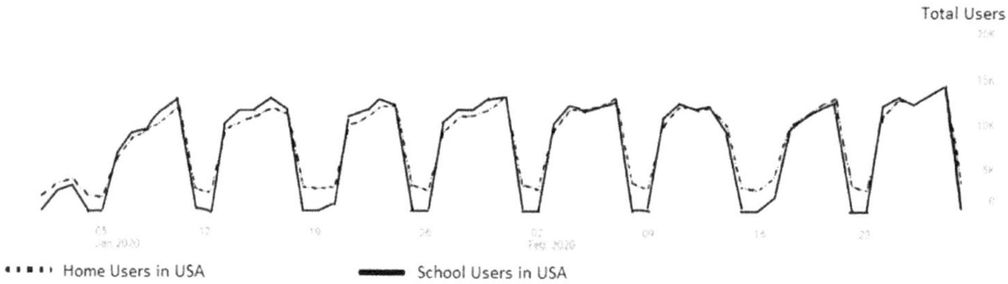

Figure 6. Japan ScratchJr Users: January 1, 2020 to February 30, 2020. Dotted Line= Home users, Solid line= School users
Source: IGI, 2021

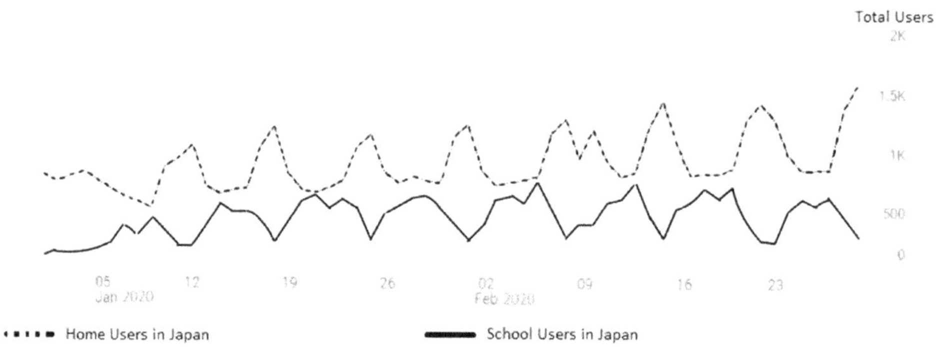

Figure 7. Japan ScratchJr Users: August 1, 2020 to March 30, 2021. Dotted Line= Home Users, Solid Line= School Users
Source: IGI, 2021

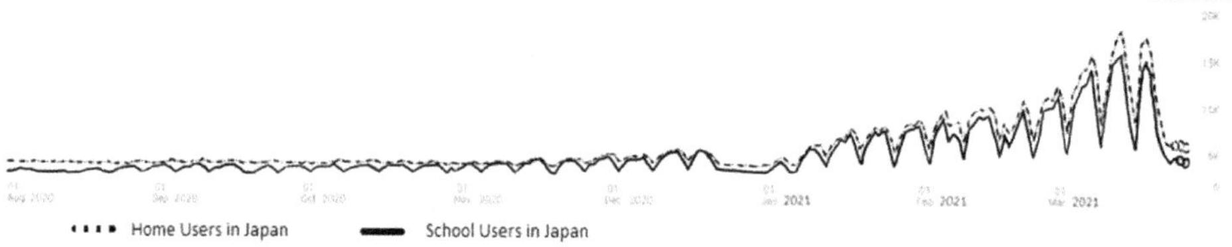

Figure 8. The percentage of projects shared increased significantly compared 2019 to 2020
Source: IGI, 2021

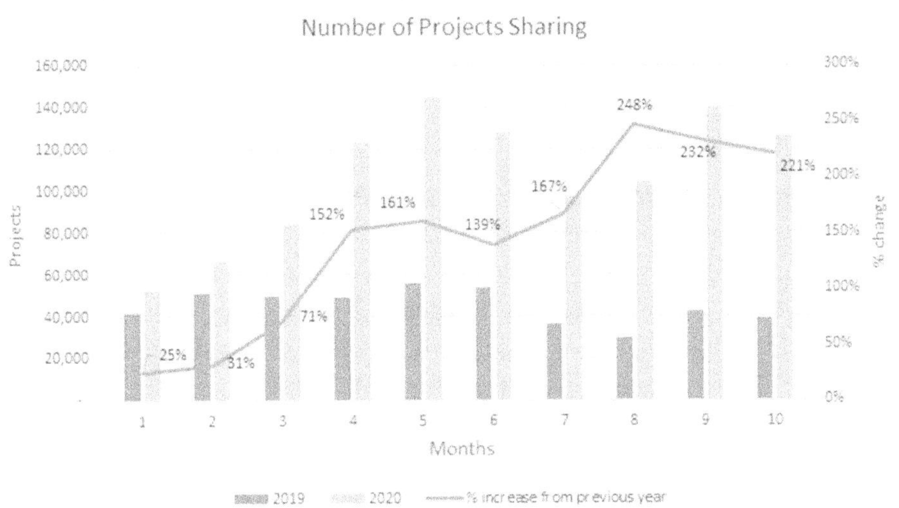

Different policies for computer science education in early childhood may also affect the usage pattern in the US and Japan. Under the Coding for All policy released in 2016, it is mandatory for schools to teach computer science to students from kindergarten on (The Whitehouse, 2016). Many US schools have incorporated ScratchJr into their curriculum, such as the K-2 CS curriculum in San Francisco, CA (*K-2 Computer Science Curriculum, n.d.*). In Japan, mandatory computer science classes were planned to be implemented in primary school starting in 2020 (Bocconi et al., 2016). It is unclear if the policy was fully implemented in 2020; however, there was a usage spike in Japan starting in 2021, which may imply that the policy was implemented. Additionally, in Figure 7, the usage pattern at home and school in Japan became more similar to the usage trend in the US, where the usage pattern follows the school periods (higher usage on the weekdays).

Change in ScratchJr Usage Pattern From COVID-19

We wanted to study how the COVID-19 pandemic impacted ScratchJr usage across the world. In 2020, due to COVID-19, 1.7 billion children were affected by the school closure (Gouedard & Pont, 2020). The consequences from the COVID-19 pandemic can clearly be seen in Figure 3, with the number of ScratchJr sessions decreased from the previous years, especially in the first half of 2020 when the pandemic started.

School closures strongly and negatively impacted the number of ScratchJr sessions. However, the analytics show that children globally shared their ScratchJr projects with others (e.g., via email) more than ever during social distancing. Particularly, in Figure 8, the number of ScratchJr projects shared increased by more than 100% after April 2020 compared to the same months in 2019. The percentage of projects shared skyrocketed over 200% after August 2020, which might be due to remote learning where students had to send projects to teachers.

What Can We Learn From the Usage Patterns?

By collecting usage data through Google Analytics, we are able to see how and when children are coding. While this data does not give us insights into the thought processes of the children, it does give us an additional viewpoint onto how children are coding. Different studies found that a type of Learning Analytics, or the type of coding blocks used may imply children's coding understanding level (Emerson et al., 2020; Price & Price-Mohr, 2018). Emerson et al. (2020) identified common misconceptions in introductory programming by analyzing how students used coding blocks. The students that lacked conceptual knowledge used fewer types of coding blocks on average compared to the other groups of students. Furthermore, the students that had a disorganized programming style had longer sequences with more errors. A different study by Price and Price-Mohr (2018) reported that students that were expert programmers (college students) spent less time and were able to code more intentionally than novices (elementary students). We cannot assess CT from just these measures alone, but we do gain a more holistic view of the child's knowledge, which ultimately allows us to make more accurate inferences about their mental processes.

A Learning Analytics study on ScratchJr analyzed young children's block usage types: 1) Coding duration; 2) Block complexity; 3) Block category; 4) Block consistency (Unahalekhaka & Bers, in press). The researchers found differences in how children in the U.S. used coding blocks at home compared to at school. Particularly, children at home used more advanced and more diverse coding blocks than children at school. The more advanced coding blocks for children ages 5-7 include control blocks such as repeat a command or trigger blocks such as if-then commands. Although the Learning Analytics alone cannot measure children's CT, some of ScratchJr coding blocks can imply CT concepts. The control coding blocks (repeat loop) align with the "control structure" concept, which is one of the seven computational thinking concepts for early childhood (Bers, 2020).

Results from Unahalekhaka and Bers (In Press) also suggested that children at home most likely had a more exploratory style of coding due to more freedom in playing with ScratchJr. Children at home also spent more time using the paint editor, a feature that supports character and background aesthetic customization. In contrast, children at school may have to follow a fixed curriculum that is focusing on using certain coding blocks. Therefore, they spent less time decorating or painting the characters and longer time with the actual coding. Furthermore, children at school used more similar coding blocks complexity level across days (Unahalekhaka & Bers, in press). This study implied that young children may need different computer science learning pedagogies across formal and informal settings to target CT centric actions. For example, teachers may give more free play time during each lesson, where children can code as they wish. Furthermore, parents may also want to provide more step-by-step scaffolding before children can code freely. A study by Strawhacker et al. (2018) reported that the teaching style such as having flexible teaching plans with open-ended coding time is beneficial for children's learning.

In this section, we illustrated how Google Analytics can report a diverse ScratchJr usage pattern across countries, circumstances, and learning settings. However, usage patterns at the aggregated level alone cannot tap into understanding an individual child's CT. To do so, we may need a different analytics tool to collect individual student's data. With more individualized learning data, we can then compare them to individual's CT scores like the measures from Grover et al. (2017). In the next section, we will expand on the limitations of Google Analytic as a tool to understand children's learning.

LIMITATIONS OF GOOGLE ANALYTICS AS A LEARNING ANALYTICS TOOL

While Google Analytics as a whole has immense potential for education, it is not able to offer the level of specificity required for educators to get useful information on students learning. For example, Google Analytics can only localize data to the city, which is far from highlighting an individual student. Most cities have multiple schools, which have multiple classrooms, which have upwards of 20 students using learning tools, such as ScratchJr, at a time. To assess a child's knowledge, particularly in early childhood, it is rarely accurate to look at big numbers and averages. Such reports often represent aggregate views of student and school data devoid of any strategic or tactical judgements. The inability to narrow down our focal point on analytics means that we can never look at just one child's work, which is hugely limiting when trying to assess something as variable as computational thinking.

Assessing computational thinking requires not only a close look at each individual student, but also a report on that child's learning process and style. Current Learning Analytics tools are very limited in types of measures that can be collected. Google Analytics can show us how a child is using the app in terms of what blocks are used and what features are employed, but there is no current way to measure the thoughts behind block and feature choice. To draw conclusions about computational thinking takes a fair bit of speculation. Without more specialized measures, drawing conclusions about thought process and reasoning is not possible.

FUTURE DIRECTIONS

Google Analytics offers huge amounts of information, but the lack of specificity and adaptation to the field of education prevents most current Learning Analytics tools from reaching their potential in the type of information that can be shown about a student. Therefore, we hope that future development in this field includes finding a way to collect individualized student data without breaching privacy obligations of students. We also hope that new, more qualitative measures can be incorporated into the data collection. Determining the specific measures that could indicate thought processes like computational thinking requires further research.

One potential promising subfield of Learning Analytics is Multi Modal-Learning Analytics (MMLA). While Learning Analytics has become more widely used in higher education, there is still lack of research on how, or if, Learning Analytics can be applied to understand learning development in the younger population. MMLA does not require the screen time interaction, instead, it collects physiological data such as speech cues, eye gazes, facial expression, and heart rate (Oviatt et al., 2018). Some researchers claimed that MMLA can assess a learner's engagement with the task, in contrast with "regular" Learning Analytics that can only track for usage pattern (e.g., number of clicks, session duration) (Crescenzi-Lanna, 2020). This could be a potential direction to take Learning Analytics for assessing thinking patterns in early childhood.

CONCLUSION

Learning Analytics, the sub field of Data Analytics that pertains to education, is becoming increasingly utilized by educators and researchers. It provides more information about students than has ever been possible, and therefore holds huge potential for improving teaching and understanding students better. As discussed, however, there are many aspects of Data Analytics that do not yet translate well to Learning Analytics. These drawbacks, though, are an effect of a newly emerging field that has not grown to its full potential yet. With further research and development, Learning Analytics could be an incredibly useful tool for understanding computational thinking and learning processes in general.

REFERENCES

Baker, R. S., & Siemens, G. (2014). Educational data mining and learning analytics. In R. K. Sawyer (Ed.), *Cambridge handbook of the learning sciences* (2nd ed., pp. 253–274). Cambridge University Press. doi:10.1017/CBO9781139519526.016

Berland, M., Baker, R. S., & Blikstein, P. (2014). Educational Data Mining and Learning Analytics: Applications to Constructionist Research. *Technology, Knowledge and Learning*, *19*(1–2), 205–220. doi:10.100710758-014-9223-7

Bers, M. U. (2020). *Coding as a Playground: Programming and Computational Thinking in the Early Childhood Classroom*. Routledge. doi:10.4324/9781003022602

Bocconi, S., Chioccariello, A., Dettori, G., Ferrari, A., & Engelhardt, K. (2016). *Developing computational thinking in compulsory education: Implications for policy and practice*. Publications Office. https://data.europa.eu/doi/10.2791/792158

Brunner, C., Fasca, C., Heinze, J., Honey, M., Light, D., Mandinach, E., et al. (2005). *Linking data and learning: The Grow Network study*. Academic Press.

Cope. (2016). *Big Data Comes to School: Implications for Learning, Assessment, and Research*. University of Illinois. https://journals.sagepub.com/doi/pdf/10.1177/2332858416641907

Crescenzi-Lanna, L. (2020). Multimodal Learning Analytics research with young children: A systematic review. *British Journal of Educational Technology*, *51*(5), 1485–1504. doi:10.1111/bjet.12959

Emerson, A., Smith, A., Rodriguez, F. J., Wiebe, E. N., Mott, B. W., Boyer, K. E., & Lester, J. C. (2020). Cluster-Based Analysis of Novice Coding Misconceptions in Block-Based Programming. *Proceedings of the 51st ACM Technical Symposium on Computer Science Education*, 825–831. 10.1145/3328778.3366924

FACT SHEET: President Obama Announces Computer Science For All Initiative. (2016, January 30). Whitehouse.Gov. https://obamawhitehouse.archives.gov/the-press-office/2016/01/30/fact-sheet-president-obama-announces-computer-science-all-initiative-0

Fischer, C., Pardos, Z. A., Baker, R. S., Williams, J. J., Smyth, P., Yu, R., Slater, S., Baker, R., & Warschauer, M. (2020). Mining Big Data in Education: Affordances and Challenges. *Review of Research in Education*, *44*(1), 130–160. doi:10.3102/0091732X20903304

Gašević, D., Dawson, S., & Siemens, G. (2015). Let's not forget: Learning analytics are about learning. *TechTrends*, *59*(1), 64–71. doi:10.100711528-014-0822-x

Ginsburg, H., & Baroody, A. (2003). *TEMA-3 examiners manual*. Pro-Ed.

Gouëdard, P., Pont, B., & Viennet, R. (2020). *Education responses to COVID-19: shaping an implementation strategy*. OECD Education Working Papers, No. 224.

Grover, S., Basu, S., Bienkowski, M., Eagle, M., Diana, N., & Stamper, J. (2017). A Framework for Using Hypothesis-Driven Approaches to Support Data-Driven Learning Analytics in Measuring Computational Thinking in Block-Based Programming Environments. *ACM Transactions on Computing Education*, *17*(3), 1–25. doi:10.1145/3105910

Ifenthaler, D., & Yau, J. Y.-K. (2020). Utilising learning analytics to support study success in higher education: A systematic review. *Educational Technology Research and Development*, *68*(4), 1961–1990. doi:10.100711423-020-09788-z

K-2 Computer Science Curriculum—Orange. (n.d.). Retrieved April 10, 2021, from https://sites.google.com/sfusd.edu/k-2cs/orange

Kent, J. (2020). *4 Emerging Strategies to Advance Big Data Analytics in Healthcare*. HealthITAnalytics. https://healthitanalytics.com/news/4-emerging-strategies-to-advance-big-data-analytics-in-healthcare

Leidl, K. D., Bers, M. U., & Mihm, C. (2017). Programming with ScratchJr: A review of the first year of user analytics. *Proceedings of the International Conference on Computational Thinking Education*.

Medal, A. (2017). *How Big Data Analytics Is Solving Big Advertiser Problems*. Entrepreneur. https://www.entrepreneur.com/article/293678

NECRL. (2004). *Using Data to Guide School Improvement*. Learning Point Associates. North Central Regional Educational Laboratory. https://files.eric.ed.gov/fulltext/ED518630.pdf

Oviatt, S. (2018, October). Ten Opportunities and challenges for advancing student-centered multimodal learning analytics. In *Proceedings of the 20th ACM International Conference on Multimodal Interaction* (pp. 87-94). ACM.

Papert, S. (1980). *Mindstorms: Children, computers, and powerful ideas*. Basic Books, Inc.

Piaget, J. (1952). The origins of intelligence in children *No. 5* (Vol. 8). International Universities Press.

Price, C. B., & Price-Mohr, R. M. (2018). An Evaluation of Primary School Children Coding Using a Text-Based Language (Java). *Computers in the Schools, 35*(4), 284–301.

Riley-Ayers. (2018). *Excerpt from Spotlight on Young Children: Observation and Assessment.* Naeyc. https://www.naeyc.org/resources/pubs/books/excerpt-from-spotlight-observation-assessment

Roberts, J. D., Chung, G. K. W. K., & Parks, C. B. (2016). Supporting children's progress through the PBS KIDS learning analytics platform. *Journal of Children and Media, 10*(2), 257–266.

Strawhacker, A., Lee, M., & Bers, M. U. (2018). Teaching tools, teachers' rules: Exploring the impact of teaching styles on young children's programming knowledge in ScratchJr. *International Journal of Technology and Design Education, 28*(2), 347–376.

Thornton-Lang. (2012) *Observation as a formal assessment tool in early childhood classrooms: A professional development module.* University of Northern Iowa. https://scholarworks.unit.edu/cgi/viewcontent.cgi?article=1238&context=grp

Trends in Digital Learning: Students' Views on Innovative Classroom Models. (2014). *Project Tomorrow.* https://tomorrow.org/speakup/2014_OnlineLearningReport.html

Unahalekhaka, A., & Bers, M. U. (in press). Taking Coding Home: Analysis of ScratchJr Usage in Home and School Settings. *Educational Technology Research and Development.*

Vatavu, R. D., Cramariuc, G., & Schipor, D. M. (2015). Touch interaction for children aged 3 to 6 years: Experimental findings and relationship to motor skills. *International Journal of Human-Computer Studies, 74*, 54–76.

Vygotsky, L. S. (1978). *Mind in society: The Development of higher psychological processes.* Harvard University Press.

Zweig. (2015). *Data collection and use in early childhood education programs: Evidence from the Northeast Region.* Regional Educational Laboratory. https:/files.eric.ed.gov/fulltext/ED555737.pdf

ADDITIONAL READING

Agus, R., & Mohamad Samuri, S. (2018). Learning Analytics Contribution in Education and Child Development: A Review on Learning Analytics. *Asian Journal of Assessment in Teaching and Learning*, *8*, 36–47. doi:10.37134/ajatel.vol8.4.2018

Alonso-Fernández, C., Calvo-Morata, A., Freire, M., Martínez-Ortiz, I., & Fernández-Manjón, B. (2019). Applications of data science to game learning analytics data: A systematic literature review. *Computers & Education*, *141*, 103612. doi:10.1016/j.compedu.2019.103612

Blikstein, P., Worsley, M., Piech, C., Sahami, M., Cooper, S., & Koller, D. (2014). Programming Pluralism: Using Learning Analytics to Detect Patterns in the Learning of Computer Programming. *Journal of the Learning Sciences*, *23*(4), 561–599. doi:10.1080/10508406.2014.954750

Chen, G., Rolim, V., Mello, R. F., & Gašević, D. (2020). Let's shine together!: A comparative study between learning analytics and educational data mining. *Proceedings of the Tenth International Conference on Learning Analytics & Knowledge*, 544–553. 10.1145/3375462.3375500

Irizarry, R. A. (2020). The Role of Academia in Data Science Education. *Harvard Data Science Review, 2*(1).

Liu, M. C., & Huang, Y. M. (2017). The use of data science for education: The case of social-emotional learning. *Smart Learn. Environ.*, *4*(1), 1. doi:10.118640561-016-0040-4

Milicevic, A., Ivanovic, M., & Budimac, Z. (2017). Data science in education: Big data and learning analytics. *Computer Applications in Engineering Education*, *25*. Advance online publication. doi:10.1002/cae.21844

Rodríguez, A. O. R., Riaño, M. A., García, P. A. G., Marín, C. E. M., Crespo, R. G., & Wu, X. (2020). Emotional characterization of children through a learning environment using learning analytics and AR-Sandbox. *Journal of Ambient Intelligence and Humanized Computing*, *11*(11), 5353–5367. doi:10.100712652-020-01887-2

Wang, L., Sy, A., Liu, L., & Piech, C. (2017). Learning to Represent Student Knowledge on Programming Exercises Using Deep Learning. In *International Educational Data Mining Society*. International Educational Data Mining Society. https://eric.ed.gov/?id=ED596596

KEY TERMS AND DEFINITIONS

COVID-19: An ongoing global pandemic of coronavirus disease 2019 (COVID-19) caused by severe acute respiratory syndrome coronavirus 2 (SARS-CoV-2).

Early Childhood: Period of time between birth and age eight.

Google Analytics: A web data analytics platform by Google that tracks website and application traffics.

Learner Interactions: An action a student takes on an online learning platform. Actions can include number of clicks, when an app is opened or closed, what pages of a site were opened, etc.

Learning Analytics: The process in collecting and analyzing data from learners to better understand and optimize their learning processes.

Multi-Modal Learning Analytics: A sub field of Learning Analytics that collects and analyzes natural human signals.

ScratchJr: A free block-based programming application for young children.

Usage Patterns: A user's behavioral patterns on a website, application, or electronic device.

Compilation of References

Abelson, H., & DiSessa, A. (1981). Turtle geometry: The computer as a medium for exploring mathematics (The MIT press series in artificial intelligence). Cambridge, MA: MIT Press.

AERA. (2015). (2015, March 5). *Study: Little Evidence That Executive Function Interventions Boost Student Achievement* [Press release]. Retrieved from https://www.aera.net/Newsroom/News-Releases-and-Statements/Study-Little-EvidenceThat-Executive-Function-Interventions-Boost-Student-Achievement

Albo-Canals, J., Martelo, A. B., Relkin, E., Hannon, D., Heerink, M., Heinemann, M., Leidl, K., & Bers, M. U. (2018). A Pilot Study of the KIBO Robot in Children with Severe ASD. *International Journal of Social Robotics, 10*(3), 371–383. doi:10.100712369-018-0479-2

Aldemir, J., & Kermani, H. (2017). Integrated STEM curriculum: Improving educational outcomes for Head Start children. *Early Child Development and Care, 187*(11), 1694–1706. doi:10.1080/03004430.2016.1185102

Allan, W., Coulter, B., Denner, J., Erickson, J., Lee, I., Malyn-Smith, J., & Martin, F. (2010). *Computational thinking for youth.* White Paper for the ITEST Small Working Group on Computational Thinking (CT).

Amelink, C. T., & Creamer, E. G. (2010). Gender differences in elements of the undergraduate experience that influence satisfaction with the engineering major and the intent to pursue engineering as a career. *Journal of Engineering Education, 99*(1), 81–92. doi:10.1002/j.2168-9830.2010.tb01044.x

American Psychological Association. (2015). Guidelines for psychological practice with transgender and gender nonconforming people. *The American Psychologist, 70*(9), 832–864. doi:10.1037/a0039906 PMID:26653312

American Psychological Association. (2021). *APA Resolution on Gender Identity Change Efforts.* American Psychological Association. Retrieved from: https://www.apa.org/about/policy/resolution-gender-identity-change-efforts.pdf

American Speech-Language-Hearing Association. (n.d.). *Practice Portal: Clinical Topics: Selective Mutism.* American Speech-Language-Hearing Association. Retrieved February 15, 2020, from https://www.asha.org/Practice-Portal/Clinical-Topics/Selective-Mutism/#collapse_8

Angevine, C., Cator, K., Roschelle, J., Thomas, S. A., Waite, C., & Weisgrau, J. (2017). *Computational Thinking for a Computational World.* Academic Press.

An, S., Tinajero, J., Tillman, D., & Kim, S. (2019). Preservice Teachers' Development of Literacy-Themed Mathematics Instruction for Early Childhood Classrooms. *International Journal of Early Childhood*, *51*(1), 41–57. doi:10.100713158-019-00232-9

Antle, A. N., & Wise, A. F. (2013). Getting down to details: Using theories of cognition and learning to inform tangible user interface design. *Interacting with Computers*, *25*(1), 1–20. doi:10.1093/iwc/iws007

Applebee, A. N., Langer, J. A., & Mullis, I. V. S. (1986). The Writing Report Card: Writing Achievement in American Schools. Princeton, NJ: Educational Testing Service; Washington, DC: Office of Educational Research and Improvement.

Arctic Apples. (2020). Retrieved from: https://www.arcticapples.com/

Arfé, B., Vardanega, T., Montuori, C., & Lavanga, M. (2019). Coding in Primary Grades Boosts Children's Executive Functions. *Frontiers in Psychology*, *10*, 2713. doi:10.3389/fpsyg.2019.02713 PMID:31920786

Bada, S. O., & Olusegun, S. (2015). Constructivism learning theory: A paradigm for teaching and learning. *Journal of Research & Method in Education*, *5*(6), 66–70.

Baker, R. S., & Siemens, G. (2014). Educational data mining and learning analytics. In R. K. Sawyer (Ed.), *Cambridge handbook of the learning sciences* (2nd ed., pp. 253–274). Cambridge University Press. doi:10.1017/CBO9781139519526.016

Bakhtin, M. M. (1981). *The dialogic imagination: Four essays* (M. Holquist, Ed. & Trans.). University of Texas Press.

Barron, B., Martin, C. K., Takeuchi, L., & Fithian, R. (2009). Parents as Learning Partners in the Development of Technological Fluency. *International Journal of Learning and Media*, *1*(2), 55–77. doi:10.1162/ijlm.2009.0021

Barrouillet, P., & Lecas, J. (1999). Mental Models in Conditional Reasoning and Working Memory. *Thinking & Reasoning*, *5*(4), 289–302. doi:10.1080/135467899393940

Barr, V., & Stephenson, C. (2011). Bringing computational thinking to K-12. *ACM Inroads*, *2*(1), 48–54. doi:10.1145/1929887.1929905

Basu, S., Mustafaraj, E., & Rich, K. (2016). CIRCL Primer: Computational Thinking. In *CIRCL Primer Series*. Retrieved from https://circlcenter.org/computational-thinking

Basu, S. (2019). Using Rubrics Integrating Design and Coding to Assess Middle School Students' Open-ended Block-based Programming Projects. *Proceedings of the 50th ACM Technical Symposium on Computer Science Education*, 1211–1217. 10.1145/3287324.3287412

Basu, S., Biswas, G., Sengupta, P., Dickes, A., Kinnebrew, J. S., & Clark, D. (2016). Identifying middle school students' challenges in computational thinking-based science learning. *Research and Practice in Technology Enhanced Learning*, *11*(1), 13. doi:10.118641039-016-0036-2 PMID:30613246

Baumeister, R. F., & Vohs, K. D. (2004). *Handbook of self-regulation: Research, theory, and applications*. Guilford Press.

Beals, L., & Bers, M. (2006). Robotic Technologies: When Parents Put Their Learning Ahead of their Child's. *Journal of Interactive Learning Research*, *17*(4), 341–366.

Bell, T., & Vahrenhold, J. (2018). CS Unplugged—How Is It Used, and Does It Work? In H.-J. Böckenhauer, D. Komm, & W. Unger (Eds.), Adventures Between Lower Bounds and Higher Altitudes: Essays Dedicated to Juraj Hromkovič on the Occasion of His 60th Birthday. doi:10.1007/978-3-319-98355-4_29

Bell, T., & Vahrenhold, J. (2018). CS unplugged—How is it used, and does it work? In H.-J. Böckenhauer, D. Komm, & W. Unger (Eds.), Adventures between lower bounds and higher altitudes: Essays dedicated to Juraj Hromkovič on the occasion of his 60th birthday. Springer International Publishing.

Bell, T., & Lodi, M. (2019). Constructing Computational Thinking Without Using Computers. *Constructivist foundations*, Vrije Universiteit Brussel. *Constructionism and Computational Thinking*, *14*(3), 342–351.

Beltagui, A., Sesis, A., & Stylos, N. (2021). A bricolage perspective on democratizing innovation: The case of 3D printing in makerspaces. *Technological Forecasting and Social Change*, *163*, 120453. doi:10.1016/j.techfore.2020.120453

Berland, M., Baker, R. S., & Blikstein, P. (2014). Educational Data Mining and Learning Analytics: Applications to Constructionist Research. *Technology, Knowledge and Learning*, *19*(1–2), 205–220. doi:10.100710758-014-9223-7

Berninger, V. W., Abbott, R. D., Vermeulen, K., Ogier, S., Brooksher, R., Zook, D., & Lemos, Z. (2002). Comparison of Faster and Slower Responders to Early Intervention in Reading: Differentiating Features of Their Language Profile. *Learning Disability Quarterly*, *25*(1), 59–76. doi:10.2307/1511191

Bers, M. U. (2018b). Coding, Playgrounds and Literacy in Early Childhood Education: The Development of KIBO Robotics and ScratchJr. *IEEE Global Engineering Education Conference (EDUCON)*, 2100. 10.1109/EDUCON.2018.8363498

Bers, M. U. (2019). Coding as another language: a pedagogical approach for teaching computer science in early childhood. *Journal of Computers in Education*, *6*(4), 499-528.

Bers, M. U. (2020). Playgrounds and Microworlds: Learning to Code in Early Childhood. In Designing Constructionist Futures: The Art, Theory, and Practices of Learning Designs. Academic Press.

Bers, M. (2007). Project Inter-Actions: A multigenerational robotic learning environment. *Journal of Science and Technology Education, 16*(6), 537–552. doi:10.100710956-007-9074-2

Bers, M. (2008). *Blocks to robots: Learning with technology in the early childhood classroom.* Teachers College Press.

Bers, M. U. (2008). *Blocks to Robots: Learning with Technology in the Early Childhood Classroom.* Teachers College Press.

Bers, M. U. (2012). *Designing Digital Experiences for Positive Youth Development: From playpen to playground.* Oxford University Press. doi:10.1093/acprof:oso/9780199757022.001.0001

Bers, M. U. (2017). The Seymour Test: Powerful Ideas in early childhood education. *International Journal of Child-Computer Interaction, 14*, 10–14. doi:10.1016/j.ijcci.2017.06.004

Bers, M. U. (2018). *Coding as a Playground: Programming and Computational Thinking in the Early Childhood Classroom.* Routledge Press.

Bers, M. U. (2018). *Coding as a playground: Programming and computational thinking in the early childhood classroom.* Routledge.

Bers, M. U. (2018). *Coding as a playground: programming and computational thinking in the early childhood classroom.* Routledge., doi:10.4324/9781315398945

Bers, M. U. (2018a). *Coding as a playground: Computational thinking and programming in early childhood.* Routledge.

Bers, M. U. (2019). Coding as Another Language: "Why Computer Science in Early Childhood Should Not Be STEM. In C. Donohue (Ed.), *Key Issues in Technology and Early Childhood.* Routledge. doi:10.4324/9780429457425-11

Bers, M. U. (2019). Coding as another language: A pedagogical approach for teaching computer science in early childhood. *Journal of Computers in Education*, 6(4), 499–528. doi:10.100740692-019-00147-3

Bers, M. U. (2022). *Beyond Coding: How Children Learn Human Values through Programming.* The MIT Press.

Bers, M. U., & Resnick, M. (2015). *The Official ScratchJr Book: Help your Kids Learn to Code.* No Starch Press.

Bers, M., New, B., & Boudreau, L. (2004). Teaching and learning when no one is expert: Children and parents explore technology. *Journal of Early Childhood Research and Practice*, 6(2).

Black, J., Brodie, J., Curzon, P., Myketiak, C., McOwan, P. W., & Meagher, L. R. (2013). Making computing interesting to school students: Teachers' perspectives. In *Proceedings of the 18th ACM Conference on Innovation and Technology in Computer Science Education.* Association for Computing Machinery. 10.1145/2462476.2466519

Blair, C. (2002). School readiness: Integrating cognition and emotion in a neurobiological conceptualization of child functioning at school entry. *The American Psychologist*, *57*(2), 111–127. doi:10.1037/0003-066X.57.2.111 PMID:11899554

Blair, C., & Razza, R. P. (2007). Relating effortful control, executive function, and false belief understanding to emerging math and literacy ability in kindergarten. *Child Development*, *78*(2), 647–663. doi:10.1111/j.1467-8624.2007.01019.x PMID:17381795

Blikstein, P. (2013). Digital Fabrication and 'Making' in Education: The Democratization of Invention. In J. Walter-Herrmann & C. Büching (Eds.), *FabLabs: Of Machines, Makers and Inventors*. Transcript Publishers. doi:10.14361/transcript.9783839423820.203

Bocconi, S., Chioccariello, A., Dettori, G., Ferrari, A., & Engelhardt, K. (2016). *Developing computational thinking in compulsory education: Implications for policy and practice*. Publications Office. https://data.europa.eu/doi/10.2791/792158

Bouck, E. C., & Yadav, A. (2020). Providing Access and Opportunity for Computational Thinking and Computer Science to Support Mathematics for Students With Disabilities. *Journal of Special Education Technology*. Advance online publication. doi:10.1177/0162643420978564

Bowman, B., Donovan, S., & Burns, M. (2001). Eager to learn: Educating our preschoolers. Washington, DC: National Academy Press.

Brackmann, C. P., Román-González, M., Robles, G., Moreno-León, J., Casali, A., & Barone, D. (2017). Development of computational thinking skills through unplugged activities in primary school. *Proceedings of the 12th Workshop on Primary and Secondary Computing Education*, 65-72. 10.1145/3137065.3137069

Bredekamp, S. (1987). *Developmentally appropriate practice in early childhood pro- grams serving children from birth through age 8*. National Association for the Education of Young Children.

Brennan, K., & Resnick, M. (2012). Using artifact-based interviews to study the development of computational thinking in interactive media design [Paper presentation]. The meeting of the American Educational Research Association, Vancouver, BC, Canada.

Brennan, K., & Resnick, M. (2012, April). New frameworks for studying and assessing the development of computational thinking. In *Proceedings of the 2012 annual meeting of the American educational research association* (*Vol. 1*, p. 25). Academic Press.

Brennan, K., Haduong, P., & Veno, E. (2020). *Assessing Creativity in Computing Classrooms*. Creative Computing Lab.

Brennan, K., & Resnick, M. (2012). Using artifact-based interviews to study the development of computational thinking in interactive media design. *Annual Meeting of the American Educational Research Association (AERA)*.

Bresler, L. (Ed.). (2007). *International handbook of research in arts education* (Vol. 16). Springer Science & Business Media. doi:10.1007/978-1-4020-3052-9

Brunner, C., Fasca, C., Heinze, J., Honey, M., Light, D., Mandinach, E., et al. (2005). *Linking data and learning: The Grow Network study*. Academic Press.

Bull, R., & Scerif, G. (2001). Executive functioning as a predictor of children's mathematics ability: Inhibition, switching, and working memory. *Developmental Psychology*, *19*(3), 273–293. PMID:11758669

Bureau of Labor Statistics. (2020). *Labor Force Statistics from the Current Population Survey*. Retrieved from: https://www.bls.gov/cps/cpsaat11.htm

Burke, Q., & Kafai, Y. B. (2010). Programming & storytelling: Opportunities for learning about coding & composition. *Proceedings of the 9th International Conference on Interaction Design and Children*. 10.1145/1810543.1810611

Bybee, R. W. (2010). Advancing STEM education: A 2020 vision. *Technology and Engineering Teacher*, *70*(1), 30.

Bybee, R. W. (2014). NGSS and the next generation of science teachers. *Journal of Science Teacher Education*, *25*(2), 211–221. doi:10.100710972-014-9381-4

Calabrese Barton, A., & Tan, E. (2018). A longitudinal study of equity-oriented STEM-rich making among youth from historically marginalized communities. *American Educational Research Journal*, *55*(4), 761–800. doi:10.3102/0002831218758668

Calabrese Barton, A., & Tan, E. (2019). Designing for rightful presence in STEM: The role of making present practices. *Journal of the Learning Sciences*, *28*(4-5), 616–658. doi:10.1080/10508406.2019.1591411

Caldwell, H., & Smith, N. (2016). *Teaching computing unplugged in primary schools: Exploring primary computing through practical activities away from the computer*. Learning Matters.

California State Board of Education. (2013). *California Common Core State Standards: English Language Arts & Literacy in History/Social Studies, Science, and Technical Subjects*. Retrieved from California Department of Education: https://www.cde.ca.gov/

Callanan, M., Cervantes, C., & Loomis, M. (2011). Informal learning. *Wiley Interdisciplinary Reviews: Cognitive Science*, *2*(6), 646–655. doi:10.1002/wcs.143 PMID:26302414

Cameron, C., McClelland, M. M., Jewkes, A., Connor, C., Farris, C., & Morrison, F. (2008). Touch your toes! Developing a direct measure of behavioral regulation in early childhood. *Early Childhood Research Quarterly*, *23*(2), 141–158. doi:10.1016/j.ecresq.2007.01.004

Campana, K., Haines, C., Kociubuk, J., & Langsam, P. (2020). Making the Connection: Computational thinking and early learning for young children and their families. *Public Libraries*, *59*(4).

Campbell, L. O., Heller, S., & Goodman, B. (2018, March). Fostering computational thinking and student engagement in the literacy classroom through pop-up makerspaces. In *Society for Information Technology & Teacher Education International Conference* (pp. 3750-3754). Association for the Advancement of Computing in Education (AACE).

Catterall, J. S., & Waldorf, L. (1999). Chicago Arts Partnerships in Education: Summary evaluation. In E. B. Fiske (Ed.), *Champions of change: The impact of the arts on learning* (pp. 47–62). Arts Education Partnership. Retrieved from https://artsedge.kennedy-center.org/champions/pdfs/ChampsReport.pdf

Center for the Developing Child. (n.d.). *A Guide to Executive Function.* Retrieved from https://developingchild.harvard.edu/guide/a-guide-to-executive-function/

Chambers, J. (2015). *Inside Singapore's plans for robots in pre-schools.* GovInsider. Retrieved from: https:// govinsider.asia/smart-gov/exclusive-singapore-puts-robots-in-pre-schools/

Chappell, C., Dabholkar, S., Dilley, C., Heiland, M., Huang, A., Kuldell, N., Kurman, M., Legault, J., Scheifele, L., Scholze, A., Takara, C., & Tuck, E. (2021, April 8-12). *The BioMaker Ecosystem: Technologies, Spaces and Curricula for K-12 Making with Biology.* American Educational Research Association 98th Virtual Annual Meeting.

Chase, M., Son, E. H., & Steiner, S. (2014). Sequencing and Graphic Novels with Primary-Grade Students. *The Reading Teacher, 67*(6), 435–443. doi:10.1002/trtr.1242

Chemaly, S. (2016, March 16). The problem with a technology revolution designed primarily for men. *Quartz.* Retrieved from https://qz.com/640302/why-is-so-much-of-our-new-technology-designed-primarily-for-men/

Chen, G., Shen, J., Barth-Cohen, L., Jiang, S., Huang, X., & Eltoukhy, M. (2017). Assessing elementary students' computational thinking in everyday reasoning and robotics programming. *Computers & Education, 109*, 162–175. doi:10.1016/j.compedu.2017.03.001

Chen, Y. F., & Martin, M. A. (2000). Using Performance Assessment and Portfolio Assessment Together in the Elementary Classroom. *Reading Improvement, 37*(1), 32–38.

Clarke, S., Resnick, L. B., & Rose, C. P. (2015). *Dialogic instruction: A new frontier.* Academic Press.

Clark, L. S. (2011). Parental Mediation Theory for the Digital Age. *Communication Theory, 21*(4), 323–343. doi:10.1111/j.1468-2885.2011.01391.x

Clements, D. (1999). The Future of Educational Computing Research: The Case of Computer Programming. In C. Hoyles & R. Noss (Eds.), Learning mathematics and Logo. Academic Press.

Clements, D. H., Battista, M. T., & Sarama, J. (2001). LOGO and Geometry. Journal for Research in Mathematics Education Monograph Series, 10.

Clements, D.H., Sarama, J., Unlu, F., Layzer, C. (2012, March). *The efficacy of an intervention synthesizing scaffolding designed to promote self- regulation with an early mathematics curriculum: Effects on executive function.* Presentation at Society for Research on Educational Effectiveness (SREE), Washington, DC.

Clements, D. H. (2007). Curriculum Research: Toward a Framework for "Research-based Curricula". *Journal for Research in Mathematics Education, 38*(1), 35–70.

Clements, D. H., & Gullo, D. F. (1984). Effects of Computer Programming on Young Children's Cognition. *Journal of Educational Psychology, 76*(6), 1051–1058. doi:10.1037/0022-0663.76.6.1051

Clements, D. H., & Sarama, J. (2004). Learning trajectories in mathematics education. *Mathematical Thinking and Learning, 6*(2), 81–89. doi:10.120715327833mtl0602_1

Clough, M. P., & Olson, J. K. (2016). Connecting science and engineering practices: a cautionary perspective. In L. A. Annetta & J. Minogue (Eds.), *Connecting Science and Engineering Education Practices in Meaningful Ways: Building Bridges* (pp. 373–385). Springer. doi:10.1007/978-3-319-16399-4_15

Code.org. (2013). *Steve Jobs on Computer Science.* Academic Press.

Code.org. (2018). *2018 Annual Report.* https://code.org/files/annual-report-2018.pdf

Code.org. (2019). Retrieved from https://code.org/

Code.org. (2020). *Leaders and Trendsetters Agree More Students Should Learn Computer Science.* https://code.org/promote

Code.org. (2020a, April 15). *CS helps students outperform in school, college, and workplace.* codeorg.medium.com

Code.org. (2020b). *CS helps students outperform in school, college, and workplace.* Retrieved from codeorg.medium.com

Code.org. (2021a). *Code.org Statistics.* Retrieved from Code.org: code.org/statistics

Code.org. (2021b). *Why Computer Science?* Retrieved from code.org: code.org/promote

Committee on STEM Education, National Science & Technology Council, the White House. (2018). *Charting a course for success: America's strategy for STEM education.* https://www.whitehouse.gov/wp-content/uploads/2018/12/STEM-Education-Strategic-Plan-2018.pdf

Congress, U. (1975). The Individuals with Disabilities Education Act–IDEA.

Connell, S. L., Lauricella, A. R., & Wartella, E. (2015). Parental Co-Use of Media Technology with their Young Children in the USA. *Journal of Children and Media, 9*(1), 5–21. doi:10.1080/17482798.2015.997440

Cope. (2016). *Big Data Comes to School: Implications for Learning, Assessment, and Research.* University of Illinois. https://journals.sagepub.com/doi/pdf/10.1177/2332858416641907

Copple, C., & Bredekamp, S. (2009). *Developmentally appropriate practice in early childhood programs serving children from birth through age 8*. National Association for the Education of Young Children.

Corbett, C., & Hill, C. (2015). *Solving the equation: the variables for women's success in engineering and computing*. The American Association of University Women.

Cordes, C., & Miller, E. (2000). *Fool's gold: A critical look at computers in childhood.* Academic Press.

Creative Computing Lab. (n.d.). *Assessing Development of Computational Practices.* https://scratched.gse.harvard.edu/ct/assessing.html

Crescenzi-Lanna, L. (2020). Multimodal Learning Analytics research with young children: A systematic review. *British Journal of Educational Technology*, *51*(5), 1485–1504. doi:10.1111/bjet.12959

CSTA & ISTE. (2011). *Operational Definition of Computational Thinking for K-12 Education.* http://www.iste.org/docs/pdfs/Operational-Definition-of-Computational-Thinking.pdf

Cumbers, J. (2019). New This Ski Season: A Jacket Brewed Like Spider's Silk. *Forbes Magazine Online*. Retrieved from: https://www.forbes.com/sites/johncumbers/2019/08/28/new-this-ski-season-a-jacket-brewed-from-spider-silk/#2788fa63561e

Cunha, F., & Heckman, J. (2007). The Technology of Skill Formation. *The American Economic Review*, *97*(2), 31–47. doi:10.1257/aer.97.2.31

Curzon, P., McOwan, P. W., Plant, N., & Meagher, L. R. (2014). Introducing teachers to computational thinking using unplugged storytelling. *Proceedings of the 9th workshop in primary and secondary computing education*, 89-92. 10.1145/2670757.2670767

Dagiene, V., & Stupurienė, G. (2016). Bebras–a sustainable community building model for the concept based learning of informatics and computational thinking. *Informatics in Education, 15*(1), 25–44. . doi:10.15388/infedu.2016.02

Dalbey, J., & Linn, M. C. (1985). The demands and requirements of computer programming: A literature review. *Journal of Educational Computing Research*, *1*(3), 253–274. doi:10.2190/BC76-8479-YM0X-7FUA

de Strulle, A., & Shen, C. (n.d.). *STEM + Computing K-12 Education (STEM+C).* https://wwwnsf.gov/funding/pgm_summ.jsp?pims_id=505006

deafkidscode.org. (n.d.). *Our Story*. Retrieved February 15, 2020, from https://www.deafkidscode.org/our-story

del Olmo-Muñoz, J., Cózar-Gutiérrez, R., & González-Calero, J. A. (2020). Computational thinking through unplugged activities in early years of Primary Education. *Computers & Education, 150*, 103832. doi:10.1016/j.compedu.2020.103832

Delacruz, S. (2020). Starting From Scratch (Jr.): Integrating Code Literacy in the Primary Grades. *The Reading Teacher, 73*(6), 805–811. doi:10.1002/trtr.1909

Denner, J., Werner, L., & Ortiz, E. (2012). Computer games created by middle school girls: Can they be used to measure understanding of computer science concepts? *Computers & Education, 58*(1), 240–249. doi:10.1016/j.compedu.2011.08.006

DevTech Research Group. (2018). *General Assessment Templates*. https://sites.tufts.edu/devtech/files/2018/03/GeneralAssessments.pdf

DevTech Research Group. (2018). *Our treasure: A KIBO coding curriculum for emergent readers*. Tufts University.

Diamond, A. (2002). Normal development of prefrontal cortex from birth to young adulthood: Cognitive functions, anatomy, and biochemistry. In D. T. Stuss & R. T. Knight (Eds.), *Principles of frontal lobe function* (pp. 466–503). Oxford University Press. doi:10.1093/acprof:oso/9780195134971.003.0029

Digital News Asia. (2015). *IDA launches $1.5m pilot to roll out tech toys for preschoolers*. Retrieved from: https://www.digitalnewsasia.com/digital-economy/ida-launches-pilot-to-roll-out-tech-toys-forpreschoolers

DiSessa, A. A. (2000). *Changing minds: Computers, learning, and literacy*. MIT Press. doi:10.7551/mitpress/1786.001.0001

Doerschuk, P., Liu, J., & Mann, J. (2007). Pilot summer camps in computing for middle school girls. *ACM SIGCSE Bulletin, 39*(3), 4–8. doi:10.1145/1269900.1268789

Dong, C., & Xu, Q. (2020). Pre-service early childhood teachers' attitudes and intentions: Young children's use of ICT. *Journal of Early Childhood Teacher Education*, 1–16. doi:10.1080/10901027.2020.1726843

Doudna, J. (2015, September). *How CRISPR lets us edit our DNA* [Video file]. Retrieved from: www.ted.com/talks/jennifer_doudna_we_can_now_edit_our_dna_but_let_s_do_it_wisely#t-686789

Dougherty, D. (2012). The maker movement. *Innovations: Technology, Governance, Globalization, 7*(3), 11–14. doi:10.1162/INOV_a_00135

Duhaime-Ross, A. (2014, September 25). Apple promised an expansive health app, so why can't I track menstruation? *The Verge*. Retrieved from https://www.theverge.com/2014/9/25/6844021/apple-promised-an-expansive-health-app-so-why-cant-i-track

Dweck, C. S. (2008). *Mindset: The new psychology of success*. Random House Digital, Inc.

Ehsan, H., Ohland, C., Dandridge, T., & Cardella, M. (2018). Computing for the Critters: Exploring Computational Thinking of Children in an Informal Learning Setting. *Proceedings of IEEE Frontiers in Education Conference*. 10.1109/FIE.2018.8659268

Elkin, M., Sullivan, A., & Bers, M. U. (2016). Programming with the KIBO Robotics Kit in Preschool Classrooms. *Computers in the Schools*, *33*(3), 169–186. doi:10.1080/07380569.2016.1216251

Emerson, A., Smith, A., Rodriguez, F. J., Wiebe, E. N., Mott, B. W., Boyer, K. E., & Lester, J. C. (2020). Cluster-Based Analysis of Novice Coding Misconceptions in Block-Based Programming. *Proceedings of the 51st ACM Technical Symposium on Computer Science Education*, 825–831. 10.1145/3328778.3366924

Erdmann, K. A., & Hertel, S. (2019). Self-regulation and co-regulation in early childhood – development, assessment and supporting factors. *Metacognition and Learning*, *14*(3), 229–238. doi:10.100711409-019-09211-w

Erdoğan, S., & Baran, G. (2009). A study on the effect of mathematics teaching provided through drama on the mathematics ability of six-year-old children. *Eurasia Journal of Mathematics, Science & Technology Education, 5*(1), 79–85. Retrieved from https://www.ejmste.com/v5n1/EURASIA_v5v1_SErdogan.pdf

Erete, S., Martin, C. K., & Pinkard, N. (2017). Digital Youth Divas: A program model for increasing knowledge, confidence, and perceptions of fit in STEM amongst black and brown middle school girls. In Moving students of color from consumers to producers of technology (pp. 152-173). IGI Global. doi:10.4018/978-1-5225-2005-4.ch008

Espy, K. A., McDiarmid, M. M., Cwik, M. F., Stalets, M. M., Hamby, A., & Senn, T. E. (2004). The contribution of executive functions to emergent mathematic skills in preschool children. *Developmental Neuropsychology*, *26*(1), 465–486. doi:10.120715326942dn2601_6 PMID:15276905

FACT SHEET: President Obama Announces Computer Science For All Initiative. (2016, January 30). Whitehouse.Gov. https://obamawhitehouse.archives.gov/the-press-office/2016/01/30/fact-sheet-president-obama-announces-computer-science-all-initiative-0

Fayer, S., Lacey, A., & Watson, A. (2017). *BLS Spotlight on Statistics: STEM Occupations - Past, Present, and Future.* https://hdl.handle.net/1813/79240

Fayer, S., Lacey, A., & Watson, A. (2017). *BLS Spotlight on Statistics: STEM Occupations-Past, Present, and Future*. U.S. Department of Labor, Bureau of Labor Statistics.

Fedorenko, E., Ivanova, A., Dhamala, R., & Bers, M. U. (2019). The Language of Programming: A Cognitive Perspective. *Trends in Cognitive Sciences*, *23*(7), 525–528. doi:10.1016/j.tics.2019.04.010 PMID:31153775

Finders, J. K., McClelland, M. M., Geldhof, G. J., Rothwell, D. W., & Hatfield, B. E. (2021). Explaining achievement gaps in kindergarten and third grade: The role of self-regulation and executive function skills. *Early Childhood Research Quarterly*, *54*, 72–85. doi:10.1016/j.ecresq.2020.07.008

Fischer, C., Pardos, Z. A., Baker, R. S., Williams, J. J., Smyth, P., Yu, R., Slater, S., Baker, R., & Warschauer, M. (2020). Mining Big Data in Education: Affordances and Challenges. *Review of Research in Education*, *44*(1), 130–160. doi:10.3102/0091732X20903304

Fitzgerald, J., & Markham, L. R. (1987). Teaching children about revision in writing. *Cognition and Instruction*, *4*(1), 3–24. doi:10.12071532690xci0401_1

Flannery, L. P., Silverman, B., Kazakoff, E. R., Bers, M. U., Bontá, P., & Resnick, M. (2013). Designing ScratchJr: support for early childhood learning through computer programming. In *Proceedings of the 12th International Conference on Interaction Design and Children*. New York, NY: Association for Computing Machinery. 10.1145/2485760.2485785

Flavell, J. H., Miller, P. H., & Miller, S. A. (1993). *Cognitive development* (3rd ed.). Prentice Hall.

Fraillon, J., Ainley, J., Schulz, W., Duckworth, D., & Friedman, T. (2018). *International Computer and Information Literacy Study*. ICILS 2018: Technical Report.

Fraillon, J., Ainley, J., Schulz, W., Friedman, T., & Duckworth, D. (2020). *Preparing for life in a digital world: IEA International computer and information literacy study 2018 international report* (Vol. 297). Springer Nature. doi:10.1007/978-3-030-38781-5

Fridin, M. (2014). Storytelling by a kindergarten social assistive robot: A tool for constructive learning in preschool education. *Computers & Education*, *70*, 53–64. doi:10.1016/j.compedu.2013.07.043

Funke, A., & Geldreich, K. (2017). Gender Differences in Scratch Programs of Primary School Children. *Proceedings of the 12th Workshop on Primary and Secondary Computing Education*, 57–64. 10.1145/3137065.3137067

Gadanidis, G. (2017). Five affordances of computational thinking to support elementary mathematics education. *Journal of Computers in Mathematics and Science Teaching*, *36*(2), 143–151.

Gage, N. A., Lierheimer, K. S., & Goran, L. G. (2012). Characteristics of Students With High-Incidence Disabilities Broadly Defined. *Journal of Disability Policy Studies*, *23*(3), 168–178. doi:10.1177/1044207311425385

Gal-Ezer, J., & Stephenson, C. (2009). *The current state of computer science in US high schools: a report from two national surveys*. Retrieved from Computer Science Teachers Association website, https://csta.acm.org/Research/sub/ Projects/ResearchFiles/StateofCSEDHighSchool.pdf

Garon, N., Bryson, S. E., & Smith, I. M. (2008). Executive Function in Preschoolers: A Review Using an Integrative Framework. *Psychological Bulletin*, *134*(1), 31–60. doi:10.1037/0033-2909.134.1.31 PMID:18193994

Garrett, J. L. (2008). STEM: The 21st century sputnik. *Kappa Delta Pi Record, 44*(4), 152–153. doi:10.1080/00228958.2008.10516514

Gašević, D., Dawson, S., & Siemens, G. (2015). Let's not forget: Learning analytics are about learning. *TechTrends, 59*(1), 64–71. doi:10.100711528-014-0822-x

Gee, J. P. (2007). What Video Games Have to Teach Us About Learning and Literacy. *Cyberpsychology & Behavior, 12*(1).

Gestsdottir, S., & Lerner, R. M. (2008). Positive development in adolescence: The development and role of intentional self-regulation. *Human Development, 51*(3), 202–224. doi:10.1159/000135757

Ginsburg, H., & Baroody, A. (2003). *TEMA-3 examiners manual*. Pro-Ed.

Gioia, I., & Guy, K. (2000). *Behavior Rating Inventory of Executive Function*. Psychological Assessment Resources.

Goddu, M. K., Lombrozo, T., & Gopnik, A. (2020). Transformations and Transfer: Preschool Children Understand Abstract Relations and Reason Analogically in a Causal Task. *Child Development, 91*(6), 1898–1915. doi:10.1111/cdev.13412 PMID:32880903

Goldstein, J., & Flake, J. K. (2016). Towards a framework for the validation of early childhood assessment systems. *Educational Assessment, Evaluation and Accountability, 28*(3), 273–293. doi:10.100711092-015-9231-8

González-González, C. S., & Arias, L. G. A. (2019). Maker movement in education: maker mindset and makerspaces. In J. L. Jurado, C. A. Collazos, y L. F. Muñoz (Eds.), Ingeniería colaborativa, aplicaciones y usos desde la perspectiva de la Interacción Humano-Computador [Collaborative engineering, applications and uses from the perspective of Human-Computer Interaction]. Editorial: Universidad San Buenaventura de Cali. Colombia.

Goswami, U. (2001). Early phonological development and the acquisition of literacy. Handbook of Early Literacy Research, 111-125.

Gouëdard, P., Pont, B., & Viennet, R. (2020). *Education responses to COVID-19: shaping an implementation strategy*. OECD Education Working Papers, No. 224.

Govind, M. (2019). *Families That Code Together Learn Together: Exploring family-oriented programming in early childhood with ScratchJr and KIBO Robotics* [Unpublished master's thesis]. Tufts University, Medford, MA, United States.

Govind, M., & Bers, M. U. (2019). *Parents Don't Need to Be Coding Experts, Just Willing to Learn With Their Children*. EdSurge. https://www.edsurge.com/news/2019-12-11-parents-don-t-need-to-be-coding-experts-just-willing-to-learn-with-their-children

Govind, M., & Bers, M. U. (2020). Family Coding Days: Engaging Children and Parents in Creative Coding and Robotics. Proceedings of Connected Learning Summit.

Govind, M., Relkin, E., & Bers, M. U. (2020). Engaging Children and Parents to Code Together Using the ScratchJr App. *Visitor Studies*, *23*(1), 46–65. doi:10.1080/10645578.2020.1732184

Gravel, B. E., Bers, M. U., Rogers, C., & Danahy, E. (2018). *Making engineering playful in schools*. The LEGO Foundation.

Gropen, J., Clark-Chiarelli, N., Hoisington, C., & Ehrlich, S. B. (2011). The importance of executive function in early science education. *Child Development Perspectives*, *5*(4), 298–304. doi:10.1111/j.1750-8606.2011.00201.x

Grover, S. (2017). Assessing Algorithmic and Computational Thinking in K-12: Lessons from a Middle School Classroom. In Emerging Research, Practice, and Policy on Computational Thinking (pp. 269-288). Springer International.

Grover, S., Basu, S., Bienkowski, M., Eagle, M., Diana, N., & Stamper, J. (2017). A Framework for Using Hypothesis-Driven Approaches to Support Data-Driven Learning Analytics in Measuring Computational Thinking in Block-Based Programming Environments. *ACM Transactions on Computing Education*, *17*(3), 1–25. doi:10.1145/3105910

Grover, S., Cooper, S., & Pea, R. (2014). Assessing computational learning in K-12. In *Proceedings of the 2014 conference on Innovation & technology in computer science education* (pp. 57-62). ACM. 10.1145/2591708.2591713

Grover, S., & Pea, R. (2013). Computational Thinking in K–12: A Review of the State of the Field. *Educational Researcher*, *42*(1), 38–43. doi:10.3102/0013189X12463051

Guzdial, M. (2008). Education: Paving the way for computational thinking. *Communications of the ACM*, *51*(8), 25–27. doi:10.1145/1378704.1378713

Guzdial, M., & Morrison, B. (2016). Seeking to making computing education as available as mathematics or science education. *Communications of the ACM*, *59*(11), 31–33. doi:10.1145/3000612

Handsfield, L. (2016). *Literacy Theory as Practice: Connecting Theory and Instruction in K–12 Classrooms*. Teachers College Press.

Hartman, S. L., & Dani, D. (2020). Full STEAM Ahead: Creating Interdisciplinary Informal Learning Opportunities for Early Childhood Teacher Candidates. *Journal of STEM Teacher Education*, *54*(1), 3. doi:10.30707/JSTE54.1/MNCB7975

Hassenfeld, Z. R., & Bers, M. U. (2020). Debugging the Writing Process: Lessons From a Comparison of Students' Coding and Writing Practices. *The Reading Teacher*, *73*(6), 735–746. doi:10.1002/trtr.1885

Hassenfeld, Z. R., Govind, M., de Ruiter, L. E., & Bers, M. U. (2020). If You Can Program, You Can Write: Learning Introductory Programming Across Literacy Levels. *Journal of Information Technology Education*, *19*, 65–85. doi:10.28945/4509

Hatton, E. (1989). Lévi-Strauss's bricolage and theorizing teachers' work. *Anthropology & Education Quarterly*, *20*(2), 74–96. doi:10.1525/aeq.1989.20.2.05x0841i

Heckman, J., & Masterov, D. (2007). The Productivity Argument for Investing in Young Children. *Review of Agricultural Economics*, *29*(3), 446–493. doi:10.1111/j.1467-9353.2007.00359.x

Hein, G. (1991). *Constructivist learning theory.* Institute for Inquiry. http://www. exploratorium. edu/ifi/resources/constructivistlearning.html

Henderson, A. (2020, July 21). *So Why Is There An "A" In STEAM?* [Blog post]. Retrieved from https://amt-lab.org/blog/2020/5/so-why-is-there-an-a-in-steam

Hendricks, C. C., Alemdar, M., & Ogletree, T. W. (2012). *The impact of participation in VEX robotics competition on middle and high school students' interest in pursuing STEM studies and STEM-related careers.* Paper presented at the ASEE Annual Conference, San Antonio, TX. Retrieved from https://peer.asee.org/22069

Hermans, F., & Aivaloglou, E. (2017). To scratch or not to scratch?: A controlled experiment comparing plugged first and unplugged first programming lessons. In *Proceedings of the 12th Workshop on Primary and Secondary Computing Education.* Association for Computing Machinery. 10.1145/3137065.3137072

Hill, C., Corbett, C., & St Rose, A. (2010). Why so few? Women in science, technology, engineering, and mathematics. American Association of University Women.

Hogan, T. P., Catts, H. W., & Little, T. D. (2005). The Relationship between Phonological Awareness and Reading: Implications for the Assessment of Phonological Awareness. *Language, Speech, and Hearing Services in Schools*, *36*(4), 285–293. doi:10.1044/0161-1461(2005/029) PMID:16389701

Horn, M. S., Crouser, R. J., & Bers, M. U. (2012). Tangible interaction and learning: The case for a hybrid approach. *Personal and Ubiquitous Computing*, *16*(4), 379–389. doi:10.100700779-011-0404-2

Huang, W., & Looi, C. K. (2020). A critical review of literature on "unplugged" pedagogies in K-12 computer science and computational thinking education. *Computer Science Education*, 1–29.

Hubwieser, P., Armoni, M., Giannakos, M. N., & Mittermeir, R. T. (2014). Perspectives and Visions of Computer Science Education in Primary and Secondary (K-12) Schools. *ACM Transactions on Computing Education, 14*(2).

Ifenthaler, D., & Yau, J. Y.-K. (2020). Utilising learning analytics to support study success in higher education: A systematic review. *Educational Technology Research and Development*, *68*(4), 1961–1990. doi:10.100711423-020-09788-z

Ingram, D., & Riedel, E. (2003). *What does arts integration do for students?* University of Minnesota, Center for Applied Research and Educational Improvement.

International Literacy Association. (2021). *Teaching with Tech.* https://www.literacyworldwide.org/blog/digital-literacies/teaching-with-tech

International Society for Technology in Education (ISTE) & The Computer Science Teachers Association (CSTA). (2011). *CT leadership toolkit.* Retrieved from https://cdn.iste.org/www-root/2020-10/ISTE_CT_Leadership_Toolkit_booklet.pdf?_ga=2.15251892.309077916.1613247518-1278422219.1611941118

Israel, M., Wherfel, Q. M., Pearson, J., Shehab, S., & Tapia, T. (2015). Empowering K–12 Students With Disabilities to Learn Computational Thinking and Computer Programming. *Teaching Exceptional Children*, *48*(1), 45–53. doi:10.1177/0040059915594790

Ito, M., Gutiérrez, K., Livingstone, S., Penuel, B., Rhodes, J., Salen, K., Schor, J., Sefton-Green, J., & Watkins, S. C. (2013). *Connected Learning: An Agenda for Research and Design.* Digital Media and Learning Research Hub.

Iwata, M., Pitkänen, K., Laru, J., & Mäkitalo, K. (2020). Exploring potentials and challenges to develop twenty-first century skills and computational thinking in K-12 maker education. In *Frontiers in Education, 5(87), 1-16.* doi:10.3389/feduc.2020.00087

Jacob, S. R., & Warschauer, M. (2018). Computational thinking and literacy. *Journal of Computer Science Integration*, *1*(1). Advance online publication. doi:10.26716/jcsi.2018.01.1.1

Jacobson, L. (2016). The Codemakers: J is for Javascript. *School Library Journal*, *62*(4).

Janveau-Brennan, G., & Markovits, H. (1999). The Development of Reasoning with Causal Conditionals. *Developmental Psychology*, *35*(4), 904–911. doi:10.1037/0012-1649.35.4.904 PMID:10442860

Jaramillo, J. M., Rendón, M. I., Muñoz, L., Weis, M., & Trommsdorff, G. (2017). Children's self-regulation in cultural contexts: The role of parental socialization theories, goals, and practices. *Frontiers in Psychology*, *8*, 923. doi:10.3389/fpsyg.2017.00923 PMID:28634460

Jenkins, T. (2002). *On the difficulty of learning to program.* https://www.psy.gla.ac.uk/~steve/localed/jenkins.html

Jones, C. D., Clark, S. K., & Reutzel, D. (2012). Enhancing Alphabet Knowledge Instruction: Research Implications and Practical Strategies for Early Childhood Educators. *Early Childhood Education*, *41*(2), 81–89. doi:10.100710643-012-0534-9

Jones, K. S. (2003). What is an affordance? *Ecological Psychology*, *15*(2), 107–114. doi:10.1207/S15326969ECO1502_1

Jungert, T., Hubbard, K., Dedic, H., & Rosenfield, S. (2018). Systemizing and the gender gap: Examining academic achievement and perseverance in STEM. *European Journal of Psychology of Education*, 479–500.

K-12 Computer Science Framework Steering Committee. (2016). *K–12 computer science framework.* https://k12cs.org

K-12 Computer Science Framework Steering Committee. (2016). *K-12 computer science framework.* https://k12cs.org/

K–12 Computer Science Framework. (2016). *K-12 CS Framework.* http://www.k12cs.org

K-2 Computer Science Curriculum—Orange. (n.d.). Retrieved April 10, 2021, from https://sites.google.com/sfusd.edu/k-2cs/orange

Kafai, Y. B., & Walker, J. T. (2020). Twenty things to make with biology. Proceedings of Constructionism, 598-606.

Kafai, Y., & Margolis, J. (2014, October 7). Why the 'coding for all' movement is more than a boutique reform. *Washington Post.* Retrieved from https:// www.washingtonpost.com/news/answer-sheet/wp/2014/10/17/whythe-coding-for-all-movement-is-more-than-a-boutique-reform

Kafai, Y. B., Fields, D. A., & Searle, K. A. (2014). Electronic Textiles as Disruptive Designs: Supporting and Challenging Maker Activities in Schools. *Harvard Educational Review, 84*(4), 532–557. doi:10.17763/haer.84.4.46m7372370214783

Kafai, Y. B., & Resnick, M. (1996). *Constructionism in practice: Designing, thinking, and learning in a digital world.* Erlbaum.

Kafai, Y., Telhan, O., Hogan, K., Lui, D., Anderson, E., Walker, J. T., & Hanna, S. (2017, June). Growing designs with biomakerlab in high school classrooms. *Proceedings of the 2017 Conference on Interaction Design and Children*, 503-508. 10.1145/3078072.3084316

Kaldor, T. (2017). *The T in STEM: Creating Play-Based Experiences That Support Children's Learning of Coding and Higher Order Thinking.* Retrieved from https://www.naeyc.org/resources/blog/creating-play-based-experiences

Kamps, D., Abbott, M., Greenwood, C., Wills, H., Veerkamp, M., & Kaufman, J. (2008). Effects of small-group reading instruction and curriculum differences for students most at risk in kindergarten: Two-year results for secondary- and tertiary-level interventions. *Journal of Learning Disabilities, 41*(2), 101–114. doi:10.1177/0022219407313412 PMID:18354931

Karpiński, Z., Di Pietro, G., & Biagi, F. (2021). *Computational thinking, socioeconomic gaps, and policy implications.* IEA Compass: Briefs in Education Series (12). Retrieved from: https://www.iea.nl/publications/series-journals/iea-compass-briefs-education-series/january-2021-computational

Kazakoff, E. R., & Bers, M. (2012). Programming in a robotics context in the kindergarten classroom: The impact on sequencing skills. *Journal of Educational Multimedia and Hypermedia, 21*(4), 371–391.

Kazakoff, E. R., & Bers, M. U. (2014). Put your robot in, Put your robot out: Sequencing through programming robots in early childhood. *Journal of Educational Computing Research, 50*(4), 553–573. doi:10.2190/EC.50.4.f

Kazakoff, E., Sullivan, A., & Bers, M. U. (2013). The effect of a classroom-based intensive robotics and programming workshop on sequencing ability in early childhood. *Early Childhood Education Journal, 41*(4), 245–255. doi:10.100710643-012-0554-5

Kent, J. (2020). *4 Emerging Strategies to Advance Big Data Analytics in Healthcare.* HealthITAnalytics. https://healthitanalytics.com/news/4-emerging-strategies-to-advance-big-data-analytics-in-healthcare

Kewalramani, S., Palaiologou, I., & Dardanou, M. (2016). Children's Engineering Design Thinking Processes: The Magic of the ROBOTS and the Power of BLOCKS (Electronics). *Eurasia Journal of Mathematics, Science and Technology Education, 16*(3). Advance online publication. doi:10.29333/ejmste/113247

Kim, Y. E., Edouard, K., Alderfer, K., & Smith, B. K. (2018). *Making culture: A national study of education makerspaces.* Drexel University.

Kingsbury, G. G., & Weiss, D. J. (1983). A comparison of IRT-based adaptive mastery testing and a sequential mastery testing procedure. In *New horizons in testing* (pp. 257–283). Academic Press. doi:10.1016/B978-0-12-742780-5.50024-X

Knight, V. F., Wright, J., & DeFreese, A. (2019). Teaching Robotics Coding to a Student with ASD and Severe Problem Behavior. *Journal of Autism and Developmental Disorders, 49*(6), 2632–2636. doi:10.100710803-019-03888-3 PMID:30734176

Koretz, D., McCaffrey, D. F., Klein, S. P., Bell, R. M., & Stecher, B. M. (1992). *The Reliability of Scores from the 1992 Vermont Portfolio Assessment Program.* Academic Press.

Kramer, J. (2007). Is abstraction the key to computing? *Communications of the ACM, 50*(4), 36–42.

Kuhl, P. K., Lim, S. S., Guerriero, S., & van Damme, D. (2019). How stereotypes shape children's STEM identity and learning. In *Developing Minds in the Digital Age: Towards a Science of Learning for 21st Century Education.* OECD Publishing. doi:10.1787/43e5bb4c-en

Kuhn, D., Nash, S. C., & Brucken, L. (1978). Sex role concepts of two- and three-year-olds. *Child Development, 49*(2), 445–451. doi:10.2307/1128709 PMID:679779

Lachapelle, C. P., & Cunningham, C. M. (2007, March). Engineering is elementary: Children's changing understandings of science and engineering. *ASEE Annual Conference & Exposition*, 33.

Ladner, R. E., & Israel, M. (2016). For all" in" computer science for all. *Communications of the ACM, 59*(9), 26–28. doi:10.1145/2971329

Lakind, A., Willett, R., & Halverson, E. R. (2019). Democratizing the maker movement: A case study of one public library system's makerspace program. *Reference and User Services Quarterly, 58*(4), 234–245. doi:10.5860/rusq.58.4.7150

Lave, J., & Wenger, E. (1991). *Situated learning: legitimate peripheral participation.* Cambridge University Press. doi:10.1017/CBO9780511815355

Lee, I., Martin, F., Denner, J., Coulter, B., Allan, W., Erickson, J., Malyn-Smith, J., & Werner, L. (2011). Computational thinking for youth in practice. *ACM Inroads*, *2*(1), 32–37. doi:10.1145/1929887.1929902

Lee, K., Sullivan, A., & Bers, M. U. (2013). Collaboration by design: Using robotics to foster social interaction in kindergarten. *Computers in the Schools*, *30*(3), 271–281.

Leidl, K. D., Bers, M. U., & Mihm, C. (2017). Programming with ScratchJr: A review of the first year of user analytics. *Proceedings of the International Conference on Computational Thinking Education.*

Lester, J. C., Rowe, J. P., & Mott, B. W. (2013). *Narrative-centered learning environments: A story-centric approach to educational games. Emerging Technologies for the Classroom. 223-237.*

Leung, S. K. (2020). Teachers' belief-and-practice gap in implementing early visual arts curriculum in Hong Kong. *Journal of Curriculum Studies*, *52*(6), 857–869. doi:10.1080/00220272.2020.1795271

Lewin-Bizan, S. G., & Urban, J. B. (Eds.). Thriving in childhood and adolescence: The role of self-regulation processes. New Directions for Child and Adolescent Development, 133, 29–44.

Littleton, K., & Howe, C. (2010). *Educational Dialogues: Understanding and Promoting Productive Interaction*. Routledge. doi:10.4324/9780203863510

Li, Y., Schoenfeld, A. H., diSessa, A. A., Graesser, A. C., Benson, L. C., English, L. D., & Duschl, R. A. (2020). Computational Thinking Is More about Thinking than Computing. *Journal for STEM Education Research*, *3*(1), 1–18. doi:10.100741979-020-00030-2 PMID:32838129

Lockwood, J., & Mooney, A. (2018). Computational thinking in education: Where does it fit? A systematic literary review. *International Journal of Computer Science Education in Schools*, *2*(1), 41–60.

Lockwood, J., & Mooney, A. (2018). Computational Thinking in education: Where does it fit? A systematic literary review. *International Journal of Computer Science Education in Schools*, *2*(1), 41–60. doi:10.21585/ijcses.v2i1.26

Lopez, S. J., & Louis, M. C. (2009). The Principles of Strengths-Based Education. *Journal of College and Character*, *10*(4). Advance online publication. doi:10.2202/1940-1639.1041

Ludwig, M., & Song, M. (2016). *Evaluation of professional development in the use of arts-integrated activities with mathematics content: Findings from the evaluation of the Wolf Trap Arts in education model development and dissemination grant.* American Institutes for Research. Retrieved from https://education.wolftrap.org/sites/default/files/Full%20WT%20AEMDD%20Report_Final_Jan-2015updated%20with%20date%2Bappendix.pdf

Lye, S. Y., & Koh, J. H. L. (2014). Review on teaching and learning of computational thinking through programming: What is next for K-12? *Computers in Human Behavior*, *41*, 51–61. doi:10.1016/j.chb.2014.09.012

Madill, H., Campbell, R. G., Cullen, D. M., Armour, M. A., Einsiedel, A. A., Ciccocioppo, A. L., & Coffin, W. L. (2007). Developing career commitment in STEM-related fields: Myth versus reality. In R. J. Burke, M. C. Mattis, & E. Elgar (Eds.), *Women and Minorities in Science, Technology, Engineering and Mathematics: Upping the Numbers* (pp. 210–244). Edward Elgar Publishing.

Manabe, H., Kanemune, S., Namiki, M., & Nakano, Y. (2011). CS unplugged assisted by digital materials for handicapped people at schools. In *Proceedings of the 5th International Conference on Informatics in Schools: Situation, Evolution and Perspectives*. Springer-Verlag. 10.1007/978-3-642-24722-4_8

Mantzicopoulos, P., & Patrick, H. (2011). Reading picture books and learning science: Engaging young children with informational text. *Theory into Practice*, *50*(4), 269–276. doi:10.1080/00405841.2011.607372

Margolis, J., Estrella, R., Goode, J., Holme, J. J., & Nao, K. (2017). *Stuck in the shallow end: Education, race, and computing*. MIT Press.

Markert, L. R. (1996). Gender related to success in science and technology. *The Journal of Technology Studies*, *22*(2), 21–29.

Marsh, J., Wood, E., Chesworth, L., Nisha, B., Nutbrown, B., & Olney, B. (2019). Makerspaces in early childhood education: Principles of pedagogy and practice. *Mind, Culture, and Activity*, *26*(3), 221–233. doi:10.1080/10749039.2019.1655651

Masoumi, D. (2020). Situating ICT in early childhood teacher education. *Education and Information Technologies*, 1–18.

Massachusetts Department of Elementary and Secondary Education. (2017). *English Language Arts and Literacy.* Retrieved from Massachusetts Department of Education: https://www.doe.mass.edu/

Maureen, I. Y., van der Meij, H., & de Jong, T. (2020). Enhancing Storytelling Activities to Support Early (Digital) Literacy Development in Early Childhood Education. *International Journal of Early Childhood*, *52*(1), 55–76. doi:10.100713158-020-00263-7

McClelland, M. M., Ponitz, C. C., Messersmith, E., & Tominey, S. (2010). Self-regulation: The integration of cognition and emotion. In The Handbook of Life-Span Development. Vol. 1: Cognition, Neuroscience, Methods (pp. 509–553). Hoboken, NJ: Wiley.

McClelland, M. M., & Cameron, C. E. (2011). Self-regulation and academic achievement in elementary school children. *New Directions for Child and Adolescent Development*, *2011*(133), 29–44. doi:10.1002/cd.302 PMID:21898897

McClelland, M. M., & Cameron, C. E. (2012). Self-regulation in early childhood: Improving conceptual clarity and developing ecologically valid measures. *Child Development Perspectives*, *6*(2), 136–142. doi:10.1111/j.1750-8606.2011.00191.x

McClelland, M. M., Cameron, C. E., Connor, C. M., Farris, C. L., Jewkes, A. M., & Morrison, F. J. (2007). Links between behavioral regulation and preschoolers' literacy, vocabulary and math skills. *Developmental Psychology*, *43*(4), 947–959. doi:10.1037/0012-1649.43.4.947 PMID:17605527

McKown, C., & Weinstein, R. S. (2003). The development and consequences of stereotype-consciousness in middle childhood. *Child Development*, *74*(2), 498–515. doi:10.1111/1467-8624.7402012 PMID:12705569

McLennan, D. P. (2017). Creating coding stories and games. *Teaching Young Children*, *10*(3). 18-21. Retrieved October 02, 2019 from https://www.naeyc.org/resources/pubs/tyc/feb2017/creating-coding-stories-and-games

McMillan, J. H. (2013). *Classroom assessment: Principles and practice for effective instruction* (6th ed.). Pearson/Allyn and Bacon.

Medal, A. (2017). *How Big Data Analytics Is Solving Big Advertiser Problems.* Entrepreneur. https://www.entrepreneur.com/article/293678

Metin, S. (2020). Activity-based unplugged coding during the preschool period. *International Journal of Technology and Design Education*, 1–17.

Metz, S. S. (2007). Attracting the engineering of 2020 today. In R. Burke & M. Mattis (Eds.), *Women and Minorities in Science, Technology, Engineering and Mathematics: Upping the Numbers* (pp. 184–209). Edward Elgar Publishing. doi:10.4337/9781847206879.00018

Miller, C. C. (2017). *Tech's Damaging Myth of the Loner Genius Nerd.* https://www.nytimes.com/2017/08/12/upshot/techs-damaging-myth-of-the-loner-genius-nerd.html

Milto, E., Portsmore, M., McCormick, M., Watkins, J., & Hynes, M. (2020). *Novel Engineering, K–8: An Integrated Approach to Engineering and Literacy.* NSTA Press.

Miner, A. S., Milstein, A., Schueller, S., Hegde, R., Mangurian, C., & Linos, E. (2016). Smartphone-based conversational agents and responses to questions about mental health, interpersonal violence, and physical health. *JAMA Internal Medicine*, *176*(5), 619–625. doi:10.1001/jamainternmed.2016.0400 PMID:26974260

Mioduser, D., Levy, S. T., & Talis, V. (2009). Episodes to scripts to rules: Concrete-abstractions in kindergarten children's explanations of a robot's behavior. *International Journal of Technology and Design Education*, *19*(1), 15–36. doi:10.100710798-007-9040-6

Mischel, W., Shoda, Y., & Rodriguez, M. L. (1989). Delay of gratification in children. *Science*, *244*(4907), 933–938. doi:10.1126cience.2658056 PMID:2658056

Modan, N. (2019, September 11). *33 states adopted 57 computer science ed policies since 2018.* K-12 Dive. https://www.educationdive.com/news/33-states-adopted-57-computer-science-ed-policies-since-2018/562530/

Moll, L., Amanti, C., Neff, D., & González, N. (2005). Funds of knowledge for teaching: Using a qualitative approach to connect homes and classrooms. In Funds of Knowledge: Theorizing Practices in Households, Communities, and Classrooms (pp. 71-88). Lawrence Erlbaum Associates.

Monhardt, L., & Monhardt, R. (2006). Creating a context for the learning of science process skills through picture books. *Early Childhood Education Journal*, *34*(1), 67–71. doi:10.100710643-006-0108-9

Moreno-León, J., & Robles, G. (2015). Dr. Scratch: A Web Tool to Automatically Evaluate Scratch Projects. *Proceedings of the Workshop in Primary and Secondary Computing Education*, 132–133. 10.1145/2818314.2818338

Moreno-LeÓn, J., Robles, G., & Roman-Gonzalez, M. (2020). Towards Data-Driven Learning Paths to Develop Computational Thinking with Scratch. *IEEE Transactions on Emerging Topics in Computing*, *8*(1), 193–205. doi:10.1109/TETC.2017.2734818

Moruzzi, C. (2017, November). Creative AI: Music composition programs as an extension of the composer's mind. In *3rd Conference on" Philosophy and Theory of Artificial Intelligence*. Springer.

Movellan, J., Eckhardt, M., Virnes, M., & Rodriguez, A. (2009). Sociable robot improves toddler vocabulary skills. *Proceedings of the 4th ACM/IEEE International Conference on Human Robot Interaction*. 10.1145/1514095.1514189

Moyer, K., & Gilmer, B. V. H. (1953). The Concept of Attention Spans in Children. *The Elementary School Journal*, *54*(1), 464–466. doi:10.1086/458623

Mulker Greenfader, C. (2019). What is the role of executive function in the school readiness of Latino students? *Early Childhood Research Quarterly*, *49*(4), 93–108. doi:10.1016/j.ecresq.2019.02.011

Mullis, I. V., & Martin, M. O. (2019). *PIRLS 2021 Assessment Frameworks*. International Association for the Evaluation of Educational Achievement. Retrieved from https://eric.ed.gov/?id=ED606056

Munoz, R., Villarroel, R., Barcelos, T. S., Riquelme, F., Quezada, A., & Bustos-Valenzuela, P. (2018). Developing Computational Thinking Skills in Adolescents With Autism Spectrum Disorder Through Digital Game Programming. *IEEE Access: Practical Innovations, Open Solutions*, *6*, 63880–63889. doi:10.1109/ACCESS.2018.2877417

Muro, M., Liu, S., Whiton, J., & Kulkarni, S. (2017). *Digitalization and the American workforce*. Brookings Institute.

Naik, G. R. (Ed.). (2012). *Applied Biological Engineering: Principles and Practice*. BoD–Books on Demand. doi:10.5772/2101

National Association for the Education of Young Children (NAEYC) & Fred Rogers Center for Early Learning and Children's Media. (2012). *Technology and Interactive Media as Tools in Early Childhood Programs Serving Children from Birth through Age 8*. https://www.naeyc.org/files/naeyc/file/positions/PS_technology_WEB2.pdf

National Center for Education Statistics. (2021). *Digest of Education Statistics: 2019*. U.S. Department of Education. https://nces.ed.gov/programs/digest/d19/

National Governors Association Center for Best Practices, Council of Chief State School Officers. (2010a). *Common Core State Standards: English Language Arts Standards: Writing, Grade 1*. Washington, DC: National Governors Association Center for Best Practices, Council of Chief State School Officers. Retrieved from Common Core State Standards Initiative: http://www.corestandards.org/

National Governors Association Center for Best Practices, Council of Chief State School Officers. (2010b). Common Core State Standards: Mathematics Standards: Number & Operations in Base Ten, Grade 1. National Governors Association Center for Best Practices, Council of Chief State School Officers.

National Research Council. (2000). *From neurons to neighborhoods: The science of early childhood development*. U.S. National Research Council.

National Research Council. (2011). *Report of a Workshop of Pedagogical Aspects of Computational Thinking*. National Academy Press.

National Research Council. (2011). *Report of a workshop on the pedagogical aspects of computational thinking*. National Academies Press.

National Research Council. (2012). *A Framework for K-12 Science Education: Practices, Crosscutting Concepts, and Core Ideas. Committee on a Conceptual Framework for New K-12 Science Education Standards. Board on Science Education, Division of Behavioral and Social Sciences and Education*. The National Academies Press.

National Science Foundation. (2017). *Women, Minorities, and Persons with Disabilities in Science and Engineering: 2017*. Special Report NSF 17-310. Available at www.nsf.gov/statistics/wmpd/

Nebeker, F. (2002). Golden accomplishments in biomedical engineering. *IEEE Engineering in Medicine and Biology Magazine*, *21*(3), 17–47. doi:10.1109/MEMB.2002.1016851 PMID:12119874

NECRL. (2004). *Using Data to Guide School Improvement*. Learning Point Associates. North Central Regional Educational Laboratory. https://files.eric.ed.gov/fulltext/ED518630.pdf

Neumann, M. (2017). Parent scaffolding of young children's use of touch screen tablets. *Early Child Development and Care*, *188*(12), 1654–1664. doi:10.1080/03004430.2016.1278215

NGSS Lead States. (2013). *Next Generation Science Standards: For States By States*. Author.

NGSS Lead States. (2013). *Next Generation Science Standards: For States, By States*. The National Academies Press.

Nystrand, M. (1997). *Opening Dialogue: Understanding the Dynamics of Language and Learning in the English Classroom*. Teachers College Press.

O'Quin, K., & Besemer, S. P. (1989). The development, reliability, and validity of the revised creative product semantic scale. *Creativity Research Journal*, *2*(4), 267–278. doi:10.1080/10400418909534323

Olds, A. R. (2001). *Child care design guide*. McGraw-Hill.

Oppenheimer, T. (2003). *The flickering mind: The false promise of technology in the classroom, and how learning can be saved*. Random House Incorporated.

Ostroff, W. L. (2016). *Cultivating curiosity in K-12 classrooms: How to promote and sustain deep learning*. ASCD.

Oviatt, S. (2018, October). Ten Opportunities and challenges for advancing student-centered multimodal learning analytics. In *Proceedings of the 20th ACM International Conference on Multimodal Interaction* (pp. 87-94). ACM.

Pane, J. F., & Myers, B. A. (2001). The impact of human-centered features on the usability of a programming system for children. *Proceedings of CHI EA'*02.

Papert, S. (2002). Hard fun. *Bangor Daily News*, 2.

Papert, S. (1980). *Mindstorms: Children, computers and powerful ideas*. Basic Books.

Papert, S. (1980). *Mindstorms: Children, Computers and Powerful Ideas*. Basic Books.

Papert, S. (1980). *Mindstorms: Children, computers, and powerful ideas*. Basic Books, Inc.

Papert, S. (1980). *Mindstorms: Children, Computers, and Powerful Ideas*. Basic Books.

Papert, S. (1980). *Mindstorms: Computers, children, and powerful ideas*. Basic Books.

Papert, S. (1987). Computer Criticism vs. Technocentric Thinking. *Educational Researcher*, *16*(1).

Papert, S. (1993). *The children's machine: Rethinking school in the age of the computer*. Basic Books.

Papert, S. (2005). You can't think about thinking without thinking about thinking about something. *Contemporary Issues in Technology & Teacher Education*, *5*(3), 366–367.

Paris, A. H., & Paris, S. G. (2003). Assessing narrative comprehension in young children. *Reading Research Quarterly*, *38*(1), 36–76. doi:10.1598/RRQ.38.1.3

Pea, R. D., & Kurland, D. M. (1984). On the cognitive effects of learning computer programming. *New Ideas in Psychology*, *2*, 137–168.

Pearce, J., & Borba, S. (2017). *What Is Family Code Night?* https://www.naesp.org/blog/what-familycode-night

Pei, C., Weintrop, D., & Wilensky, U. (2018). Cultivating computational thinking practices and mathematical habits of mind in lattice land. *Mathematical Thinking and Learning*, *20*(1), 75–89.

Peppler, K. A., & Warschauer, M. (2012). Uncovering Literacies, Disrupting Stereotypes: Examining the (Dis)Abilities of a Child Learning to Computer Program and Read. *International Journal of Learning and Media*, *3*(3), 15–41. doi:10.1162/IJLM_a_00073

Pérez-Marín, M., Hijón-Neira, R., Bacelo, A., & Pizarro, C. (2018). Can computational thinking be improved by using a methodology based on metaphors and Scratch to teach computer programming to children? *Computers in Human Behavior*.

Perlis, A. J. (1962). The computer in the university. In M. Greenberger (Ed.), *Computers and the world of the future* (pp. 180–219). MIT Press.

Petre, M., & Price, B. (2004). Using robotics to motivate 'back door' learning. *Education and Information Technologies*, *9*(2), 147–158. doi:.0000027927.78380.60 doi:10.1023/B:EAIT

Piaget, J. (1952). The origins of intelligence in children *No. 5* (Vol. 8). International Universities Press.

Piaget, J. (1952). *The origins of intelligence in children*. International Universities Press.

Piaget, J. (1971). Developmental stages and developmental processes. In D. R. Green, M. P. Ford, & G. B. Flamer (Eds.), *Measurement and Piaget* (pp. 172–188). McGraw-Hill.

Pierson, E., Momoh, L., & Hupert, N. (2015). *Summative Evaluation Report for the Be A Scientist!* Project's Family Science Program. https://iridescentlearning.org/wp-content/uploads/2014/01/BAS-2015-Eval-FINAL-3.pdf

Pivetti, M., Di Battista, S., Agatolio, F., Simaku, B., Moro, M., & Menegatti, E. (2020). Educational Robotics for children with neurodevelopmental disorders: A systematic review. *Heliyon*, *6*(10), e05160. doi:10.1016/j.heliyon.2020.e05160 PMID:33072917

Plucker, J. A., Beghetto, R. A., & Dow, G. T. (2004). Why Isn't Creativity More Important to Educational Psychologists? Potentials, Pitfalls, and Future Directions in Creativity Research. *Educational Psychologist*, *39*(2), 83–96. doi:10.120715326985ep3902_1

Ponitz, C. C., McClelland, M. M., Matthews, J. S., & Morrison, F. J. (2009). A structured observation of behavioral self-regulation and its contribution to kindergarten outcomes. *Developmental Psychology*, *45*(3), 605–619. doi:10.1037/a0015365 PMID:19413419

Portelance, D. J., & Bers, M. U. (2015). Code and Tell: Assessing young children's learning of computational thinking using peer video interviews with ScratchJr. *Proceedings of the 14th International Conference on Interaction Design and Children (IDC '15)*. 10.1145/2771839.2771894

Portelance, D. J., Strawhacker, A., & Bers, M. U. (2015). Constructing the ScratchJr programming language in the early childhood classroom. *International Journal of Technology and Design Education*, •••, 1–16.

Prensky, M. (2001). Digital Natives, Digital Immigrants Part 2: Do They Really Think Differently? *On the Horizon*, *9*(6), 1–6. doi:10.1108/10748120110424843

Price, C. B., & Price-Mohr, R. M. (2018). An Evaluation of Primary School Children Coding Using a Text-Based Language (Java). *Computers in the Schools*, *35*(4), 284–301.

Promise, D. (2017). *Computational Thinking for a Computational World.* Retrieved from https://digitalpromise.org/wp-content/uploads/2017/12/dp-comp-thinking-v1r5.pdf

Przybylski, A. K., & Weinstein, N. (2019). Digital Screen Time Limits and Young Children's Psychological Well-Being: Evidence From a Population-Based Study. *Child Development*, *90*(1), e56–e65. doi:10.1111/cdev.13007 PMID:29235663

Pugnali, A., Sullivan, A., & Bers, M. U. (2017). The Impact of User Interface on Young Children's Computational Thinking. *Journal of Information Technology Education: Innovations in Practice*, *16*, 172–193. doi:10.28945/3768

RAND Reading Study Group. (2002). *Reading for Understanding, toward an R&D Program in Reading Comprehension*. RAND.

Relkin, E. (2018). *Assessing young children's computational thinking abilities* (Master's thesis). Retrieved from ProQuest Dissertations and Theses database. (UMI No. 10813994)

Relkin, E. (2018). *Assessing Young Children's Computational Thinking Abilities* (Masters Thesis). Tufts University.

Relkin, E., & Bers, M. U. (2020). *Exploring the Relationship Among Coding, Computational Thinking, and Problem Solving in Early Elementary School Students* [Symposium]. Annual Meeting of the American Educational Research Association (AERA), San Francisco, CA.

Relkin, E., & Bers, M. (2021). *TechCheck-K: A Measure of Computational Thinking for Kindergarten Children. In 2021 IEEE Global Engineering Education Conference (EDUCON).* IEEE. Retrieved from https://sites.tufts.edu/devtech/files/2021/05/1487.pdf

Relkin, E., & Bers, M. U. (2019). Designing an Assessment of Computational Thinking Abilities for Young Children. In L. E. Cohen & S. Waite-Stupiansky (Eds.), *STEM for Early Childhood Learners: How Science, Technology, Engineering and Mathematics Strengthen Learning* (pp. 85–98). Routledge. doi:10.4324/9780429453755-5

Relkin, E., & Bers, M. U. (2019). *Designing an assessment of computational thinking abilities for young children. In STEM for Early Childhood Learners: How Science, Technology, Engineering and Mathematics Strengthen Learning*. Routledge.

Relkin, E., de Ruiter, L. E., & Bers, M. U. (2021). Learning to Code and the Acquisition of Computational Thinking by Young Children. *Computers & Education*, *169*, 104222. Advance online publication. doi:10.1016/j.compedu.2021.104222

Relkin, E., de Ruiter, L., & Bers, M. U. (2020). TechCheck: Development and Validation of an Unplugged Assessment of Computational Thinking in Early Childhood Education. *Journal of Science Education and Technology*, *29*(4), 482–498. doi:10.100710956-020-09831-x

Relkin, E., Govind, M., Tsiang, J., & Bers, M. (2020). How Parents Support Children's Informal Learning Experiences with Robots. *Journal of Research in STEM Education*, *6*(1), 39–51. doi:10.51355/jstem.2020.87

Resnick, L. B., Asterhan, C. S. C., & Clarke, S. (2018). Next Generation Research in Dialogic Learning. In G. E. Hall, L. F. Quinn & D. M. Gollnick (Eds.), Wiley Handbook of Teaching and Learning (pp. 338-323). Wiley-Blackwell.

Resnick, M. (2017). *Lifelong Kindergarten: Cultivating Creativity through Projects, Passion, Peers, and Play*. MIT Press.

Resnick, M., Maloney, J., Monroy-Hernández, A., Rusk, N., Eastmond, E., Brennan, K., ... Kafai, Y. (2009). Scratch: Programming for all. *Communications of the ACM*, *52*(11), 60–67.

Resnick, M., Maloney, J., Monroy-Hernandez, A., Rusk, N., Eastmond, E., Brennan, K., Millner, A., Rosenbaum, E., Silver, J., Silverman, B., & Kafai, Y. (2009). Scratch: Programming for Everyone. *Communications of the ACM*, *52*(11), 60–67. doi:10.1145/1592761.1592779

Resnick, M., & Siegel, D. (2015). A Different Approach to Coding. *International Journal of People-Oriented Programming*, *4*(1), 1–4.

Resnick, M., & Silverman, B. (2005). Some reflections on designing construction kits for kids. *Proceeding of the 2005 Conference on Interaction Design and Children - IDC '05*, 117–122. 10.1145/1109540.1109556

Rideout, V. J. (2014). *Learning at home: Families' educational media use in America.* A report of the Families and Media Project. The Joan Ganz Cooney Center at Sesame Workshop.

Riley-Ayers. (2018). *Excerpt from Spotlight on Young Children: Observation and Assessment.* Naeyc. https://www.naeyc.org/resources/pubs/books/excerpt-from-spotlight-observation-assessment

Robelen, E. W. (2011). STEAM: Experts make case for adding arts to STEM. *Education Week*, *31*(13), 8.

Roberts, J. D., Chung, G. K. W. K., & Parks, C. B. (2016). Supporting children's progress through the PBS KIDS learning analytics platform. *Journal of Children and Media*, *10*(2), 257–266.

Robertson, J., Gray, S., Toye, M., & Booth, J. N. (2020). The relationship between executive functions and computational thinking. *International Journal of Computer Science Education in Schools*, *3*(4), 35–49. doi:10.21585/ijcses.v3i4.76

Rogoff, B. (1999). Cognition as a collaborative process. In Handbook of child psychology. New York: Wiley.

Rogoff, B., Paradise, R., Arauz, R. M., Correa-Chávez, M., & Angelillo, C. (2003). Firsthand learning through intent participation. *Annual Review of Psychology*, 54. PMID:12499516

Román-González, M., Moreno-León, J., & Robles, G. (2019). Combining Assessment Tools for a Comprehensive Evaluation of Computational Thinking Interventions. In *Computational Thinking Education* (pp. 79–98). Springer. doi:10.1007/978-981-13-6528-7_6

Román-González, M., Pérez-González, J., & Jiménez-Fernández, C. (2017). Which cognitive abilities underlie computational thinking? Criterion validity of the Computational Thinking Test. *Computers in Human Behavior, 72*, 678–691. doi:10.1016/j.chb.2016.08.047

Roque, R. (2016). Family Creative Learning: Designing Structures to Engage Kids and Parents as Computational Creators. In K. Peppler, Y. Kafai, & E. Halverson (Eds.), *Makeology in K-12, Higher, and Informal Education*. Routledge.

Roque, R., Lin, K., & Liuzzi, R. (2014). Engaging Parents as Creative Learning Partners in Computing. *Exploring the Material Conditions of Learning, 2*, 687–688.

Rose, S. P., Habgood, M. P. J., & Jay, T. (2017). An Exploration of the Role of Visual Programming Tools in the Development of Young Children's Computational Thinking. *The Electronic Journal of e-Learning, 15*(4), 297-309.

Rothbart, M. K. (2007). Temperament, development, and personality. *Current Directions in Psychological Science, 16*(4), 207–212. doi:10.1111/j.1467-8721.2007.00505.x

Rothbart, M. K., Sheese, B. E., & Posner, M. I. (2007). Executive attention and effortful control: Linking temperament, brain networks, and genes. *Child Development Perspectives, 1*(1), 2–7. doi:10.1111/j.1750-8606.2007.00002.x

Rubinstein, A., & Chor, B. (2014). Computational thinking in life science education. *PLoS Computational Biology, 10*(11), e1003897. doi:10.1371/journal.pcbi.1003897 PMID:25411839

Rumelhart, D. E. (1994). Toward an interactive model of reading. In R. B. Ruddell, M. R. Ruddell, & H. Singer (Eds.), *Theoretical models and processes of reading* (pp. 864–894). International Reading Association.

Rusk, N., Berg, R., & Resnick, M. (2005). *Rethinking robotics: Engaging girls in creative engineering*. Proposal to the National Science Foundation, Cambridge. Retrieved from https://www.media.mit.edu/publications/rethinking-robotics-engaging-girls-in-creative-engineering-2/

Ryan, E. G. (2013, November 8). Smartphones are made for giant man-hands. *Jezebel*. Retrieved from https://jezebel.com/smartphones-are-made-for-giant-man-hands-1461122433

Sameroff, A. J., & Haith, M. M. (1996). *The Five to Seven Year Shift: The Age of Reason and Responsibility*. The University of London.

Sameroff, A. J., & Haith, M. M. (1996). *The Five to seven year shift: The age of reason and responsibility*. University of Chicago Press.

Sanders, M. E. (2008). Stem, stem education, stemmania. *Technology Teacher*.

Sano, A. (2019, March 27). *Coding will be mandatory in Japan's primary schools from 2020*. Nikkei Asia. https://asia.nikkei.com/Economy/Coding-will-be-mandatory-in-Japan-s-primary-schools-from-2020#:~:text=TOKYO%20%2D%2D%20Computer%20programming%20will,highly%20sought%20information%20technology%20skills

Sattler, J. M. (2014). *Foundations of behavioral, social and clinical assessment of children*. Jerome M. Sattler, Publisher, Incorporated.

Saxena, A., Lo, C. K., Hew, K. F., & Wong, G. K. W. (2020). Designing Unplugged and Plugged Activities to Cultivate Computational Thinking: An Exploratory Study in Early Childhood Education. *The Asia-Pacific Education Researcher*, *29*(1), 55–66. doi:10.100740299-019-00478-w

Scherer, R., Siddiq, F., & Sánchez Viveros, B. (2018). The cognitive benefits of learning computer programming: A meta-analysis of transfer effects. *Journal of Educational Psychology*, *111*(5), 764–792. doi:10.1037/edu0000314

Schunk, D. H., & Zimmerman, B. J. (1997). Social origins of self-regulatory competence. *Educational Psychologist*, *32*(4), 195–208. doi:10.120715326985ep3204_1

Scott, K., Sheridan, K., & Clark, K. (2014). Culturally Responsive Computing: A theory revisited. *Learning, Media and Technology*, *40*(4), 1–25.

Seiter, L., & Foreman, B. (2013). Modeling the learning progressions of computational thinking of primary grade students. *Proceedings of the Ninth Annual International ACM Conference on International Computing Education Research - ICER '13*, 59. 10.1145/2493394.2493403

Seow, P., Looi, C. K., How, M. L., Wadhwa, B., & Wu, L. K. (2019). Educational policy and implementation of computational thinking and programming: Case study of Singapore. In *Computational thinking education* (pp. 345–361). Springer.

Serafini, F., & Gee, E. (2017). *Remixing multiliteracies: Theory and practice from New London to new times*. Teachers College Press.

Sheffield, R. S., Koul, R., Blackley, S., Fitriani, E., Rahmawati, Y., & Resek, D. (2018). Transnational examination of STEM education. *International Journal of Innovation in Science and Mathematics Education (formerly CAL-laborate International), 26*(8).

Shute, V. J., Sun, C., & Asbell-Clarke, J. (2017). Demystifying computational thinking. *Educational Research Review*, *22*, 142–158. doi:10.1016/j.edurev.2017.09.003

Signorella, M. L., Bigler, R. S., & Liben, L. S. (1993). Developmental differences in children's gender schemata about others: A meta-analytic review. *Developmental Review*, *13*(2), 147–183. doi:10.1006/drev.1993.1007

Smith, M. (1982). *Creators not consumers: Rediscovering social education*. NAYC.

Smith, R., Snow, P., Serry, T., & Hammond, L. (2020). The Role of Background Knowledge in Reading Comprehension: A Critical Review. *Reading Psychology*, *42*(3).

Snodgrass, M. R., Israel, M., & Reese, G. C. (2016). Instructional supports for students with disabilities in K-5 computing: Findings from a cross-case analysis. *Computers & Education, 100*, 1–17. doi:10.1016/j.compedu.2016.04.011

Snow, C. E., Van Hemel, S. B., & Committee on Developmental Outcomes Assessments for Young Children. (2008). *Early childhood assessment: Why, what, and how*. Washington, DC: National Academies Press.

Spencer, S. J., Steele, C. M., & Quinn, D. M. (1999). Stereotype threat and women's math performance. *Journal of Experimental Social Psychology, 35*(1), 4–28. doi:10.1006/jesp.1998.1373

Springer, K., & Keil, F. (1991). Early Differentiation of Causal Mechanisms Appropriate to Biological and Nonbiological Kinds. *Child Development, 62*(4), 767–781. doi:10.2307/1131176 PMID:1935342

Steele, C. M. (1997). A threat in the air: How stereotypes shape intellectual identity and performance. *The American Psychologist, 52*(6), 613–629. doi:10.1037/0003-066X.52.6.613 PMID:9174398

Steele, C. M. (1999). Thin ice: "Stereotype threat" and black college students. *Atlantic Monthly, 284*(2), 44–47, 50–54.

Steele, C. M., & Aronson, J. (1995). Stereotype threat and the intellectual test performance of African-Americans. *Journal of Personality and Social Psychology, 69*(5), 797–811. doi:10.1037/0022-3514.69.5.797 PMID:7473032

STEM Education Act of 2015, House of Representatives 1020, 114th Congress. (2015). https://www.congress.gov/bill/114th-congress/house-bill/1020

Strawhacker, A. & Bers, M. U. (2018b). Promoting Positive Technological Development in a Kindergarten Makerspace: A Qualitative Case Study. *European Journal of STEM Education, 3*(3), 9. doi:10.20897/ejsteme/3869

Strawhacker, A., Verish, C., Shaer, O., & Bers, M. U. (2020a, April). Debugging as Inquiry in Early Childhood: A case study using the CRISPEE prototype. *Computational Thinking for Science Learning. Symposium. Annual Meeting of the American Educational Research Association (AERA).*

Strawhacker, A. L., & Bers, M. U. (2015). "I want my robot to look for food": Comparing children's programming comprehension using tangible, graphical, and hybrid user interfaces. *International Journal of Technology and Design Education, 25*(3), 293–319.

Strawhacker, A. L., Lee, M. S. C., & Bers, M. U. (2017). Teaching tools, teachers' rules: Exploring the impact of teaching styles on young children's programming knowledge in ScratchJr. *International Journal of Technology and Design Education*. Advance online publication. doi:10.100710798-017-9400-9

Strawhacker, A., & Bers, M. U. (2018). What they learn when they learn coding: Investigating cognitive domains and computer programming knowledge in young children. *Educational Technology Research and Development*. Advance online publication. doi:10.100711423-018-9622-x

Strawhacker, A., & Bers, M. U. (2018a). Makerspaces for early childhood education (principles of space redesign) & Maker values of early childhood educators, organizing a grassroots space. In B. E. Gravel, M. U. Bers, C. Rogers, & E. Danahy (Eds.), *Making engineering playful in schools* (pp. 18–29). The LEGO Foundation.

Strawhacker, A., & Sullivan, A. (2021). Computational Expression: How dramatic arts support computational thinking in young children. In M. U. Bers (Ed.), *Computational thinking and coding in early childhood*. IGI Global.

Strawhacker, A., Verish, C., Shaer, O., & Bers, M. (2020c). Young children's learning of bioengineering with CRISPEE: A developmentally appropriate tangible user interface. *Journal of Science Education and Technology*, *29*(3), 319–339. doi:10.100710956-020-09817-9

Strawhacker, A., Verish, C., Shaer, O., & Bers, M. U. (2020b). Designing with Genes in Early Childhood: An exploratory user study of the tangible CRISPEE technology. *International Journal of Child-Computer Interaction*, *26*, 26. doi:10.1016/j.ijcci.2020.100212

Strong-Wilson, T., & Ellis, J. (2007). Children and place: Reggio Emilia's environment as third teacher. *Theory into Practice*, *46*(1), 40–47. doi:10.1080/00405840709336547

Sullivan, A. (2016). *Breaking the STEM Stereotype: Investigating the Use of Robotics to Change Young Children's Gender Stereotypes About Technology and Engineering* (Doctoral Dissertation). Tufts University, Medford, MA.

Sullivan, A., Elkin, M., & Bers, M. U. (2015). KIBO Robot Demo: Engaging young children in programming and engineering. In *Proceedings of the 14th International Conference on Interaction Design and Children (IDC '15)*. ACM.

Sullivan, A. (2019). *Breaking the STEM stereotype: reaching girls in early childhood*. Rowman & Littlefield.

Sullivan, A. (2019). *Breaking the STEM Stereotype: Reaching Girls in Early Childhood*. Rowman & Littlefield.

Sullivan, A. (2020). *STEM Tools, Games, and Products to Engage Girls in Pre-K through Early Elementary School. Technological Horizons in Education.*

Sullivan, A. A. (2019). *Breaking the STEM stereotype: Reaching girls in early childhood*. Rowman & Littlefield Publishers.

Sullivan, A. A., Bers, M. U., & Mihm, C. (2017). Imagining, Playing, and Coding with KIBO: Using Robotics to Foster Computational Thinking in Young Children. *Proceedings of the International Conference on Computational Thinking.*

Sullivan, A., & Bers, M. U. (2015). Robotics in the early childhood classroom: Learning outcomes from an 8-week robotics curriculum in pre-kindergarten through second grade. *International Journal of Technology and Design Education*. Advance online publication. doi:10.100710798-015-9304-5

Sullivan, A., & Bers, M. U. (2016). Girls, boys, and bots: Gender differences in young children's performance on robotics and programming tasks. *Journal of Information Technology Education: Innovations in Practice, 15*, 145–165. doi:10.28945/3547

Sullivan, A., & Bers, M. U. (2017). Dancing robots: Integrating art, music, and robotics in Singapore's early childhood centers. *International Journal of Technology and Design Education*. Advance online publication. doi:10.100710798-017-9397-0

Sullivan, A., & Bers, M. U. (2018). Investigating the use of robotics to increase girls' interest in engineering during early elementary school. *International Journal of Technology and Design Education, 29*(5), 1033–1051. doi:10.100710798-018-9483-y

Sullivan, A., & Bers, M. U. (2018b). The Impact of Teacher Gender on Girls' Performance on Programming Tasks in Early Elementary School. *Journal of Information Technology Education: Innovations in Practice, 17*, 153–162. doi:10.28945/4082

Sullivan, A., & Bers, M. U. (2019). VEX Robotics Competitions: Gender differences in student attitudes and experiences. *Journal of Information Technology Education, 18*, 97–112. doi:10.28945/4193

Sullivan, A., & Bers, M. U. (Manuscript submitted for publication). Increasing female representation on VEX robotics competition teams: Results from a three-year study. *International Journal of Technology and Design Education.*

Sullivan, A., Elkin, M., & Bers, M. U. (2015). KIBO Robot Demo: Engaging young children in programming and engineering. *Proceedings of the 14th International Conference on Interaction Design and Children (IDC '15).* 10.1145/2771839.2771868

Swartz, M. I., & Crowley, K. (2004). Parent Beliefs about Teaching and Learning in a Children's Museum. *Visitor Studies, 7*(2), 5–16.

Takeuchi, L., & Stevens, R. (2011). *The New Coviewing: Designing for Learning through Joint Media Engagement.* The Joan Ganz Cooney Center at Sesame Workshop.

Tang, X., Yin, Y., Lin, Q., Hadad, R., & Zhai, X. (2020). Assessing computational thinking: A systematic review of empirical studies. *Computers & Education, 148*, 103798. doi:10.1016/j.compedu.2019.103798

Taylor, M. S. (2018). Computer Programming With Pre-K Through First-Grade Students With Intellectual Disabilities. *The Journal of Special Education, 52*(2), 78–88. doi:10.1177/0022466918761120

Taylor, M. S., Vasquez, E., & Donehower, C. (2017). Computer Programming with Early Elementary Students with Down Syndrome. *Journal of Special Education Technology, 32*(3), 149–159. doi:10.1177/0162643417704439

The Computer Science Teachers Association (CSTA). (2021). *K-12 CS Education Glossary.* https://www.csteachers.org/page/glossary

The Condition of Education: Students with Disabilities. (2020). National Center of Education Statistics. https://nces.ed.gov/programs/coe/indicator_cgg.asp

The Toy Association. (2019). *STEM/STEAM Formula for Success*. https://www.toyassociation.org/ta/research/reports/stem-steam/toys/research-and-data/reports/stem-steam.aspx?hkey=6e80262f-1fea-4b37-a5e2-9679ec26f048

Thies, R., & Vahrenhold, J. (2013). *On plugging unplugged into CS classes*. doi:10.1145/2445196.2445303

Thies, R., & Vahrenhold, J. (2012). Reflections on Outreach Programs in CS Classes: Learning Objectives for" Unplugged" Activities. *Proceedings of the 43rd ACM technical symposium on Computer Science Education*, 487-492. 10.1145/2157136.2157281

Thornton-Lang. (2012) *Observation as a formal assessment tool in early childhood classrooms: A professional development module*. University of Northern Iowa. https://scholarworks.unit.edu/cgi/viewcontent.cgi?article=1238&context=grp

Toikkanen, T., & Leinonen, T. (2017). The coding ABC MOOC: Experiences from a coding and computational thinking MOOC for Finnish primary school teachers. In *Emerging research, practice, and policy on computational thinking* (pp. 239–248). Springer International Publishing.

Trends in Digital Learning: Students' Views on Innovative Classroom Models. (2014). *Project Tomorrow*. https://tomorrow.org/speakup/2014_OnlineLearningReport.html

Tucker, A., McCowan, D., Deek, F., Stephenson, C., Jones, J., & Verno, A. (2006). *A model curriculum for K–12 computer science: Report of the ACM K–12 task force curriculum committee* (2nd ed.). Association for Computing Machinery.

Tufekci, Z. (2013, November 4). It's a man's phone. *Medium.* Retrieved from https://medium.com/technology-and-society/its-a-mans-phone-a26c6bee1b69

U.S. Bureau of Labor Statistics. (2021). *Labor Force Statistics from the Current Population Survey CPS CPS Program Links.* Author.

Unahalekhaka, A., & Bers, M. U. (in press). Taking Coding Home: Analysis of ScratchJr Usage in Home and School Settings. *Educational Technology Research and Development.*

Upadhyaya, B., McGill, M. M., & Decker, A. (2020). A Longitudinal Analysis of K-12 Computing Education Research in the United States: Implications and Recommendations for Change. *Proceedings of the 51st ACM Technical Symposium on Computer Science Education*, 605-611. 10.1145/3328778.3366809

Vakil, S. (2018). Ethics, identity, and political vision: Toward a justice-centered approach to equity in computer science education. *Harvard Educational Review*, *88*(1), 26–52. doi:10.17763/1943-5045-88.1.26

Vatavu, R. D., Cramariuc, G., & Schipor, D. M. (2015). Touch interaction for children aged 3 to 6 years: Experimental findings and relationship to motor skills. *International Journal of Human-Computer Studies*, *74*, 54–76.

Vee, A. (2013). Understanding computer programming as a literacy. *Literacy in Composition Studies*, *1*(2), 42–64.

Vee, A. (2017). *Coding Literacy: How Computer Programming Is Changing Writing*. The MIT Press. doi:10.7551/mitpress/10655.001.0001

Venville, G., Gribble, S. J., & Donovan, J. (2005). An exploration of young children's understandings of genetics concepts from ontological and epistemological perspectives. *Science Education*, *89*(4), 614–633. doi:10.1002ce.20061

Verish, C., Strawhacker, A., Bers, M. U., & Shaer, O. (2018). CRISPEE: A Tangible Gene Editing Platform for Early Childhood. *Proceedings of the Twelfth International Conference on Tangible, Embedded and Embodied Interaction (TEI)*. 10.1145/3173225.3173277

Viana, A. G., Beidel, D. C., & Rabian, B. (2009). Selective mutism: A review and integration of the last 15 years. *Clinical Psychology Review*, *29*(1), 57–67. doi:10.1016/j.cpr.2008.09.009 PMID:18986742

Vogel, S., Hoadley, C., Castillo, A. R., & Ascenzi-Moreno, L. (2020). Languages, literacies, and literate programming: Can we use the latest theories on how bilingual people learn to help us teach computational literacies? *Computer Science Education*, *30*(4), 420–443. doi:10.1080/08993408.2020.1751525

Vogel, S., Santo, R., & Ching, D. (2017, March). Visions of computer science education: Unpacking arguments for and projected impacts of CS4All initiatives. In *Proceedings of the 2017 ACM SIGCSE Technical Symposium on Computer Science Education* (pp. 609-614). ACM.

Vohs, K. D., & Baumeister, R. F. (2004). Understanding self-regulation: An Introduction. In R. F. Baumeister & K. D. Vohs (Eds.), *Handbook of Self-Regulation: Research, theory, and applications* (pp. 1–9). Guilford Press.

von Wangenheim, C. G., Hauck, J. C. R., Demetrio, M. F., Pelle, R., da Cruz Alves, N., Barbosa, H., & Azevedo, L. F. (2018). CodeMaster—Automatic Assessment and Grading of App Inventor and Snap! Programs. *Informatics in Education*, *17*(1), 117–150. doi:10.15388/infedu.2018.08

Vossoughi, S., & Vakil, S. (2018). Toward what ends? A critical analysis of militarism, equity, and STEM education. In *Education at war* (pp. 117–140). Fordham University Press. doi:10.2307/j.ctt2204pqp.9

Vygotsky, L. S. (1987). Thinking and speech (N. Minick, Trans.). In R. W. Rieber & A. S. Carton (Eds.), The collected works of L. S. Vygotsky (Vol. 1., pp. 39-285). New York: Plenum Press. (Original work published 1934)

Vygotsky, L. (2012). *Thought and language*. MIT Press.

Vygotsky, L. S. (1978). *Mind in Society*. Harvard University Press.

Vygotsky, L. S. (1978). *Mind in society: The Development of higher psychological processes*. Harvard University Press.

Wadsworth, B. J. (1996). *Piaget's theory of cognitive and affective development: Foundations of constructivism*. Longman Publishing.

Walker, J., & Strawhacker, A. (Co-chairs). (2021, April 8-12). The Biomaker Ecosystem: Technologies, Spaces and Curriculum for K-12 Making with Biology [Symposium]. *American Educational Research Association* (Virtual Conference).

Wang, J., & Hejazi Moghadam, S. (2017, March). Diversity barriers in K-12 computer science education: structural and social. *Proceedings of the 2017 ACM SIGCSE Technical Symposium on Computer Science Education*, 615-620. 10.1145/3017680.3017734

Werner, L., Denner, J., & Campe, S. (2014). Using computer game programming to teach computational thinking skills. *Learning, Education And Games, 37*. Retrieved from https://dl.acm.org/citation.cfm?id=2811150

Werner, L., Denner, J., Campe, S., & Kawamoto, D. C. (2012). The fairy performance assessment: measuring computational thinking in middle school. *Proceedings of the 43rd ACM Technical Symposium on Computer Science Education*, 215–220. 10.1145/2157136.2157200

Westlund, J., & Breazeal, C. (2015). The Interplay of Robot Language Level with Children's Language Learning During Storytelling. In *Proceedings of the Tenth Annual ACM/IEEE International Conference on Human-Robot Interaction Extended Abstracts*. ACM. 10.1145/2701973.2701989

Willis, A. I., & Harris, V. (2000). Political acts: Literacy learning and teaching. *Reading Research Quarterly*, *35*(1), 72–88. doi:10.1598/RRQ.35.1.6

Wilson, A., Hainey, T., & Connolly, T. M. (2013). Using Scratch with Primary School Children: An Evaluation of Games Constructed to Gauge Understanding of Programming Concepts. *International Journal of Game-Based Learning*, *3*(1), 93–109. doi:10.4018/ijgbl.2013010107

Wilson, C., Sudol, L. A., Stephenson, C., & Stehlik, M. (2010). *Running on empty: The failure to teach K-12 computer science in the digital age*. The Association for Computing Machinery and the Computer Science Teachers Association.

Wilson-Lopez, A., Larsen, V., & Gregory, S. (2017). Reading and Engineering: Elementary Students' Co-Application of Comprehension Strategies and Engineering Design Processes. *Journal of Pre-College Engineering Education Research*, *6*(2), 39–57. doi:10.7771/2157-9288.1116

Wing, J. (2011). *Research notebook: Computational thinking—What and why?* https://www.cs.cmu.edu/link/research-notebookcomputational-thinking-what-and-why

Wing, J. (2011). Research notebook: Computational thinking—What and why? *The Link Magazine*. Retrieved from https://www.cs.cmu.edu/link/research-notebookcomputational-thinking-what-and-why

Wing, J. (2011). Research notebook: computational thinking—What and why? *The Link Magazine*. Retrieved from: https://www.cs.cmu.edu/link/research-notebookcomputational-thinking-what-and-why

Wing, J. M. (2006). Computational Thinking. *CACM Viewpoint*, 33-35. http://www.cs.cmu.edu/afs/cs/usr/wing/www/publications/Wing06.pdf

Wing, J. M. (2008). Computational thinking and thinking about computing. *Philosophical Transactions of the Royal Society A: Mathematical, Physical and Engineering Sciences, 366*(1881), 3717–3725.

Wing, J. (2006). *Computational thinking. Communications of Advancing Computing Machinery, 49 (3), 33-36*. Association for Computing Machinery.

Wing, J. M. (2006). Computational thinking. *CACM Viewpoint*, *49*(3), 33–35. doi:10.1145/1118178.1118215

Wing, J. M. (2006). Computational thinking. *Communications of the ACM, 49*(3), 33-35. through programming robots in early childhood. *Journal of Educational Computing Research, 50*(4), 553–573.

Witherspoon, E. B., Schunn, C. D., Higashi, R. M., & Baehr, E. C. (2016). Gender, interest, and prior experience shape opportunities to learn programming in robotics competitions. *International Journal of STEM Education*, *3*(1), 18. doi:10.118640594-016-0052-1

Wittgenstein, L. (1997). Philosophical Investigations (2nd ed.). Cambridge: Blackwell.

Wohl, B., Porter, B., & Clinch, S. (2015). Teaching computer science to 5–7 year-olds: An initial study with scratch, cubelets and unplugged computing. *Proceedings of the Workshop in Primary and Secondary Computing Education*, 55–60. 10.1145/2818314.2818340

Wortham, S. C. (2006). *Early childhood curriculum: Developmental bases for learning and teaching*. Kevin M.

Yadav, A., Good, J., Voogt, J., & Fisser, P. (2017). Computational thinking as an emerging competence domain. In Technical and vocational education and training (Vol. 23, pp. 1051–1067). doi:10.1007/978-3-319-41713-4_49

Yadav, A., Hong, H., & Stephenson, C. (2016). Computational Thinking for All: Pedagogical Approaches to Embedding 21st Century Problem Solving in K-12 Classrooms. *TechTrends*, *60*(6), 565–568. doi:10.100711528-016-0087-7

Yager, R. E. (1996). Meaning of STS for science teachers. *Science/technology/Society: as reform in science education*, 16-24.

Yelland, N. (2005). Mindstorms or a storm in a teacup? A review of research with Logo. *International Journal of Mathematical Education in Science and Technology*, *26*(6), 853–869. doi:10.1080/0020739950260607

Yelland, N., & Masters, J. (2007). Rethinking scaffolding in the information age. *Computers & Education, 48*(3), 362–382. doi:10.1016/j.compedu.2005.01.010

Zapata-Cáceres, M., Martín-Barroso, E., & Román-González, M. (2020). Computational Thinking Test for Beginners: Design and Content Validation. In *2020 IEEE Global Engineering Education Conference (EDUCON)* (pp. 1905-1914). IEEE. 10.1109/EDUCON45650.2020.9125368

Zeidler, D. L., Herman, B. C., Clough, M. P., Olson, J. K., Kahn, S., & Newton, M. (2016). Humanitas emptor: Reconsidering recent trends and policy in science teacher education. *Journal of Science Teacher Education, 27*(5), 465–476. doi:10.100710972-016-9481-4

Zelazo, P. D., Carter, A., Reznick, J. S., & Frye, D. (1997). Early development of executive function: A problem solving framework. *Review of General Psychology, 1*(2), 198–226. doi:10.1037/1089-2680.1.2.198

Zweben, S., & Bizrot, B. (2015). *2014 Taulbee survey.* Retrieved from the Computing Research Association website, https://cra.org/wp-content/uploads/2015/06/2014-Taulbee-Surv

Zweig. (2015). *Data collection and use in early childhood education programs: Evidence from the Northeast Region.* Regional Educational Laboratory. https:/files.eric.ed.gov/fulltext/ED555737.pdf

About the Contributors

Jessica Blake-West is a master's student at Tufts University, studying Human Factors Engineering, and the Lab Manager of DevTech Research Group. She received her B.S. in Cognitive Neuroscience at Brown University in 2020. While studying at Brown, she worked extensively with children through internships at Women and Infants' Hospital in Providence Rhode Island, Brown's Swearer Center, and DevTech Research Group. Her studies at Brown focused on developmental neuroscience and psychology, as well as computer science and education studies. After her internship at DevTech in 2019, Jessica took a great interest in educational technology and teaching computational thinking, and decided to return to DevTech post-graduation as the Lab Manager. Now at DevTech, her work is primarily focused on developing the Coding Stages Assessment and ScratchJr administration, which includes using Google Analytics to assess ScratchJr usage.

Laura de Ruiter is a Research Assistant Professor at the DevTech Research Group at the Eliot-Pearson Department of Child Study and Human Development at Tufts University. She studies language acquisition and cognitive development in young children. Her interests include understanding how children map mental representations and language, and uncovering the cognitive underpinnings of developmental computer science. Laura uses predominantly quantitative research methods, including experiments, corpus analyses and psychometric assessments. She received her Ph.D. from the Max Planck Institute for Psycholinguistics in Nijmegen (The Netherlands), and completed a postdoctoral fellowship at the ESRC International Research Centre for Language and Communicative Development (LuCiD) at the University of Manchester (UK).

Madhu Govind is a doctoral student in the Eliot-Pearson Department of Child Study and Human Development at Tufts University and a graduate researcher at the DevTech Research Group. She received her B.S. in Child Studies and Neuroscience at Vanderbilt University and her M.A. at Tufts University in Child Study and Human Development. Her interests broadly encompass the ways in which innovative education technologies and pedagogies can promote children's learning and creative expression. Madhu's research interests are shaped by personal life experiences and over a decade of teaching experience as an after-school tutor, public middle school special educator, and math department chair. Her current work focuses on PK-2 teachers' relationships with and attitudes toward coding and robotics education for children. Madhu's doctoral research is generously supported by the Tufts Provost Fellowship and the Evans Literacy Fellowship.

Ziva R. Hassenfeld is the Jack, Joseph and Morton Mandel Assistant Professor in Jewish Education at Brandeis University. She studies reading comprehension from a sociocultural perspective, focusing on how children develop interpretations of the Hebrew Bible as a case of student reading development. She uses a variety of qualitative methods including ethnographic observation, stimulated recall interviewing, and think-aloud interviewing. These investigations connect her to the worlds of biblical hermeneutics, both contemporary and rabbinic, as well as literary theory and criticism.

Libby Hunt holds a master's degree from the Eliot-Pearson Department of Child Study and Human Development at Tufts University, with a concentration in 21st Century Literacies: Media and Technology. She received a B.A. in English Literature from Wheaton College (MA). Libby's primary research interests include the impacts of media on children's social-emotional development, how educational media can promote learning, and media literacy. During her time at Tufts, Libby worked as a research assistant for the DevTech Research Group and the Children's Television Project.

Tess Levinson is a Ph.D. Student at the Eliot-Pearson Department of Child Study and Human Development at Tufts University and is a member of the DevTech Research Group. Tess received her B.S. in Cognitive Studies and Disability Studies at Vanderbilt University, where she became interested in the interaction between social-emotional development and academic learning. Prior to Tufts, she was a research coordinator at the Perelman School of Medicine at the University of Pennsylvania, coordinating a research grant on the neural correlates of motivation in adolescents at risk for psychosis. Her current research interests include the intersection of social and emotional learning and computer science learning in young children, the neurocognitive processes associated with learning to code, and inclusive coding and robotics educational environments for children with disabilities.

Claudia Mihm is currently a Master's student at Harvard Graduate School of Education, studying Technology, Innovation and Education. She attended Tufts

University for her undergraduate degree, majoring in Computer Science and Child Study and Human Development. Throughout her time at Tufts, she worked as an undergraduate research assistant in the DevTech research group, primarily supporting research around ScratchJr and the KIBO robotics kit, and creating a guide to help teachers support their students through the transition from ScratchJr to Scratch. Over the past 5 years, Claudia has taught computational thinking and computer science to young learners in informal settings, and is primarily interested in leveraging computational thinking approaches and technological tools to create engaging learning experiences across subjects.

Elizabeth Kazakoff Myers, Ph.D., is currently Director of Education Research and Evaluation at WGBH Educational Foundation where she manages formative and summative research activities in children's educational media. She has an extensive background in the development and evaluation of K-12 STEM and EdTech resources with a particular focus on the intersection of new technologies and child development. Dr. Myers has held positions in and led research projects across academic, industry, government, and non-profit settings. Dr. Myers earned a B.S in Psychology from Rensselaer Polytechnic Institute, M.Ed. in Psychological Studies from Cambridge College, and a Ph.D. in Child Study and Human Development from Tufts University.

XuanKhanh Nguyen is an undergraduate student in the Computer Science Department at Tufts University and an undergraduate research assistant at the DevTech Research Group. She will receive her B.S. at Tufts University with a major in Data Science. XuanKhanh's goal of being a Data Scientist to transform the education system and focus on innovative educational technologies. Her current capstone project focuses on studying the effectiveness and relevance of Computer Science Standards to prepare students for college, career, and life. At DevTech, XuanKhanh's research work focuses on ScratchJr's data analytics using Google Analytics and Machine Learning models.

Emily Relkin, M.A., is a Ph.D. student at the Eliot-Pearson Department of Child Study and Human Development at Tufts University and is a member of the DevTech Research Group. Emily received her B.A. from Muhlenberg College in Psychology and her M.A. from Tufts University in Child Study and Human Development. Her research focuses on understanding and assessing the development of computational thinking abilities in young children. She developed and validated TechCheck, a novel unplugged computational thinking assessment for 5-9-year-olds that is being used in research and educational settings.

Amanda Strawhacker, Ph.D., is the Associate Director of the Early Childhood Technology (ECT) Graduate Certificate Program at Tufts University's Eliot-Pearson Department of Child Study and Human Development. Her work involves teaching, developing curriculum, and professional development around educational technology. Prior to her role at ECT, Amanda was a Ph.D. student at the DevTech Research Group. She has contributed to the research and development of several technologies including the ScratchJr programming app, the KIBO robotics kit, the Early Childhood Makerspace at Tufts, and most recently the CRISPEE bioengineering kit. Amanda is a two-time winner of the Eliot-Pearson Research-Practice Integration Award, and was a speaker with TEDxYouth@BeaconStreet. Her research interests include engaging children in playful learning about computational thinking, biology, and ethics, and supporting in-service educators and adults in fostering children's early STEM experiences.

Amanda Sullivan is a research consultant, educator, and author who broadly focuses on the impact of new technologies and media on young children. Her research specifically explores strategies for breaking gender stereotypes and engaging girls in STEM & STEAM. Amanda holds a Master's and Ph.D. in Child Development from Tufts University where she worked with the Developmental Technologies (DevTech) Research Group. She is the co-creator of theScratchJr Coding Cards: Creative Coding Activities for Children Ages 5-7, published by No Starch Press, and author of the book Breaking the STEM Stereotype: Reaching Girls in Early Childhood,published by Rowman & Littlefield. Her work has been featured in GeekWire, WIRED magazine, the New York Times, and more. As a former drama teacher, Amanda is an advocate for STEAM education and integrating the arts with technology. She has over a decade of classroom experience teaching early childhood and elementary school robotics, coding, drama, film production, and more. Amanda is a Lecturer in the Early Childhood Technology (ECT) graduate certificate program at Tufts University and an Associate Faculty member in the College of Doctoral Studies at the University of Phoenix.

Apittha Unahalekhaka is a doctoral student in the Eliot-Pearson Department of Child Study and Human Development at Tufts University and a graduate researcher at the DevTech Research Group. She received her B.S. at University of Toronto with a double major in Neuroscience and Economics, M.M.S. at Duke University, and Ed.M. at Harvard Graduate School of Education. Her research interests are data science for education and socio-emotional development in early childhood with technological learning tools. Prior to graduate schools, she was a management consultant at Gallup, external relations associate at Teach For Thailand, and an intern at The United Nations High Commissioner for Refugees (UNHCR). Her current research work focuses on analyzing ScratchJr data analytics and how young children's collaboration affect the quality of their coding projects.

Miki Vizner, M.A., is passionate about creating technologies that allow young children to grow through play and self-expression. He strives to create technologies that empower, not pacify. He's spent a decade collecting skills that make him as comfortable working with a classroom full of kindergartners as he is in manufacturing facilities across Asia. Now, he is a mechanical engineer developing integrated power systems for developing countries. Before, he spent two years building robots and rapid prototyping tools for young children as part of the DevTech research group during a master's program in child studies and human development at Elliot Pearson. He also tended bar. Before that, he spent two and a half years in Rwanda, developing a center for vulnerable high schoolers to learn science by doing. He is interested in preoperational logic, making absurd objects, talking to young children about big ideas, and being outside.

Index

E

F

G

H

I

J

K

L

M

P

R

S

T

U

W

www.ingramcontent.com/pod-product-compliance
Lightning Source LLC
LaVergne TN
LVHW080603300926
849162LV00005B/1038
* 9 7 8 1 7 9 9 8 7 3 0 8 2 *